Understanding and Solving
Environmental Problems in the 21st Century

Understanding and Solving Environmental Problems in the 21st Century
Toward a new, integrated hard problem science

Edited by

Robert Costanza
*University of Maryland Institute for Ecological Economics,
Solomons, MD, USA*

and

Sven Erik Jørgensen
*Royal Danish School of Pharmacy,
Department of General Chemistry, Copenhagen, Denmark*

2002

ELSEVIER

Amsterdam • Boston • London • New York • Oxford • Paris
San Diego • San Francisco • Singapore • Sydney • Tokyo

ELSEVIER SCIENCE Ltd
The Boulevard, Langford Lane
Kidlington, Oxford OX5 1GB, UK

© 2002 Elsevier Science Ltd. All rights reserved.

This work is protected under copyright by Elsevier Science, and the following terms and conditions apply to its use:

Photocopying

Single photocopies of single chapters may be made for personal use as allowed by national copyright laws. Permission of the Publisher and payment of a fee is required for all other photocopying, including multiple or systematic copying, copying for advertising or promotional purposes, resale, and all forms of document delivery. Special rates are available for educational institutions that wish to make photocopies for non-profit educational classroom use.

Permissions may be sought directly from Elsevier Science Global Rights Department, PO Box 800, Oxford OX5 1DX, UK; phone: (+44) 1865 843830, fax: (+44) 1865 853333, e-mail: permissions@elsevier.co.uk. You may also contact Global Rights directly through Elsevier's home page (http://www.elsevier.com), by selecting 'Obtaining Permissions'.

In the USA, users may clear permissions and make payments through the Copyright Clearance Center, Inc., 222 Rosewood Drive, Danvers, MA 01923, USA; phone: (+1) 978 7508400, fax: (+1) 978 7504744, and in the UK through the Copyright Licensing Agency Rapid Clearance Service (CLARCS), 90 Tottenham Court Road, London W1P 0LP, UK; phone: (+44) 207 631 5555; fax: (+44) 207 631 5500. Other countries may have a local reprographic rights agency for payments.

Derivative Works

Tables of contents may be reproduced for internal circulation, but permission of Elsevier Science is required for external resale or distribution of such material. Permission of the Publisher is required for all other derivative works, including compilations and translations.

Electronic Storage or Usage

Permission of the Publisher is required to store or use electronically any material contained in this work, including any chapter or part of a chapter.

Except as outlined above, no part of this work may be reproduced, stored in a retrieval system or transmitted in any form or by any means, electronic, mechanical, photocopying, recording or otherwise, without prior written permission of the Publisher.

Address permissions requests to: Elsevier Science Global Rights Department, at the mail, fax and e-mail addresses noted above.

Notice

No responsibility is assumed by the Publisher for any injury and/or damage to persons or property as a matter of products liability, negligence or otherwise, or from any use or operation of any methods, products, instructions or ideas contained in the material herein. Because of rapid advances in the medical sciences, in particular, independent verification of diagnoses and drug dosages should be made.

First edition 2002

Library of Congress Cataloging in Publication Data
A catalog record from the Library of Congress has been applied for.

British Library Cataloguing in Publication Data
A catalogue record from the British Library has been applied for.

ISBN: 0-08-044111-4

⊗ The paper used in this publication meets the requirements of ANSI/NISO Z39.48-1992 (Permanence of Paper).
Printed in The Netherlands.

Preface

This book is intended to serve a number of purposes and audiences, including:
(1) to act as a "state of the art" assessment of integrated environmental science and its relation to real world problem solving, aimed at the academic community, and
(2) to act as a sourcebook for managers, policy makers, and the informed public on both the state of the science and the state of consensus among scientists on key environmental issues.

Background and process

This book is a product of the 2nd EcoSummit, held in Halifax, Nova Scotia, June 18–22, 2000. The aim of the EcoSummit was to encourage integration of both the natural and social sciences with the policy and decisionmaking community, for the purpose of developing a deeper understanding of complex environmental problems. This understanding is a necessary basis for sustainable solutions to environmental problems.

The EcoSummit centered around the following six themes:
(1) Integrated modeling and assessment.
(2) Complex, adaptive, hierarchical systems.
(3) Ecosystem services.
(4) Science and decisionmaking.
(5) Ecosystem health and human health.
(6) Quality of life and the distribution of wealth and resources.

These themes were seen as necessary in order to reach the goal of understanding and solving environmental problems in the 21st century. Each theme agenda was developed in advance of the summit, and a working group on each of the six themes was held during the course of the summit.

The structure and process of the EcoSummit was different than most scientific meetings. It was a "summit" in that all delegates actively participated in one or more working groups charged with developing the chapters in this book. A plenary session was held the first morning where each theme chair presented a background talk and posed the key questions that formed the basis of the working group agendas. On each subsequent day there was another plenary session, at which the rapporteurs summarized their working group's progress and outcomes.

This book is an attempt to capture the "consensus" of the meeting. Consensus in this context is taken to mean the "sense of the meeting". Every individual at the meeting will probably not agree with every word in the following pages, but the vast majority will agree with the vast majority of what is said. In that sense, this book represents (to a fair approximation) the collective view of the participants, rather than the individual view of any one or small group of the participants.

Acknowledgements

We are indebted to the staff at Elsevier Science publishers, who played a central role in organizing and funding the EcoSummit. Mary Malin was the main conference organizer, aided by Julie Ingram and Gerald Dorey. We would also like to thank the other members of the organizing committee of the EcoSummit (William Mitsch, Johannes Heeb, Tony Jakeman, Anthony King, Mohi Munawar, David Rapport, and Mark Schwartz) and all of the presenters and participants (see the Participant List, p. vii). Finally, a special thanks to Amanda Walker, who did the technical editing on the manuscript, helped cajole authors to send in their revisions, and contributed in innumerable ways in the process of taking the book from first drafts to final publication.

Robert Costanza
and
Sven Erik Jørgensen

EcoSummit Participant List

Work Group: Integrated Assessment and Modeling

Put O. Ang Jr., Department of Biology, The Chinese University of Hong Kong, China
Rob M. Argent, Centre for Environmental Applied Hydrology, University of Melbourne, Australia
David Barker, Geostructures Consulting, United Kingdom
M. Bruce Beck, Warnell School of Forest Resources, University of Georgia, USA
Shui Bin, Department of Engineering and Public Policy, Carnegie Mellon University, USA
Heather Breeze, Gorsebrook Research Institute, Saint Mary's University, Canada
Tony Charles, Management Science/Environmental Science, Saint Mary's University, Canada
Peter Deadman, Faculty of Environmental Studies, University of Waterloo, Canada
Ingrid S. Eriksson, Institute of Agricultural Engineering, Swedish University of Agricultural Sciences (SLU), Sweden
Chris Fletcher, Gorsebrook Research Institute, Saint Mary's University, Canada
Anthony Friend, School of Community and Regional Planning, Canada
Philippe Girardin, INRA, Equipe Agriculture et Environment, France
Matt Hare, Swiss Federal Institute of Environmental Science & Technology, Switzerland
Graham Harris, CSIRO Land and Water, Canberra, Australia
Ralf Hoch, University of Kassel, Germany
Tony Jakeman, Centre for Resource and Environmental Studies, The Australian National University, Australia
Marco Janssen, Free University, The Netherlands
Elias Kautsky, Institute of Environmental Science and Technology, Yokohama National University, Japan
Ulrik Kautsky, Swedish Nuclear Fuel and Waste Management Co., Sweden
Linda Kumblad, Department of Systems Ecology, Stockholm University, Sweden
Guy Larocque, Natural Resources Canada, Canada

Rebecca Letcher, Centre for Resource and Environmental Studies, The Australian National University, Australia
Kevin Lim, Faculty of Environmental Studies, University of Waterloo, Canada
Silvia Maltagliati, Department of Energetics "Sergio Stecco", University of Florence, Italy
Lubos Matejicek, Institute for Environmental Studies, Charles University, Czech Republic
David Mauriello, US Environmental Protection Agency, USA
Tom Maxwell, University of Maryland, USA (as of September 2002: Gund Institute of Ecological Economics, The University of Vermont, Burlington, VT)
Shailendra Mudgal, INERIS, France
Bjorn Naeslund, Department of Systems Ecology, Stockholm University, Sweden
Watam Naito, Yokohama National University, Japan
Naoki Nakatani, Osaka Prefecture University, Japan
Dapo Odulaja, International Centre of Insect Physiology and Ecology, Kenya
Rannveig Olafsdottir, Department of Physical Geography, Lund University, Sweden
Femi Osidele, Warnell School of Forest Resources, University of Georgia, USA
Claudia Pahl-Wostl, Swiss Federal Institute of Environmental Science & Technology, Switzerland
Richard Park, Eco Modeling, USA
Paul Parker, Faculty of Environmental Studies, University of Waterloo, Canada
Dominique Pelletier, Laboratoire MAERHA, IFREMER, France
Jim Reilly, New Jersey Office of State Planning, USA
Andrea Rizzoli, Instituto Dalle Molle di Studi sull'Intelligenza Artificiale (IDSIA), Switzerland
Michelle Scoccimarro, Integrated Catchment Assessment and Management Centre, The Australian National University, Australia
Michael Sonnenshein, University of Oldenburg, Germany
Paul Sullivan, Department of Applied Mathematics, The University of Western Ontario, Canada
Parviz Tarikhi, Iranian Remote Sensing Center, Iran
Alexey Voinov, Institute for Ecological Economics, Center for Environmental Science, University of Maryland, USA (as of September 2002: Gund Institute of Ecological Economics, The University of Vermont, Burlington, VT)

Work Group: Complex Adaptive Hierarchical Systems

Simone Bastianoni, University of Siena, Siena, Italy,
Stuart R. Borrett, Institute of Ecology, University of Georgia, Athens, GA, USA

EcoSummit Participant List

Sherry Brandt-Williams, FEDP Rookery Bay National Estuarine Research Reserve, Naples, FL, USA
Jae S. Choi, Dalhousie University, Halifax, Nova Scotia, Canada
Marko Debeljak, University of Ljubljana, Slovenia
Brian D. Fath, U.S. Environmental Protection Agency, Cincinnati, OH, USA
Julio Fonseca, Institute for Marine Research, Coimbra University, Coimbra, Portugal
William E. Grant, Ecological Systems Laboratory, Department of Wildlife and Fisheries Sciences, Texas A&M University, College Station, TX, USA
Dwikorita Karnawati, Gadjah Mada University, Bulaksumur, Yogyakarta, Indonesia
João C. Marques, Institute for Marine Research, Coimbra University, Coimbra, Portugal
Anton Moser, Institute Of Biotechnology, Graz University Of Technology, Graz, Austria
Felix Müller, Ecology Center, University of Kiel, Kiel, Germany
Claudia Pahl-Wostl, EAWAG, Ueberlandstr., Duebendorf, Switzerland
Bernard C. Patten, Institute of Ecology, University of Georgia, Athens, GA, USA
Ralf Seppelt, Technical University Braunschweig, Braunschweig, Germany
Wolf H. Steinborn, Ökologie-Zentrum der Universität Kiel, Kiel, Germany
Yuri M. Svirezhev, Potsdam Institute for Climate Impact Research, Potsdam, Germany

Work Group: Ecosystem Services

Marie Adamsson, Göteborg University, Department of Applied Environmental Science, Göteborg
David Barker, Geostructures Consultant, Model Farm, Crockham Hill, Edenbridge, Kent, UK
Anja Brüll, Berlin, Germany
Belinda Campbell, Department of Biological Engineering, Dalhousie University, Halifax, Nova Scotia, Canada
Andrew Dakers, Natural Resources Engineering Consultancy and Part-time Lecturer, Environmental Management and Design Division, Lincoln University, New Zealand
Stefan Gossling, Lund, Sweden
Björn Guterstam, Global Water Partnership, Stockholm, Sweden
Bill Hart, Centre for Resources Studies, Daltech-Dalhousie University, Halifax, Nova Scotia, Canada
Johannes Heeb, International Ecological Engineering Society, Wolhusen, Switzerland
Steven Loiselle, University of Siena, Italy

Ulo Mander, Institute of Geography, University of Tartu, Tartu, Estonia
Donata Melaku Canu, CNR-ISDGM, Venice, Italy
Ralf Roggenbauer, Leitha, Austria
Michel Roux, Swiss Federal Research WSL, Birmensdorf, Switzerland
George D. Santopietro, Department of Economics, Radford University, VA, USA
Deidre Stuart, School of Natural Resource Sciences, Queensland University of Technology, Brisbane, Queensland, Australia
Mary Trudeau, Ontario, Canada
Hein D. van Bohemen, Ministry of Transport, Public Works and Water Management, Delft, Netherlands
Alan Werker, Department of Civil Engineering, University of Waterloo, Ontario, Canada

Work Group: Science and Decisionmaking

Jim Berkson, Department of Fisheries and Wildlife Sciences, Virginia Polytechnic Institute and State University, USA
Rebekah Blok, Martec Limited, Canada
Mark Borsuk, Duke University, USA
Valerie Brown, Faculty of Environmental Management & Agriculture, University of Western Sydney, Australia
Randall Bruins, US Environmental Protection Agency, USA
Kevin Cover, City of Ottawa, Canada
Virginia Dale, Environmental Sciences Division, Oak Ridge National Laboratory, USA
Jodi Dew, Virginia Polytechnic Institute and State University, USA
Carl Etnier, Agricultural University of Norway, Norway
Lucia Fanning, Dalhousie University, Canada
Francisca Felix, Delft University of Technology, The Netherlands
Mohd. Nordin Hasan, Institute for Environment and Development, University of Kebangsaan, Malaysia
Huasheng Hong, Xiamen University, China
A.W. King, Environmental Sciences Division, Oak Ridge National Laboratory, USA
Norbert Krauchi, Forest Ecosystems and Ecological Risks Division, Swiss Federal Institute for Forest, Snow and Landscape Research (WSL), Switzerland
Wolfram Krewitt, System Analysis and Technology Assessment, Institute of Technical Thermodynamics, Stuttgart, Germany
Ken Lubinsky, US Geological Survey, USA
John Olson, Villanova University, USA
Janina Onigkeit, University of Kassel, Germany
Gary Patterson, Nova Scotia, Canada

EcoSummit Participant List

Irene Peters, Swiss Federal Institute for Environmental Science and Technology (EAWAG), Switzerland
K. S. Rajan, Institute of Industrial Science, The University of Tokyo, Japan
Peter Reichert, Department of Systems Analysis, Integrated Assessment and Modelling (SIAM), Swiss Federal Institute for Environmental Science and Technology (EAWAG), Switzerland
Edward J. Rykiel Jr., Washington State University, USA
Mark Schwartz, University of California, Davis, USA
Kamala Sharma, University of Sydney, Australia
Jason Shogren, University of Wyoming, USA
Val Smith, University of Kansas, USA
Michael Sonnenschein, University of Oldenburg, Germany
Robert St-Louis, Environment Canada, Canada
Deidre Stuart, School of Natural Resource Sciences, Queensland University of Technology, Brisbane, Queensland, Australia
Ray Supalla, University of Nebraska, USA
Diederik van der Molen, RIZA, The Netherlands
Henk van Latesteijn, Scientific Council for Government Policy, The Netherlands

Work Group: Quality of Life and the Distribution of Wealth and Resources

Joshua Farley, University of Maryland Institute for Ecological Economics, Solomons, MD, USA (as of September 2002: Gund Institute of Ecological Economics, The University of Vermont, Burlington, VT)
Robert Costanza, University of Maryland Institute for Ecological Economics, Solomons, MD, USA (as of September 2002: Gund Institute of Ecological Economics, The University of Vermont, Burlington, VT)
Paul Templet, Institute for Environmental Studies, Louisiana State University, Baton Rouge, LA, USA
Michael Corson, Texas A&M University, College Station, TX, USA
Philippe Crabbe, University of Ottawa, Ottawa, Canada
Ricardo Esquivel, Colonia Mexico, Merida 97128, Yucatan, Mexico
Koyu Furusawa, Kokugakuin University, Faculty of Economics, Tokyo, Japan
William Fyfe, University of Western Ontario, Department of Earth Sciences, London, Ontario
Orie Loucks, Miami University, OH, USA
Kelly MacDonald, Environment Canada, Dartmouth, Nova Scotia, Canada
Lorna MacPhee, Saint Mary's University, Halifax, Nova Scotia, Canada
Chris Miller, Brenau University, Gainesville, GA, USA
Patricia O'Brien, University of Vermont, Burlington, VT, USA
Gary Patterson, Agriculture Canada, Truro, Nova Scotia, Canada

Jaques Ribemboim, Department of Agriculture, Director, State of Pernambuco, Recife, PE, Brazil
Lynne Scott, School of Mathematics, University of Southern Australia, Mawson Lakes, Australia
Helena Urbano, Av. Boa Viagem 328, Recife, PE, Brazil
Sara J. Wilson, GPI Atlantic, Halifax, Nova Scotia, Canada

Work Group: Ecosystem Health and Human Health

Put O. Ang Jr., Department of Biology, The Chinese University of Hong Kong, China
Dave Cote, Terra Nova National Park, Canada
Le Dien Duc, Vietnam National University, Hanoi, Vietnam
Job S. Ebenezer, Evangelical Lutheran Church, Chicago, USA
Dean Fairbanks, University of Pretoria, South Africa
Bob Ford, Fredrick Community College, USA
Rob Gordon, Nova Scotia Agricultural College, Canada
Yang Guang, Nova Scotia Agricultural College, Canada
Judith Guernsey, Dalhousie University, NS, Canada
Abduel Hadi Harman Shaa, University Kebangsaan Malaysia, Malaysia
Andrew Hamilton, Commission for Environmental Cooperation, Canada
William Hart, DalTech, Dalhousie University, Canada
Huasheng Hong, Xiamen University, China
Jennifer Hounsell, International Society for Ecosystem Health, Canada
John Howard, The University of Western Ontario, Canada
Banquin Huang, Xiamen University, China
Yanhe Huang, Fujian Agricultural University, China
Dwikorita Karnawati, Gadjah Mada University, Indonesia
Robert Lannigan, The University of Western Ontario, Canada
Sharon Lawrence, Aquatic Ecosystem Health and Management Society, Gaia Project, Canada
Yan Liu, Xiamen University, China
Diane Malley, Aquatic Ecosystem Health and Management Society, Canada
Leanne McLean, Dalhousie University, Canada
Robert McMurtry, Health Canada and The University of Western Ontario, Canada
Vincent Mercier, Environment Canada, Canada
Naoki Mori, University of Tokyo, Japan
Mohi Munawar, Aquatic Ecosystem Health and Management Society, and Fisheries and Oceans Canada
Mary Ann Naragdao, University of the Phillipines, Phillipines, and Dalhousie University, Canada

Katsuo Okamoto, National Institute of Agro-Environmental Sciences, Japan
Daniel Rainham, Dalhousie University, Canada
D.J. Rapport, University of Guelph, College Faculty of Environmental Design and Rural Development, Guelph, ON, Canada, and The University of Western Ontario, Faculty of Medicine and Dentistry, London, ON, Canada
Dieter Riedel, Health Canada
Elizabeth Rodriguez, Parque Nacional Mirador del Norte, Dominican Republic
Meenu Saraf, Gujarat University, India
Helene Savard, Sir Sandford Fleming College, Canada
Paul Schaberg, USDA Forest Service, Northeastern Research Station, Burlington, VT, USA
Neil Scott, University of New Brunswick-Saint John, Canada
Annabelle Singleton, Saint Mary's University, Canada
Risa Smith, British Columbia Ministry of Environment, Lands and Parks, Canada
Harold Taylor, The International Coalition for Land & Water Stewardship in the Red River Basin, Canada
Nguyen Thi Hoang Lien, Hanoi University of Science, Vietnam
Liette Vasseur, University of Moncton, Canada
Shihe Xing, Fujian Agriculture University, China
Hoang Xuan Co, Hanoi University of Science, Vietnam

Contents

Preface	v
EcoSummit Participant List	vii

Introduction: Understanding and Solving Environmental Problems
in the 21st Century: Toward a new, integrated "hard problem science"
R. Costanza and S.E. Jørgensen 1
References 3

Chapter 1
Integrated Assessment and Modeling – Science for Sustainability
G. Harris 5
Abstract 5
1. Introduction 6
2. The global context 6
3. Changes to the enterprise of scientific research 7
4. "Clean and green" drivers on policy and markets 10
5. The science of IA and IAM – integration and synthesis 11
6. IAM and ESM – the science of the future? 14
References 16

Chapter 2
The Potential for Integrated Assessment and Modeling to Solve
Environmental Problems: Vision, Capacity, and Direction
P. Parker, R. Letcher, A. Jakeman, with M.B. Beck, G. Harris, R.M. Argent,
M. Hare, C. Pahl-Wostl, A. Voinov, M. Janssen, P. Sullivan, M. Scoccimarro,
A. Friend, M. Sonnenshein, D. Barker, L. Matejicek, D. Odulaja,
P. Deadman, K. Lim, G. Larocque, P. Tarikhi, C. Fletcher, A. Put,
T. Maxwell, A. Charles, H. Breeze, N. Nakatani, S. Mudgal, W. Naito,
O. Osidele, I. Eriksson, U. Kautsky, E. Kautsky, B. Naeslund, L. Kumblad,
R. Park, S. Maltagliati, P. Girardin, A. Rizzoli, D. Mauriello, R. Hoch,
D. Pelletier, J. Reilly, R. Olafsdottir, S. Bin 19
Abstract 19
1. Introduction 20

2. Integration — 21
3. Visions of the future — 23
   3.1. Optimistic view — 23
   3.2. Pessimistic view — 24
4. Evolution of IAM — 25
5. IAM current position: Points of agreement — 26
6. Case studies — 28
7. Model complexity — 29
8. Validation — 29
9. Integrated Assessment — 31
10. Agent-based models — 32
11. Communication — 33
12. Values in models — 34
13. Future of IAM — 34
14. Links to other groups — 36
15. Conclusion — 37
References — 38

Chapter 3
Complex Adaptive Hierarchical Systems
B.C. Patten, B.D. Fath, J.S. Choi, S. Bastianoni, S.R. Borrett,
S. Brandt-Williams, M. Debeljak, J. Fonseca, W.E. Grant, D. Karnawati,
J.C. Marques, A. Moser, F. Müller, C. Pahl-Wostl, R. Seppelt,
W.H. Steinborn and Y.M. Svirezhev — 41
Abstract — 41
1. The new confrontation – biocomplex wholeness — 42
   1.1. Complexity — 43
   1.2. Adaptation and hierarchy — 46
2. Measuring the organizational complexity of CAHSystems — 53
   2.1. Measuring the complexity of genomes and organisms – biocomplexity — 54
   2.2. Exergy-based orientors in a natural (virgin) forest — 57
   2.3. Exergy and information of solar radiation — 62
   2.4. Emergy and exergy — 65
   2.5. Integration of orientors — 69
   2.6. Adaptation and hierarchy, again — 72
3. Systemicity — 73
   3.1. Eco-anthropic CAHSystems — 73
   3.2. A polycentric approach to integrated assessment — 75
   3.3. Eco-geological assessment towards sustainable coastal development in Yogyakarta, Indonesia – scale adjustment to observe and analyze CAHSystems — 77

4. Concluding thoughts 82
Glossary 82
References 87

Chapter 4
Complex Adaptive Hierarchical Systems
B.C. Patten, B.D. Fath, J.S. Choi, with S. Bastianoni, S.R. Borrett,
S. Brandt-Williams, M. Debeljak, J. Fonseca, W.E. Grant, D. Karnawati,
J.C. Marques, A. Moser, F. Müller, C. Pahl-Wostl, R. Seppelt,
W.H. Steinborn and Y.M. Svirezhev 95
Abstract 95
1. Introduction 95
2. About theory 96
3. About applications 97
4. About modeling 98
5. Conclusions 99
References 99

Chapter 5
Ecosystem Services, Their Use and the Role of Ecological Engineering:
State of the Art
A. Dakers 101
Abstract 101
1. Introduction 102
2. Defining ecosystem services 103
3. Humankind's relationship with the natural environment 105
   3.1. Anthropocentric or ecocentric valuing 105
   3.2. Ecosystem relationships – embeddedness 105
   3.3. Valuing ecosystem services 109
4. The use and misuse of ecosystem services 109
5. Designing and engineering to restore a sustainable relationship with ecosystems 110
   5.1. Making better use of ecosystem services 110
   5.2. Frontline projects 111
   5.3. Players involved in achieving better use of ecosystem services 111
   5.4. Engineer as designer 112
6. Ecological engineering 114
   6.1. Case Study 1: Ministry of Transport in the Netherlands 115
   6.2. Case Study 2: Oxelösund Våtmark, Sweden 116
   6.3. Case Study 3: Donaumoos – Germany 116
   6.4. Case Study 4: Kaja, Ås, Norway 116
   6.5. Case Study 5: Aremark 117

6.6. Case Study 6: Kågeröd Recycling Project ... 117
6.7. Case Study 7: Ruswil, Switzerland ... 117
6.8. Case Study 8: Calcutta Wastewater-fed Aquaculture ... 118
6.9. Case Study 9: Stensund Aquaculture Centre ... 119
6.10. Case Study 10: Water Enhancement Programme, Christchurch ... 120
6.11. Case study evaluation ... 121
7. Under-utilization of ecosystem services ... 122
8. Conclusions ... 123
Acknowledgements ... 124
References ... 124

Chapter 6
Ecosystem Services
B. Guterstam, A. Werker, with M. Adamsson, D. Barker, A. Brüll, A. Dakers, S. Gossling, J. Heeb, S. Loiselle, U. Mander, D. Melaku Canu, R. Roggenbauer, M. Roux, G.D. Santopietro, D. Stuart, M. Trudeau, H.D. van Bohemen ... 127
Abstract ... 127
1. Introduction ... 127
2. Ecosystem services ... 129
3. Key questions and common ground ... 132
4. The role of ecosystems services tomorrow ... 134
5. Conclusions ... 137
References ... 137

Chapter 7
Science and Decisionmaking
V.H. Dale ... 139
Abstract ... 139
1. Science and decisionmaking ... 140
2. Scientists' role in decisionmaking ... 140
3. Three case studies ... 141
   3.1. Mount St. Helens ... 141
   3.2. Tennessee Cedar Barrens ... 143
   3.3. The Brazilian Amazon ... 144
   3.4. Lessons learned ... 145
4. Characteristics of scientists and decisionmakers influence how they interact ... 146
5. Questions about the relationship between science and decisionmaking ... 149
Acknowledgements ... 150
References ... 151

Chapter 8
Science and Decisionmaking
E.J. Rykiel Jr., with J. Berkson, V.A. Brown, W. Krewitt, I. Peters, M. Schwartz, J. Shogren, D. Van der Molen, R. Blok, M. Borsuk, R. Bruins, K. Cover, V. Dale, J. Dew, C. Etnier, L. Fanning, F. Felix, M. Nordin Hasan, H. Hong, A.W. King, N. Krauchi, K. Lubinsky, J. Olson, J. Onigkeit, G. Patterson, K.S. Rajan, P. Reichert, K. Sharma, V. Smith, M. Sonnenschein, R. St-Louis, D. Stuart, R. Supalla and H. van Latesteijn   153
Abstract   153
1. Introduction   153
   1.1. Working definitions   153
   1.2. Multiple roles of science   154
   1.3. The role of scientists in controversial issues   155
   1.4. Scientists and activism   156
   1.5. Education of scientists   156
   1.6. Science based on holism   157
2. Increasing the effectiveness of the individual environmental scientist   158
   2.1. Pathways to involvement in decisionmaking   158
   2.2. Changing the environmental science curriculum   159
3. Case studies   160
   3.1. Integrating science and economics for environmental policymaking in Europe   161
   3.2. Lake management and demand-driven research in the Netherlands   162
4. Conclusions   163
Acknowledgements   164
References   165

Chapter 9
Ecosystem Health and Human Health
L. Vasseur, D.J. Rapport, J. Hounsell   167
Abstract   167
1. Introduction   167
2. Ecosystems, humans, and the concept of health   168
   2.1. Ecosystems   169
   2.2. Health   170
3. Ignoring the link   172
4. Climate change   175
5. Agrosystems and food production   178
6. Biodiversity and declining productive capacity   181
7. Discussion   183
Acknowledgements   185
References   185

Chapter 10
Ecosystem Health and Human Health: Healthy Planet, Healthy Living
L. Vasseur, P.G. Schaberg, J. Hounsell, with P.O. Ang Jr., D. Cote, L.D. Duc,
J.S. Ebenezer, D. Fairbanks, B. Ford, W. Fyfe, R. Gordon, Y. Guang,
J. Guernsey, A. Hadi Harman Shaa, A. Hamilton, W. Hart, H. Hong,
J. Howard, B. Huang, Y. Huang, D. Karnawati, R. Lannigan, S. Lawrence,
Z. Li, Y. Liu, D. Malley, L. McLean, R. McMurtry, V. Mercier, N. Mori,
M. Munawar, M.A. Naragdao, K. Okamoto, D. Rainham, D. Riedel,
E. Rodriguez, M. Saraf, H. Savard, N. Scott, A. Singleton, R. Smith,

H. Taylor, N.T. Hoang Lien, S. Xing and H. Xuan Co	189
Abstract	189
1. Introduction	190
2. Linkages between ecosystem health and human health	193
2.1. Air quality	193
2.2. Water resources	194
2.3. Food resources	196
2.4. Soils	197
2.5. Biodiversity	197
2.6. Other models	198
3. Sources of solutions	199
4. Priority actions	202
5. Barriers to effective action	205
5.1. Sustenance needs	205
5.2. Little connection to the land	206
5.3. Resistance to change	206
5.4. Ignorance	207
5.5. Low critical mass	207
6. Measures, indicators, or metrics of progress	208
7. Conclusions	213
Acknowledgements	217
References	217

Chapter 11
Quality of Life and the Distribution of Wealth and Resources

R. Costanza, J. Farley and P. Templet	221
Abstract	221
1. How is Quality of Life (QOL) defined?	221
2. How has Quality of Life been measured?	225
2.1. Economic income, economic welfare, and human welfare	225
2.2. Level and pattern of economic activity: gross national product	225
2.3. Sustainable economic income	228
2.4. Measuring economic welfare	231

    2.5. Assessing human welfare directly   234
3. A comparison of two approaches to fairness in the distribution of wealth
    and resources   237
    3.1. Fairness across individuals in space   237
    3.2. Fairness across individuals in time   240
    3.3. Fairness across countries in space and time   247
4. Can we measure fairness?   248
5. What is the relationship between fairness and QOL?   251
6. Principles for achieving a sustainable, fair, and high-QOL society   253
References   255

Chapter 12
Quality of Life and the Distribution of Wealth and Resources
J. Farley, R. Costanza, P. Templet, with M. Corson, Ph. Crabbé,
R. Esquivel, K. Furusawa, W. Fyfe, O. Loucks, K. MacDonald,
L. MacPhee, L. McArthur, C. Miller, P. O'Brien, G. Patterson,
J. Ribemboim and S.J. Wilson   259
Abstract   259
1. How do we define Quality of Life (QOL)?   259
    1.1. What are human needs?   260
    1.2. Satisfiers and wants   260
    1.3. Implications of our definition for improving QOL   262
    1.4. QOL and the four capitals   263
2. How can we measure QOL?   264
    2.1. Are objective measures suitable?   264
    2.2. Operationalizing human needs assessment as a measure of QOL   266
    2.3. Ecosystem services: indicators to integrate with QOL   267
    2.4. The implications of using HNA as a measure of QOL   267
3. Development of indicators of fairness in the distribution of wealth and
    resources   269
    3.1. Natural capital and market failures   270
    3.2. The elimination of poverty   273
    3.3. Maximum income level   274
    3.4. Geographical fairness   275
4. Approaches to measuring fairness   277
    4.1. Ecosystem health and functioning markets   278
    4.2. Poverties and pathologies   279
    4.3. Wealth and power   279
    4.4. A Quality of Life Gini Coefficient?   281
5. Implications of the relationship between fairness and QOL   282
    5.1. Positional wealth   283
    5.2. Income inequality as a detriment to QOL   283

5.3. Do we still need incentives to produce?	284
6. How do we achieve sustainable, fair, and high QOL?	284
6.1. Current world setting	285
6.2. Policy suggestions	289
6.3. Natural capitalism, increased efficiency, industrial ecology, and dematerialization	290
7. Conclusion	297
Appendix. The Sustainability Bill of Rights	298
References	298

Conclusions
S.E. Jørgensen and R. Costanza	303
References	307
Author Index	309
Subject Index	319

Introduction: Understanding and Solving Environmental Problems in the 21st Century: Toward a new, integrated "hard problem science"

R. Costanza and S.E. Jørgensen

Existing social, economic, and political institutions, as well as academic disciplines, evolved at a time when natural resources and ecological services were vast relative to the human presence, and human impacts were relatively small and local in comparison. We have now moved from this relatively "empty world" to a world that is relatively full (Daly, 1992). In this new "full world", human impacts are more global and far-reaching, and the emphasis must shift from addressing problems in isolation to studying whole complex systems and the dynamic interactions between the parts. Complex systems are characterized by non-linearities, autocatalysis, complex, time delayed feedback loops, emergent phenomena, and chaotic behavior (Costanza et al., 1993; Kauffman, 1993; Patten and Jørgensen, 1995; Jørgensen, 1995). This means that the whole is significantly different from the simple sum of the parts, and scaling (the transfer of understanding across spatial, temporal, and complexity scales) is a core problem. Incorporating both biophysical and social dynamics makes these problems "wickedly complex" and difficult. They are impossible to address from within the confines of any single discipline.

To address these substantial challenges, we need to develop a new integrative approach to science, education, policy, and management that transcends existing disciplinary and other boundaries. We coined the term "hard problem science" (HPS) to refer to this new approach. It implies the following set of characteristics:

- **Consilience among all the sciences:** A balanced and pluralistic kind of "leaping together", in which the natural and social sciences and the humanities all contribute equitably, is needed. HPS needs to be truly transdisciplinary and multiscale, rather than either reductionist or holistic. One's discipline will be noted much as one's place of birth is noted today – where one started on life's journey, but not what totally defines one's life.

- **A balance between synthesis and analysis:** HPS research and education need to balance analysis and synthesis to produce not just data, but knowledge and even wisdom. This will enable vastly improved links with social decisionmaking.
- **A pragmatic philosophy built on complex systems theory and modeling:** The limits of predictability of complex, adaptive, living systems need to be recognized, and a "pragmatic modeling" philosophy of science needs to be adopted. This will allow new, adaptive approaches to environmental management and better links with social decisionmaking.
- **A multiscale approach:** A multiscale approach to understanding, modeling, and managing complex, adaptive, living systems needs to be the norm, and methods for transferring knowledge across scales need to be vastly improved.
- **A consistent theory of cultural and biological co-evolution:** A consistent theory of biological and cultural co-evolution needs to evolve and increase understanding of human's place in nature and the possibilities of designing a sustainable and desirable human presence in the biosphere.
- **A recognition of the central role of envisioning in science:** Envisioning and goal setting need to be recognized as critical parts of both science and social decisionmaking. We need to create a shared vision of a desirable and sustainable future, and implement adaptive management systems at multiple scales in order to get us there.

The remainder of this book fleshes out these characteristics of HPS, around six core themes that structured discussion at the conference which formed the basis for this book (the "EcoSummit" – see the preface for more on the process). Following this introductory chapter, there are 12 additional chapters (two for each theme) plus a conclusions chapter summarizing the overall results. The first chapter in each theme is a background and overview of the theme. A first draft of this chapter was delivered by the working group chairs at the opening plenary session of the EcoSummit. It was edited based on further input from the working groups. The second chapter in each theme was drafted at the summit by each working group. It represents the general consensus of the working group (and the broader summit participants) on the issues brought up in the background chapters. Each working group consisted of 20 to 50 participants, who are listed under each consensus chapter. They communicated via email after the EcoSummit to finalize their consensus.

The six themes are:
(1) Integrated assessment and modeling (IAM)
(2) Complex, adaptive, hierarchical systems (CAHS)
(3) Ecosystem services (ES)
(4) Science and decisionmaking (SDM)
(5) Ecosystem health (EH) and human health
(6) Quality of life (QOL) and the distribution of wealth and resources

In keeping with the nature of the problems being addressed, the background and consensus chapters are complex, integrated, transdisciplinary, and interconnected with each other. Taken together, we believe that they provide the basis for understanding and solving environmental problems in the 21st century and beyond.

References

Costanza, R., Wainger, L., Folke, C. and Mäler, K.-G., 1993, Modeling complex ecological economic systems: toward an evolutionary, dynamic understanding of people and nature. BioSci. 43: 545–555.

Daly, H.E., 1992, Allocation, distribution, and scale: towards an economics that is efficient, just, and sustainable. Ecol. Econ. 6:185–193.

Jørgensen, S.E., 1995, The growth rate of zooplankton at the edge of chaos: ecological models. J. Theor. Biol. 175:13–21.

Kauffman, S., 1993, The Origins of Order: Self-Organization and Selection in Evolution (Oxford University Press, New York).

Patten, B.C. and Jørgensen, S.E., 1995, Complex Ecology: The Part–Whole Relationship in Ecosystems (Prentice Hall, Englewood Cliffs, NJ).

Background

Chapter 1

Integrated Assessment and Modeling – Science for Sustainability

G. Harris

Abstract

In response to pressure from major natural resource management (NRM) crises around the globe, the nature of environmental science is changing rapidly. From research carried out by individual scientists and a focus on publication of results in the scientific literature, there is a rapid trend towards large-scale research projects carried out by interdisciplinary teams, which are focused on achieving outcomes. Liaison with government and the community is essential if science is to make difference on the ground through the adoption of changed management and farming practices. At larger and larger spatial scales, complex interactions between science, economics, and society are being explored in order to provide a sound basis for the analysis of a range of management techniques. Economic instruments, such as the valuation of ecosystem services, are being widely employed to provide improved NRM policies and practices. Various kinds of innovative modeling techniques are being employed to summarize knowledge and to provide forecasting capabilities including various forms of dynamic and agent-based modeling. The challenge lies in finding solutions to the problems of communication between a wide range of disciplines and communities and in building models which incorporate a range of data sets and types. The "social engineering" and networking ("soft" science) required to build and sustain teams is as onerous and important as the task of building and maintaining models of various kinds (the "hard" science). Ecological science is being challenged to contribute to the development of a more sustainable human society in terms of the "triple bottom line" – ecological, economic, and social sustainability. This is a major challenge for science in the new millennium.

Understanding and Solving Environmental Problems in the 21st Century
Edited by R. Costanza and S.E. Jørgensen
© 2002 Elsevier Science Ltd. All rights reserved

1. Introduction

The way we do scientific research is changing – and changing rapidly. Those of us who have been doing environmental research for more than thirty years have seen major changes in the way research is performed. From the days of pure research carried out by individuals or small teams, we have moved into an era of massive interdisciplinary research teams tackling applied problems of global import. Governments and society now require more accountability; they require not just outputs (scientific papers), but also outcomes – science must make a difference. This is reflected in the scope and title of this chapter. Integrated assessment implies that not only is the science exemplary but that it is now being done in the context of the social and economic forces at work in society. Science is being used to generate useful knowledge and significant outcomes in a social and economic context. Knowledge is now used for prediction, for commercial gain and as a guide to action through "evidence-based" policy. Science and society are debating the needs and requirements of each other – and not always successfully.

2. The global context

As we move into the new millennium, we have begun to realize that society faces major global environmental problems: emissions of greenhouse gases, global warming, loss of biodiversity, acid rain, and perturbations in the global cycles of major nutrients to name but a few. As we begin to search for ways to respond, we are beginning to suffer from a plethora of acronyms: IA, integrated assessment; IAM, integrated assessment and modeling; ESM, earth system modeling; NRM, natural resource management; ESD, ecologically sustainable development. All these reflect the need to mobilize scientific, economic, and sociological tools to address some of the most important issues of our day (Bailey, 1997; Velinga, 1998). There is now a major focus on the "triple bottom line" or sustainability in terms of ecological, sociological, and economic factors.

Each of these lines of inquiry use the latest scientific tools (e.g., computer modeling, systems simulation, remote sensing from satellites, the World-Wide-Web, and other forms of information technology) to assemble, integrate, and synthesize data from a wide range of sources, at a wide range of scales. These data are then combined with economic and other models to attempt to assess the interactions and surprises that are likely to occur. Prediction, and the ability to try out some different "what if?" scenarios, is essential if we are to plot a course away from global disaster (Schellnhuber, 1999). There are huge intellectual challenges here, which I shall discuss.

Not the least of the challenges for the new millennium is the role that science plays in some of the political and policy debates raging over international conventions such as those on the control of greenhouse gas emissions, and on biodiversity and NRM. This is applied science and international politics being played out at a grand scale. Take, for example, the attempts through ESM to model the possible surprises and feedback in global warming caused by the increase of greenhouse gases in the atmosphere (Dowlatabadi, 1995) or the attempts to model the policy options for control of acid rain in Europe (Schopp et al., 1999). Recent reports from the United Nations and other bodies concerned with resource use and development, have warned of growing international conflict over resource use – particularly the use and extraction of trans-boundary resources like water (De Soysa and Gleditsch, 1999).

Taking stock of global resources (United Nations Development Program et al., 2000) has conclusively shown us that we are now doing significant damage to the fabric of the global ecosystem, including degrading terrestrial and aquatic systems which provide many important services. The human race has long neglected the important interactions between the human and the natural world and the critical role played by natural ecosystems in the sustainability of the planet. These "ecosystem services" as they are called, clean the air and water and provide a number of critical functions for planetary sustainability. There has been a recent attempt to place monetary value on the services (Costanza et al., 1997) and to bring them into the mainstream of economic accounting (see the special issue of the journal *Ecological Economics,* Vol. 29, 1999). It would seem that the value of these services far exceeds the global gross domestic product.

We must act – the cost of doing nothing is higher than the cost of action. But what are we to do? How do we best manage natural resources at regional scales? What are the best NRM policies in an era of rights markets and resource trading? How do we preserve biodiversity and restore natural ecosystems? How do we establish markets for ecosystem services? How do we approach ESD? These are the kinds of questions now being addressed. Horgan (1996) is wrong, this is not the end of science, it is not even the end of the beginning! To look into the future we need entirely new kinds of science and intellectual effort to integrate and synthesize data from disparate sources.

3. Changes to the enterprise of scientific research

While science was once the preserve of the privately wealthy, during the 1960s, governments around the world invested heavily in research and development (R&D) and a rising number of professionals joined the growing ranks of staff in universities and government R&D agencies. The great enterprise of science reached its zenith in the mid 1960s driven forward by the conviction that the "white hot heat of the technological revolution" (as it is was described by the British Prime Minister of the day) would lead to economic and environmental benefits such as those that science had provided to the winners of World War II. But already the signs of disillusionment were there to be seen (Dixon, 1973).

Now after such events as the atom bomb, Chernobyl, and other environmental disasters, there is much less public trust of the whole science and technology enterprise. As a result, and because of more difficult economic times and scarcer funds, governments and the public demand accountability and transparency – outcomes and adoption pathways rule. Environmental problems have become more complex, and transdisciplinary teams are required to tackle the complexities of NRM. The days of the lone boffin doing as she pleased are numbered. Science has moved into a state known as Mode II science (Gibbons et al., 1994) or "science in the context of its application".

Environmental science is now characterized by collaboration, partnerships, and strong linkages to government policy and commercial outcomes. Government policies throughout the western world now link science very closely to the whole process of innovation, not just the development of new ideas and techniques, but the application of these ideas and techniques and the generation of wealth, jobs, and global competitiveness. Increasingly, government policy is also being driven by global concerns – there is now tight linkage to international conventions on oceans policy, on biodiversity, and on global warming and greenhouse gas emissions to name just a few.

The upshot of all this is that R&D is being called on to produce systems solutions, solutions that link science to innovation and global economic policy, to global environmental concerns, and to regional development. And it is being expected to do this with fewer and fewer resources and with tighter and tighter controls on accountability and transparency, largely at the behest of those who have little understanding of the nature and culture of the scientific enterprise, and of the time and effort required to develop and maintain the necessary skills. Government policies around the globe are just as focused on turning what we already know into "useful knowledge" for industry and policymakers as they are in funding new research. Indeed, in some cases, there is a strong feeling that we need less research and more transfer of knowledge to the community. Little wonder that environmental R&D is being inexorably drawn towards Mode II science, to IA and IAM. Little wonder that the global modeling community is developing ESM as well as new global circulation models!

The much vaunted "purchaser–provider" model of science requires that the purchaser actually know something about that which he or she is purchasing. In my experience, particularly in some of the more complex areas of NRM, this is rarely the case. Many R&D managers do not understand what constitutes the "state of the art" at the time and frequently support second-rate science and suboptimal outcomes. In days of scarce resources we need to keep a very clear eye on what questions can be asked, what questions should be asked, and on priorities. These are often best understood by the scientific community, but trust has been so eroded that the purchasers will not listen to the scientific community, which is perceived to be "barrow pushing". After all, being perpetually stretched for resources the scientists always have a vested interest and are often cast in the guise of mendicant priests.

There is tension between the "hard" science of the researchers and modelers, and the "soft" science of those concerned with the community, adoption, and outcomes. The truth is, that at the turn of the millennium, we need both sets of skills. If we are going to turn around some of the evident environmental problems at regional and global scales then science and the community must come together. But the game is changing by the week as technology and society interact and new tools emerge. Thanks to the rapid development of computers, information technologies, and the Internet, we can now tackle and solve problems that would have been out of the question in the early days of my own scientific career.

Using techniques such as Geographic Information Systems, statistics, and simulation modeling, we can now model spatially distributed processes up to global scales. This was unheard of even a decade ago. On the "soft" side we are operating in a social and policy environment that is also very new. With the rise of the markets and the disappearance of the big regulatory NRM agencies, the advent of global environmental treaties and economic globalization, the rise of community concern and incentives for action, we are in a completely different world. I do not expect these trends to slow down – if anything I expect the rate of change to increase. When I hear my own staff in CSIRO complaining of the high level of social interaction and the networking "overhead" that so characterizes their work, I have to tell them that this networking is not an "overhead": it is a key part of what we do – after all we are here to make a difference.

So, in a few short years, we have gone from reductionist science and the publication of outputs (bricks?) in the scientific literature (e.g., Forscher, 1963), to useful knowledge and to dose–response models, through economic and innovation policies, to outcomes. This is a fundamental challenge to the whole enterprise of R&D. Integration, synthesis, outcomes, team work, collaboration, and partnerships are rarely included in the traditional model of career progress and the reward systems of the Academy. More and more of us now work in the environment where "the Minister wants an answer in 10 minutes!" – and if we are to influence government policy then the Minister must have her answer.

So the irony of it all is that at the turn of the millennium we are moving away from new investments in environmental R&D just at the time when the problems are getting larger and more urgent. The problems we must tackle are wickedly complex problems. These are problems of intermediate complexity, lying between science and society, showing emergence and deep structure – and including culture. These are the problems of understanding emergence in Complex Adaptive Systems (CAS) where the agents, both the natural world and ourselves, change their behavior as the problems emerge and grow. Mobbs and Dovers (1999) have assessed the peculiarly complex attributes of NRM problems – their complexity and intractability.

4. "Clean and green" drivers on policy and markets

Agriculture is a particularly good example of the wicked complexities of IA. Western agriculture has long depended on a relatively small number of crops, which have undergone long-term breeding and selection for high yields. These crops come from a relatively small number of plant types and are much less diverse than the highly diverse plant assemblages normally grown and harvested by indigenous peoples all around the globe. It is now clear that the western crops are not only demanding of water and fertilizers but they are also "leaky" in terms of water and nutrient use (Daily, 2001). Natural ecosystems show high biodiversity and are highly efficient at water use and nutrient recycling, leaking few nutrients into ground waters, streams, and rivers (Chapin et al., 2000). Because, especially in arid regions like Australia, water-use efficiency is at a premium and "leaky" western crops tend to lead to severe NRM problems like dryland salinity (Walker et al., 1999), there is increasing interest in the development of cropping systems which mimic the natural ecosystems of the region and provide similar services. There is a need for Integrated Assessments of the impact of agriculture (Bland, 1999).

One of the major drivers of human behavior and the application of IA in agriculture is the growing demand for "clean and green" or organic produce. Many of the practices of western agriculture are not sustainable and cause damage to the environment; organic farming is more sustainable (Tilman, 1998; Reganold et al., 2001). Amongst other things, agriculture has been responsible for overgrazing, increased rates of erosion, and large-scale changes in nutrient cycles, with increased nutrient loads to rivers and coastal ocean waters (Vitousek et al., 1997; Bland, 1999). Sustainable agricultural production is increasingly being demanded by consumers and is likely to figure significantly in future World Trade Organization (WTO) negotiations. Those who insist that the global economy must be seen as a "wholly owned subsidiary of the environment", if we are to live sustainably, are beginning to have an impact on the behavior of consumers. Three global trends are visible. First, the increasing demands by consumers for accreditation and environmental management systems on the part of producers. Second, the growing rejection by consumers of the corporatization of the multinational agri-business and its links with powerful government interests. This was brought into focus by the demonstrations surrounding the meeting of the WTO in Seattle in 2000. Third (and this is almost certainly a response to the previous two), the push by large multinational business into the area of sustainability and environmental stewardship.

The complexity of an IA for agriculture and other land uses is well illustrated by the recent problems of Bovine Spongiform Encephalopathy (BSE) and foot and mouth disease in the United Kingdom. Both BSE and foot and mouth disease have led to major changes in both the nature of animals farmed and their populations.

BSE caused a switch in production from cattle to sheep, while foot and mouth disease led to the elimination of large populations of grazing animals from the landscape. These changes to species composition and population densities led to alterations in grazing pressure and hence to erosion rates and water quality. More importantly foot and mouth disease led to the closure of large areas of rural UK where tourism was a much more significant source of revenue than farming. Suddenly agriculture was seen as merely one form of rural land use with high impact on the overall rural economy, on ecosystem services, and on conservation. A similar problem is found in Australia where agriculture is largely the cause of dryland salinity but is only one part of a series of rural land uses, including tourism, which require preservation of landscape function and biodiversity.

In the last few decades, the trend towards globalization of trade has been spurred on by changes in technology, the increasingly global span of financial trading, and by a number of trends in economic and environmental management. Trade protection is reducing, and more and more countries are being opened up to international flows of information, resources, and finance. Many multinational corporations now control key market sectors around the globe and there is an increasing focus on environmental management and NRM. Simultaneously, more and more governments are moving to a reliance on market economics for NRM. Economic instruments are a popular way to manage natural resources and to include the cost of ecosystem services in the overall equation (Costanza et al., 1997). This is both "let the market rule" – a more laissez-faire approach to a reduced role for government regulation in the market place – and an attempt to price ecosystem services into the true cost of production, so that, for example, users are required to pay for both the cost of supplying water and for the replacement of the necessary infrastructure. Australia has gone farther down the route of corporatizing and privatizing its water supply infrastructure than almost any other country. As a result of the success of this policy, plans are now afoot for the establishment of trading schemes for salt, nutrients, carbon, and ecosystem services (including biodiversity). A recent policy statement by the Australian Prime Minister proposes to spend $1.4B in twenty-one catchments over the next seven years to address dryland salinity using a mix of policy instruments including market economics.

5. The science of IA and IAM – integration and synthesis

Given the complexity of the problems outlined it is immediately clear that this is not the end of science (Horgan, 1996), it is not even the end of the beginning. But it is a new kind of science. The theme of sustainability and IA is inherently multi-sectoral and is characterized by technical indeterminacy and value multiplicity (Ravetz, 2000). We need both new scholarship (in the true sense of the word) and community interaction. We need a new combination of environmental R&D with social and economic research – a combination of the "hard" and "soft" approaches. In combination this is a whole new enterprise played out at regional and global scales. There is a need for those who can cross intellectual barriers, a need for "knowledge brokers". who can teach, translate jargon, and generate trust on all sides. Science must beware hubris and the "don't you worry about that" approach. A little humility goes a long way.

So just how do we integrate across disciplines and synthesize knowledge so as to produce useful outcomes? How do we do this in an environment where data sources have different types and degrees of error, where some data types from disparate disciplines are even incompatible? How do we keep the community on side and committed to change – and at the same time convince our political and economic masters to keep funding the whole enterprise? How do we do this in an environment where the technology is changing by the month, where there is total interpenetration between the information technology, the science, and the means of synthesis and delivery – and the finance gurus have decided to outsource the whole thing!? This is not rocket science; it is much more difficult!

There is a need for integration between scientists of differing disciplines, between science, economics, and society and between cultures. Even the first is difficult enough; no one should ever underestimate the difficulties of communication, even between scientists, let alone between science and the community. Add to this the difficulty of communication between cultures – both within and between countries. All this is being complicated by changing global cultures and the rise of post-modern ideas about science, values, and the nature of truth. It is indeed a brave new world.

How do we address the difficulties of combining and integrating knowledge between disciplines and the problems associated with assembling and applying models of various kinds? There are a range of approaches being used, from hard modeling and the use of supercomputers to simulate various possible futures at global scales (Dowlatabadi, 1995), to agent modeling where adaptive changes are allowed as time goes on (Janssen and de Vries, 1998; Janssen et al., 2000), to a large range of sociological tools and techniques (Syme et al., 1999).

I want to be a little provocative and make a plea for some old-fashioned scholarship in all this. While teams are needed to cross disciplines and provide solutions to complex problems (and science does require those who can manage the social interactions between scientists), nevertheless, in my experience, the real integration and synthesis often takes place between the ears of well-read individuals. Synthesis and integration requires a real in-depth knowledge and familiarity with a wide range of scientific literature. While new developments in information technologies (such as electronic journals and searchable databases) are a boon, it still requires a few people to sit down and go through the mountains of literature – through the random piles of scientific "bricks" (Forscher, 1963) – seeking the foundations of edifices and the outlines of new architects' plans. This takes time, patience, and no little skill. But time is something we rarely have and such skills are neglected and not often rewarded. If we really do want to integrate and produce new ideas and solutions, then the culture must be changed in ways to make this single-minded approach possible.

Around the world a number of new approaches to modeling and synthesis are being discussed, agent modeling, goal functions and systems simulation, the properties of CAS, and so on. All are brave attempts to seek more simple routes through the complexity we face. Not all will prove successful but anyone seeking a path through a maze will initially try a number of ways forward (Parson, 1995). Certainly there is a growing view that the complexity we face derives from the interaction (and adaptation) of many relatively simple agents (Holland, 1998) and that the complex pattern of systems behavior that we observe emerges from the underlying pandemonium (Harris, 1999). The question of debate is at what level can we represent reality, at what level can we model and predict with confidence? If the behavior of the real world is really like a CAS then there are emergent levels of abstraction that give us predictive power.

So there is a real link between IA and IAM and ESM. IA and IAM attempt to integrate across disciplines and to deal with the complexity of regional NRM problems. For example, dryland salinity in Australia where agricultural practice has led to land clearing, reduced biodiversity, "leaky" systems, and mobilization of salt in the landscape. The NRM problem is to replace the present system of agriculture with a system which is economically productive and profitable at both farm enterprise and regional scales whilst preserving and enhancing the biophysical function of the landscape and the native biodiversity. ESM, on the other hand, starts at biospheric scales with models of the global atmospheric and oceanic circulation, and of rising greenhouse gases and global warming. Added to these models are now scenarios for the effects of this on the biosphere, on biodiversity and on the social and economic impacts of global warming. This is a "top down" approach to similar problems as those listed above. Both meet at regional scales, because large-scale NRM problems must be solved at regional scales and because any implementation of policies to prevent and mitigate global warming must be implemented at regional scales.

Both types of models confront problems of representation and abstraction across different disciplines where data are usually quantitatively and qualitatively different and where perceptions of reality and importance differ also. Through all this, we must remember that there is an inverse relationship between model complexity and its probability of use. There is a real need for simple, robust heuristics, which can be used to support decisionmaking and policy development. This is a major challenge. Scientists must remember that in the real world, decisions are going to be taken come what may – the least they can do is to try to influence the decision that is taken and not have it done in a "content-free" way.

The extension of predictive modeling to realms outside quantitative environmental science is also a major challenge. In all cases, the predictions move us outside the known parameter space of both present conditions and the models themselves. This means that we are predicting out into unknown territory, environmentally, economically, and socially. There are approaches to this, using "tolerable windows" and knock out models (Schellnhuber, 1999) and other ways to constrain the scenarios, but nonetheless, this is dangerous territory. Here be monsters! Devotees of the BBC "Yes, Minister" TV series will recognize this as a "courageous" attempt to bring disparate disciplines into a common (often economic) framework and to make predictions about presently impossible economic and social behavior.

In all cases, what the modelers and the policymakers seek is advance warning of hysteresis effects – points of no return in the predictive landscape. We already know of some such effects. One particularly well-known example is the concept of the critical nutrient load for lakes and estuaries (Harris, 1997; Scheffer, 1998). Once this critical load is exceeded, nutrient enrichment produces practically irreversible changes to the aquatic ecosystem, such that recovery is made almost impossible. What the IAM and ESM modelers seek are similar points of no return at regional and global scales. Even though the greenhouse effect is producing only slow and apparently steady warming, are there surprises in store? Will there be similar points of no return arising from unexpected feedback in the biosphere? Obviously it is important to avoid such situations if at all possible.

6. IAM and ESM – the science of the future?

Integrated Assessment is therefore a way of systems thinking; a way of trying to balance the "triple bottom line" of ecology, economics, and sociology. While some argue that ESD is an oxymoron and that we are already into a regime of "managed sustainability", we nonetheless have to give it a go because we have no choice. We are attempting restoration ecology at massive scales (Schrope, 2001). This is science in a new social context – science done "softly". Scientists must recognize the absolute importance of community trust and action. After all it is not us who will make the decision, but the whole of society.

In Australia we have learned some very important lessons from the LANDCARE movement, a rural community movement to repair damaged land and to deliver NRM at local scales. But we now know that the NRM problems we face are bigger than can be tackled by local voluntarism alone. We need a new pact between science, government, and industry to tackle the problems of landscape destruction, the loss of biodiversity and ecosystem services in our rural lands (Daily, 2001). We need more people to realize that the economy really is a wholly owned subsidiary of the environment. What price are an information technology revolution and an economic boom if the air is unbreathable and the water undrinkable?

We need to develop social process and a deeper understanding of the working of the biosphere, which results in ESD. We must remember that many of our limitations arise from muddled thinking and deep cultural biases, which in many cases have a distinctly northern-hemisphere origin and bias. This is not a criticism but a timely reminder that we carry much historical cultural baggage which does not always help us face reality. As humans of particular size and life span we are better at perceiving some types of problems than others. Quite often the truth is staring us in the face but for some peculiarly human reason we do not see it. IAM and ESM are no different; they are peculiarly human and culturally determined attempts to solve a highly complex global problem with the tools and skills available at the end of the millennium. As technology and knowledge expands there will be others.

Governments around the world are debating the pros and cons of trading regimes, incentives, and other market instruments. Gone are the days of regulation as the means to the same end. We merely attempt to think through the policy implications of present knowledge but seen through a filter of the present state of science and of policy development. This is a very contingent world. What science must do is to continually explore the limits of present knowledge, and the limits to action, incentives, and policy. Science is part of society but is an agent for change. This is an uncomfortable marriage.

This is new science, new economics, and new policy for a whole new scale of NRM problems – the problems of regional sustainability and landscape management. Above all, science must remember that there are always moral and ethical aspects to human behavior. Is the use of economic instruments and rights markets the most moral and ethical way to manage the biosphere? Does the application of science itself always produce the most moral and ethical outcomes? The answer to both questions must be no. We talk of environmental management or even stewardship – I prefer the term compassion, and so would many others.

Science, scientists should remember, is the ultimate rationalist argument. We often forget that we live in a world that has largely rejected rationalism and positivism – we live in a time of radical doubt and radical certainty. Perceptions of risk and reward, of greed and fear dominate our lives. There is a need to build a new narrative, which combines the best of science with the best of human nature. This requires science to engage society. What IA, IAM, ESM and ESD are about is an attempt to build a sustainable future for us all, and that includes the best possible outcomes, the inclusion of moral and ethical considerations, as well as the "triple bottom line". IA is the key debate about global futures, it is challenging and essential. This is definitely NOT the end of science, but the science we do will be different in many important ways. It is a vision splendid for the new millennium.

References

Bailey, P.D., 1997, IEA, a new methodology for environmental policy? Environ. Impact Assess. Rev. 17:221–226.
Bland, W.L., 1999, Toward integrated assessment in agriculture. Agric. Syst. 60:157–167.
Chapin III, F.S., Zavaleta, E.S., Eviner, V.T., Naylor, R.L., Vitousek, P.M., Reynolds, H.L., Hooper, D.U., Lavorel, S., Sala, O.E., Hobbie, S.E., Mack, M.C. and Díaz, S., 2000, Consequences of changing biodiversity. Nature 405:234–242.
Costanza, R., d'Arge, R., de Groot, R., Farber, S., Grasso, M., Hannon, B.M., Limburg, K., Naeem, S., O'Neill, R.V., Paruelo, J., Raskin, R.G., Sutton, P. and van den Belt, M.J., 1997, The value of the world's ecosystem services and natural capital. Nature 387:253–260. See http://www.floriplants.com/news/article.htm.
Daily, G.C., 2001, Ecological forecasts. Nature 411:245.
De Soysa, I. and Gleditsch, N.P., 1999, To Cultivate Peace: Agriculture in a World of Conflict. PRIO report 1/99 (International Peace Research Institute, Oslo) 90 pp.
Dixon, B., 1973, What is Science For? (Pelican Books, London) 284 pp.
Dowlatabadi, H., 1995, Integrated assessment models of climate change. Energy Policy 23:289–296.
Forscher, B.K., 1963, Chaos in the brickyard. Science 142:339.
Gibbons, M., Limoges, C., Nowotny, H., Schwartzman, S., Scott, P. and Trow, M., 1994, The New Production of Knowledge (Sage Publications, London) 179 pp.
Harris, G.P., 1997, Algal biomass and biogeochemistry in catchments and aquatic ecosystems: scaling of processes, models and empirical tests. Hydrobiology 349:19–26.
Harris, G.P., 1999, This is not the end of limnology (or of science): the world may well be a lot simpler than we think. Freshw. Biol. 42:689–706.
Holland, J.H., 1998, Emergence: From Chaos to Order (Addison-Wesley, Reading, MA) 258 pp.
Horgan, J., 1996, The End of Science (Abacus, Little Brown and Co., London) 324 pp.
Janssen, M.A. and de Vries, B., 1998, The battle of perspectives: a multi-agent model with adaptive responses to climate change. Ecol. Econ. 26:43–65.
Janssen, M.A., Walker, B.H., Langridge, J. and Abel, N., 2000, An adaptive agent model for analysing coevolution of management and policies in a complex rangeland system. Ecol. Model. 131:249–268.
Mobbs, C. and Dovers, S., 1999, Social, economic, legal, policy and institutional R&D for natural resource management: issues and direction for LWRRDC. Occasional Paper 01/99 (Land and Water Research and Development Corporation, Canberra ACT).
Parson, E.A., 1995, Integrated assessment and environmental policy making. Energy Policy 23: 463–475.
Ravetz, J.R., 2000, Integrated assessment for sustainability appraisal in cities and regions. Environ. Impact Assess. Rev. 20:31–64.
Reganold, J.P., Glover, J.D., Andrews, P.K. and Hinman, H.R., 2001, Sustainability of three apple production systems. Nature 410:926–930.
Scheffer, M., 1998, Ecology of Shallow Lakes (Chapman and Hall, London) 357 pp.
Schellnhuber, H.J., 1999, "Earth system" analysis and the second Copernican revolution. Nature 402 (6761) Supplement: C19-C23.
Schopp, W., Amann, M., Cofala, J., Heyes, C. and Klimont, Z., 1999, Integrated assessment of European air pollution control strategies. Environ. Softw. Model. 14:1–9.
Schrope, M., 2001, Save our swamp. Nature 409:128–130.
Syme, G.J., Nancarrow, B.E. and McCreddin, J.A., 1999, Defining the components of fairness in the allocation of water to environmental and human uses. J. Environ. Manag. 57:51–70.
Tilman, D.G., 1998, The greening of the green revolution. Nature 396:211–212.

United Nations Development Program, United Nations Environment Program, World Bank and World Resources Institute, 2000, World Resources 2000–2001: People and Ecosystems: The Fraying Web of Life (World Resources Institute, Washington, DC) 400 pp.

Velinga, P., 1998, European forum on Integrated Environmental Assessment (EFIEA). Glob. Environ. Chang. 9:1–3.

Vitousek, P.M., Aber, J.D., Howarth, R.W., Likens, G.E., Matson, P.A., Schindler, D.W., Schlesinger, W.H. and Tilman, D.G., 1997, Human alteration of the global nitrogen cycle: sources and consequences. Ecol. Appl. 7:737–750.

Walker, G., Gilfedder, M. and Williams, J., 1999, Effectiveness of Current Farming Systems in the Control of Dryland Salinity (CSIRO Land and Water, Canberra, ACT) 16 pp.

Consensus

Chapter 2

The Potential for Integrated Assessment and Modeling to Solve Environmental Problems: Vision, Capacity, and Direction

P. Parker, R. Letcher, A. Jakeman, with M.B. Beck, G. Harris, R.M. Argent, M. Hare, C. Pahl-Wostl, A. Voinov, M. Janssen, P. Sullivan, M. Scoccimarro, A. Friend, M. Sonnenshein, D. Barker, L. Matejicek, D. Odulaja, P. Deadman, K. Lim, G. Larocque, P. Tarikhi, C. Fletcher, A. Put, T. Maxwell, A. Charles, H. Breeze, N. Nakatani, S. Mudgal, W. Naito, O. Osidele, I. Eriksson, U. Kautsky, E. Kautsky, B. Naeslund, L. Kumblad, R. Park, S. Maltagliati, P. Girardin, A. Rizzoli, D. Mauriello, R. Hoch, D. Pelletier, J. Reilly, R. Olafsdottir, S. Bin

Abstract

To understand the environmental problems of the 21st century, teams of researchers and partners must work together to build integrated tools to analyze and represent the known processes underway. Environmental problems are caused by complex interactions among physical and human systems that demand analytical skills beyond those of a single discipline. Integrated assessment and modeling (IAM) is applied at the global level (e.g., climate change models) and at the local/regional level (e.g., watershed models). The central role of human decisions and actions as drivers of environmental change is recognized in new IAM approaches. The chapter discusses various definitions of IAM and identifies five different types of integration that are needed for the effective solution of environmental problems. The future is then depicted in the form of two brief scenarios: one optimistic and one pessimistic. The evolution of IAM and its current state are then briefly reviewed and examples of recent case studies are identified. Integrated assessment is introduced as a means to respond to the need for better indicators of sustainability. The issues of complexity and validation are recognized as more complex than in traditional disciplinary approaches. Communication is identified as a central issue both internally among team members and externally with decisionmakers, stakeholders, and other scientists. Links with other research groups are recognized and points of shared interest identified. Finally, it is concluded that the process of integrated assessment and modeling is considered as important as the product for any particular project. By learning to work together and recognize the contribution of all team members and participants, it is believed that we will have a strong scientific and social basis to address the environmental problems of the 21st century.

1. Introduction

Environmental problems in the 21st century are typically complex. They are caused by complex interactions among physical and human systems that demand analytical skills beyond those of a single discipline. To understand these problems more fully, teams of researchers work together to build integrated tools to analyse and represent the known processes underway. The modeling of environmental processes has been undertaken for decades, but the need for integrated assessment and modeling (IAM) is heightened as the extent and severity of environmental problems in the 21st century worsens. The scale of IAM is not restricted to the global level as in climate change models, but includes local and regional models of environmental problems. The central role of human decisions and actions as drivers of environmental change is recognised in new IAM approaches. Earlier forms of systems modeling are being replaced with new integrated models that incorporate human components that facilitate scenario generation and decision support functions. This chapter presents the deliberations of forty-five scientists from Africa, Asia, Australia, Europe, and North America who are directly involved with and concerned about the direction of environmental research in the future. Their shared vision and advice for future directions in research, application development, and communication in IAM are offered to encourage progress in the development of tools to help address future environmental problems. However, consensus was not reached on all points in this discussion and the chapter also identifies points where opinions differed and debates are expected to continue.

The chapter first discusses various definitions of IAM and identifies five different types of integration that are needed for the effective solution of environmental problems. The future is then depicted in the form of two brief scenarios: one optimistic and one pessimistic. The evolution of IAM and its current state are then briefly reviewed and examples of recent case studies are identified. Integrated assessment is introduced as a means to respond to the need for better indicators of sustainability. The issues of complexity and validation are recognised as more complex than in traditional disciplinary approaches. Communication is identified as a central issue both internally among team members and externally with decisionmakers, stakeholders, and other scientists. Links with other research groups are recognised and points of shared interest identified. Finally, it is concluded that the process of integrated assessment and modeling is considered as important as the product for any particular project. By learning to work together and recognise the contribution of all team members and participants, it is believed that we will have a strong scientific and social basis to address the environmental problems of the 21st century.

2. Integration

Given the wide acceptance of the need for an integrated approach to environmental assessment and modeling, as well as to environmental management more generally, it may be somewhat surprising that there is no generally agreed upon definition of what constitutes integration, or more specifically, what is IAM. Risbey et al. (1996) state that the linking of mathematical representations of different components of natural and social systems in a computer simulation model is one way in which integration is undertaken. More broadly, a model is a simplification of reality. People think and communicate in terms of models. These may include
- Data models that are representations of measurements and experiments;
- Qualitative, conceptual models as verbal or visual descriptions of systems and processes;
- Quantitative numeric models that are formalizations of the qualitative models;
- Mathematical methods and models used to analyze the numeric models and to interpret the results;
- Decisionmaking models that transform the values and knowledge into actions.

Within the IAM process we attempt to integrate these various models in a transparent and interactive framework that allows for the participation of stakeholders in all the stages of the process. This framework offers a means to integrate the individual models of stakeholders at a variety of scales and it organizes the stakeholders' community by helping them communicate understanding, values, and concerns. Most important is not the unique model implementation that is developed, but the ongoing process of integrated assessment. As Risbey et al. (1996) stress, IAM is more than just a model building exercise, it is also a "methodology that can be used for gaining insight over an array of environmental problems spanning a wide variety of spatial and temporal scales".

This view of the importance of the process of IAM rather than just the model building aspect is supported more generally. Rotmans and Van Asselt (1996) provide a definition of IAM stating, "Integrated Assessment is an interdisciplinary and participatory process combining, interpreting and communicating knowledge from diverse scientific disciplines to allow a better understanding of complex phenomena". They stress the importance of integrated assessment models as frameworks to organise recent disciplinary research and note that the explicit purpose of IAM is to inform policy and to support decisionmaking. They state that IAM, as an intuitively based process, is not new and conclude that the new element in IAM is the use of integrated frameworks such as conceptual frameworks or computer based simulation models. Finally, they note the ideal state of IAM as an iterative process of investigation and recommendation, stressing the importance of communication. Communication not only of results from scientists to decisionmakers, but also of lessons learned by decisionmakers and the visions and views expressed by society, from stakeholders back to the scientist.

Margerum and Born (1995) in their discussion of Integrated Environmental Management note that while integration is a goal that is often strived for, in practice it is never truly achieved. This is an important observation for IAM, in that while the ideal of IAM may be difficult or even impossible to achieve, focusing on the process of IAM allows important lessons to be learnt. In many ways it is the process, rather than the outcome, which is of paramount importance in IAM.

To meet the need for integration or integrated scientific studies to address environmental issues, the various definitions of integration used by authors should be incorporated in a multi-dimensional form of IAM. The term 'integrated' has been used by different authors as describing various forms of integration, such as linking models with GIS (Geographic Information Systems), integrating software, or even stakeholder participation in the IAM process. At least five different types of integration can be identified within IAM. These integration types are illustrated in fig. 1. Issues are the centrepoint of integration as IAM seeks to avoid the fragmented approach traditionally adopted by science and recognises links between environmental issues and the need to include such interactions as part of the study. The integration of different issues is reinforced with the integration of multiple stakeholders as part of the research process. Conversely, models designed to depict processes at the local scale may not be appropriate to apply at a larger scale. Different disciplines may focus on different scales and this creates challenges for the integration of modules from these different sources.

The integration of different disciplines is required to gain insights into complex processes. Scale is important as IAM studies are often designed at one scale (e.g., global climate change), yet decision support is required at the local or catchment scale. Decision support requires application development where the general scientific model is embedded in an 'user-friendly' application to meet the needs of decisionmakers. The integration of models or linking of discrete modules is a common method adopted in IAM. Figure 1 shows that in IAM, a variety of *stakeholders, scales, disciplines* and *models* are integrated for the consideration of integrated environmental management issues. This is in many ways the ideal of IAM, whereas in practice one or more of these forms of integration may be ignored, often for good reason. In some cases, practitioners even argue about the necessity of models as a part of IAM, that is, about the role of non-model based integrated assessment. However, in its most comprehensive form, IAM contains all five elements of integration.

More broadly, the term 'integrated' is often used interchangeably with similar terms in the environmental management literature. Downs et al. (1991) reviewed

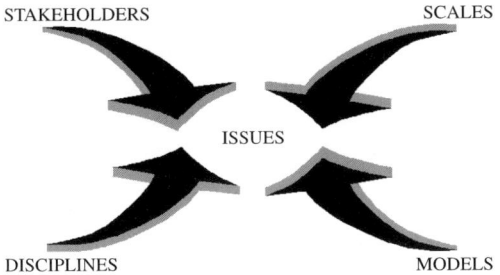

STAKEHOLDERS   SCALES

ISSUES

DISCIPLINES   MODELS

Fig. 1. Types of integration to address environmental issues

terms used when referring to Integrated River Basin Management. In particular, they found that four terms – comprehensive, integrated, ecosystem, and holistic – are used interchangeably to some extent, even though they have slightly different meanings. We accept the overlap in meanings of these terms and assert that IAM seeks to achieve multiple forms of integration in its approach to environmental issues.

3. Visions of the future

The use of IAM tools to solve environmental problems in the future could follow one of several paths. To illustrate the range in future possibilities, two brief scenarios are offered: the optimistic view and the pessimistic view.

3.1. Optimistic view

The optimistic view foresees new IAM tools being used to better manage the environment of our planet. The new tools successfully integrate insights from the natural and social sciences. The future 'knowledge-based' economy enables teams to use their brainpower to integrate models developed by scientist partners and to convert scientific models into practical application packages for use by the decisionmaker partners. The results from the modeling and testing of alternate future scenarios are used to depict scenarios that extend beyond the recent range of experience to assist in the evaluation of more extreme events or frequencies and to develop appropriate environment and resource management policies. These results are effectively communicated to politicians, decisionmakers, and community partners who then use the findings in their decisions about future actions. The overall pattern is one of integrated modeling, integrated application, integrated communication, and integrated decisionmaking. The new underpinning to this integrated approach is an integrated scientific education where the insights and skills of single disciplines are not left in isolation, but are set within the broader context of other disciplines and processes that also influence the overall pattern.

Environmental science, by being grounded in the complexity of real environmental problems, thus serves as the base for the development of a new integrated science education that incorporates rather than isolates core features. An enhanced environmental future is the expected outcome.

3.2. Pessimistic view

The pessimistic view foresees IAM tools failing to influence decisionmaking regarding the management of the Earth's environment. Fragmented, piecemeal, or single-purpose decisions prevail. The result is unintended consequences where harvest techniques reinforced with new technologies result in unsustainable harvests, population collapse, and rapid ecosystem change. The cumulative impacts of many independent decisions also create environmental problems on a scale and magnitude that were not foreseen by the individual decisionmaker. Following this pattern, major environmental catastrophes occur before society moves away from the dominant economic growth paradigm. Crisis-based reactions prove insufficient to address the complex problems that arise. In some cases, the species, ecosystem, or resource may be lost and consequently not be available for any form of assessment or management. The failure of the fragmented approach leads to the belated realization that an integrated response is needed. A shift is made to adopt an integrated approach, but valuable time has been lost and environmental problems worsened. The outcome is a recognised need for integrated approaches, but the environment has suffered increased damage and its natural capital has been eroded.

Some of the barriers to achieving the first scenario are built into our current global system and need to be addressed if the environmental damage of the second scenario is to be avoided. These barriers include
- Separation between scientists in different disciplines,
- Separation between scientists and application modelers,
- Separation between application modelers and software interface designers,
- Separation between scientists and decisionmakers,
- Separation between scientists and the community, and
- Fragmentation in education.

In each case, integration is proposed as a means to overcome the separation and fragmentation. A core question is how to teach the next generation of scientists to avoid the current fragmentation. The need is recognized for multidisciplinary education that focuses on collaboration and accepts responsibility for its vision of the future as value based. Dominant social values and welfare goals shape science so an open and honest approach is advocated to recognise these links. For example, by changing the social values, the objectives of environmental decisionmaking may change from maximum sustainable yield of species x to maintenance of species diversity in habitat y. The consensus opinion of the workshop participants is that an integrated approach is required to address environmental problems in the 21st century. The uncertainty discussed is over the timing and priority placed on developing the required integrated tools. To set the context for future development of integrated models and assessment approaches, a brief review of the evolution of IAM is provided.

4. Evolution of IAM

Integrated assessment and modeling has changed over recent decades both in terms of the approaches taken and in terms of the actors or key drivers promoting model construction. While the impact of humans on the physical environment has been recognised for many decades, the magnitude of the impact has grown rapidly since the 1950's. By the 1960's and 1970's, the magnitude of impacts was recognised as scientists, such as Forester, Meadows, and others, built Earth systems models (ESM) or conducted integrated applied systems analysis (IASA) to demonstrate the effect of continued growth, consumption, and pollution on the planet. Complex descriptions of global systems were obtained and macro prescriptions offered to avoid the predicted catastrophes.

These early models typically were driven by the scientists who built them and generally used empirical data at the global or continental scale. These models depicted aggregate patterns, yet despite the prescriptions offered, they failed to provide tools to decisionmakers who operated at the local or national level. As a result, the findings of these early models were largely ignored. Population and consumption levels continued to rise and the demands on the physical environment continued to grow. The failure of these early models to change society's actions was attributed to their lack of clear, problem-focussed objectives.

Changes in modeling approaches arose in part by changes in the financial support available. The funding priority and mechanisms of national governments changed to redirect funds to projects that were identified as important by end-users rather than supporting proposals by scientists in isolation.

Decisionmakers faced with allocation decisions and the need for policies to manage particular resources (water, forest, habitat, etc.) found the global models to be of little use and asked scientists to design models focused on their particular resource. The 1980's and 1990's witnessed the development of many models that integrated components of the resource issue faced by particular decisionmakers. The result was a broad array of models directed at the needs of particular resource managers. In this period the main drivers were decisionmakers who sponsored research directed to meet their particular needs. Scientists found themselves responding to the needs of particular clients with resources allocated to questions of the greatest financial interest rather than necessarily of the greatest scientific importance to better understand the system. More fundamentally, by focussing on the allocation of particular resources, the question of ecological sustainability was avoided and decisions could be made to allocate a particular resource in ways that met stakeholder objectives, but failed to meet broader sustainability objectives.

Many considered the emphasis to have shifted too far toward a narrow problem focus that missed important links to broader environmental and social systems. Even when the interests of multiple parties were recognised, the 'big' voices with money and power were typically heard more than the 'small' voices or interests without financial resources.

A new partnership for effective integrated assessment and modeling is called for in the future. Neither scientists nor resource managers should dominate the design and priorities of models to address environmental problems. A broader partnership is required. Examples of these broader partnerships can be found by exploring the current range of integrated models being developed. However, before identifying particular case studies, a review of the current state of IAM is provided.

5. IAM current position: Points of agreement

Effective management of environmental systems has been hampered historically by the lack of sufficient links between policymakers and scientists and by the difficulty of interpreting results from a wide variety of disciplines (Palmer, 1992; Syme et al., 1994; Park and Seaton, 1996). Integrated assessment and modeling (IAM) has been suggested as a solution to these problems, forcing closer links between decisionmakers and scientists and providing a framework for communication of results across disciplinary boundaries through to decisionmakers (Park and Seaton, 1996). More broadly, Born and Sonzogni (1995) see integrated environmental management, for which IAM is a generally applied tool, as an opportunity to broaden the purpose of management, overcoming the problems of previous approaches, which focused on only a portion of the environmental system and were implemented incrementally. Rotmans and Van Asselt (1996) see two distinct roles for IAM. First, as a way to keep track of how the pieces of the puzzle fit together and thereby offer a distinct advantage of indicating priorities for narrower disciplinary research. Second, in providing "an opportunity to develop a coherent framework for consideration of multiple objectives of decisionmaking and identification of possible policy criteria".

Rotmans and Van Asselt (1996) emphasise two main goals for IAM: that it should add value compared to insights derived from disciplinary research, and that it should provide decisionmakers with useful information. Consortium for International Earth Science Information Network (CIESIN) (1995) concludes that integrated assessment has a number of benefits including the ability to answer broad questions of how important the issue is and to assess potential responses to an issue. They suggest that integrated assessment can be used as a framework in which to structure present knowledge, helping to keep the entire problem in view and facilitating the search for possible responses. They maintain that this framework also provides a comprehensive structure for assembling, organizing, and communicating advances in knowledge as they occur and can help to structure uncertainty and sensitivity.

From our discussion we conclude that IAM in its current state has several distinctive features. IAM is currently a *problem-focused area of research*, with research often being *project-based*, and undertaken on a *demand-pull*, or *stakeholder needs* basis. IAM projects are generally undertaken to address specific sustainability or management issues, in contrast to previous systems modeling when research was often science driven and focused on providing complex systems descriptions and prescriptions for decisionmakers. IAM aims to be responsive to different groups of stakeholders, including client groups, government and policymakers, and community members and organisations. It combines the natural and social sciences to provide a broader view of both the system itself and the impediments to better management and sustainability, and seeks to enhance communication both between researchers and stakeholders, and among IAM participants.

We agree that the science behind IAM is often not new, rather the current state of IAM has been enabled by new technologies, so that in many ways it can be considered to be the combining of old areas of science and research to consider problems in new, more holistic ways. This more holistic approach to science raises additional problems, outside of those experienced in disciplinary research, which are the core issues requiring resolution and innovation in IAM research.

We consider that one of the main methodological problems in IAM is scale, and the resolution of different scales for different system components. Global system models differ greatly from local catchment models and creating links between models designed for different scales remains a major challenge. Even in catchment models, the boundaries and scale of hydrological aspects can be very different than those for socioeconomic aspects. Uncertainty and error estimation are also more challenging when modules based on different scales and different approaches are combined.

While there is no single way in which to perform IAM, the successes of individual research efforts in IAM in the past have been dependent on a number of common features. Janssen and Goldworthy (1996) discuss the importance of multidisciplinary research for natural resource management and the attributes of multidisciplinary teams and research efforts that are required for success. They emphasise the importance of the *team developing their own sets of norms and values*, aside from norms and values which each team member experiences within their own discipline, and the need for individual team members to be prepared to respect contributions from other disciplines and to view their disciplinary ability as a contribution to a joint goal. They see the problem-solving orientation of multidisciplinary teams and their shared objectives as the foundation on which these norms and values may be established. They conclude that due to the difficulty in imposing rigid scientific norms on multidisciplinary teams, there is a need for multidisciplinary teams to establish their own standards of excellence, from very basic things such as presentation, to more complex details such as module integration and error estimation. While the ideal qualifications for team members have not been well documented, Janssen and Goldworthy (1996) state that attitude, communication skills, education, and experience are all important attributes.

Another important aspect of IAM, which often affects the success of the IAM process, is the way in which stakeholder participation is managed throughout the project. Margerum and Born (1995), in their discussion of Integrated Environmental Management (IEM), note that for integration to be successful in IEM, "provisions must be made to include the fullest range of participants who accurately reflect the set of concerns of the public". This statement is equally applicable to stakeholder participation in IAM. They conclude that IEM, and we would argue IAM, "requires participants to: take a more inclusive view that considers the scope of environmental and human systems; examines interconnections; identify common goals; and selectively identify the key elements on which to focus attention". Morgan and Dowlatabadi (1996) describe a number of attributes, which they feel are the hallmarks of good integrated assessment. These include the characterisation and analysis of uncertainty as a central focus of assessments, the use of an iterative approach, and the recognition and inclusion of parts of the problem about which there is little information, by using techniques such as expert opinion, where formal models are inappropriate.

6. Case studies

The literature on IAM is growing rapidly and many studies are available. Rather than attempting to describe the many studies in detail, selected case studies, their model components, and study area are listed in table 1.

Table 1
Integrated Assessment and Modeling (IAM) case studies

Author	Date	Components	Study area
Grayson et al.	1994	Physical, biological	Latrobe River Basin, Australia
Scoccimarro et al.	1999	Hydrological, socio-economic, decision	River Basin, Thailand
Van Waveren	1999	Physical, biological	Water management, Netherlands
Voinov et al.	1999	Physical, biological, socio-economic	Patuxent River Watershed, USA

7. Model complexity

It is hard to imagine succeeding with mere simplicity in the form of models developed and employed in integrated assessment, unless this is in the use of regression relationships, for example, in one of the constituent knowledge domains (and where integrated assessment is being implemented largely without the use of models). We take it for granted, therefore, as the preceding discussion has shown, that all integrated assessments address complex situations, so that the IAMs developed for exploring such situations are necessarily complex, that is, they tend to be of a high order, with many state variables and dense interactions among those states. Furthermore, the goal of integrated assessment is not to enquire into the nature of complexity itself, in the popularly understood manner of the agenda of the Santa Fe Institute in the early 1990's (Waldrop, 1992). Rather, in the process of integrated assessment we may assemble the constituent, disciplinary parts of the overall model according to what is thought to be appropriate to the problem at hand (along the lines of what has been called "demand-side" modeling). The goal is not to end up with a model as a finished product (transferable to many like situations elsewhere), but to adapt it within the process of integrated assessment, as a vehicle of problem exploration, for instance, or as a device for communicating the relevant science to a lay audience. The resulting complexity is a blend of the complexity of the constituent disciplines (i.e., the behavior of an aquatic foodweb is not generally a simple thing) and the exigences of the stakeholders' hopes and fears for the future. For example, the investigation of the propagation of pathogens through a food web requires a greater degree of model complexity than the foodweb model itself, without even beginning to consider incorporating the simulation of human agency as it interacts with the environmental system. Complex interactions can generate counter-intuitive results and these need to be verified as valid before decisionmakers are willing to initiate prescribed actions.

Scale issues are extremely important when describing processes and in some cases, proper scaling may dramatically decrease complexity. The hierarchical/modular approach can be used so that the complexity in certain modules may not appear in higher hierarchical levels and thus simplify analysis and the interpretation of results.

So the real issues are: 1) can we comprehend what we have done with the model, sufficient to interpret its results in ways that will be understandable to the professional scientist and communicable to an audience of scientifically lay stakeholders; and 2) how valid, or trustworthy, has the model been as it has evolved throughout this process?

8. Validation

What constitutes validation of a model, without entering into what we now know is the vexed question of finding the right label for it (Oreskes, 1998), has arguably changed significantly in the past decade. At one time "history matching" and "peer review" were the two necessary and sufficient cornerstones of the process. But with the increasing difficulties of actually being able to match history and increasingly an absence of observed history to be matched (in part, because we move on to ever more extensive, more avowedly interdisciplinary problems), has come a dissatisfaction with the sufficiency of these conventional cornerstones (Beck et al., 1997; Beck and Chen, 2000).

In lay terms, the following are the essential, contemporary questions one would like to have answered in seeking to evaluate a model.

1: Has the model been constructed of approved materials, i.e., approved constituent hypotheses (in scientific terms)?
2: Does its behaviour approximate well that observed in the real system?
3: Does it work, i.e., does it fulfil its designated task, or serve its intended purpose?

We are familiar with peer review as the means of answering the first of these, and of matching history being directed at the second. The third question, of course, has always been vitally important, yet we have rarely been able to address it in a manner allowing us to discriminate between a process of model design, construction, and application deemed to be superior and another deemed inferior. Moreover, the purpose to which a model might be put may be quite varied, for instance, to provide

1: a succinctly encoded archive of contemporary knowledge for storage and retrieval;
2: a collation tool for allowing different sets of data to be viewed or examined together;
3: a tool for helping to develop an understanding of the system being managed and the types of interactions that exist between, for example, the social, economic, and biophysical sub-systems;
4: an instrument of prediction in support of decisionmaking or policy formulation;
5: a device for communicating scientific notions to and/or from a scientifically lay audience;
6: an exploratory vehicle for scenario building and the discovery of our ignorance.

The last pair of purposes (arguably of particular significance to integrated assessment) is notably not what one would normally expect of a model, at least not when considering how its design (and performance) should be evaluated. Significantly too, the terms embedded in these statements (archive, instrument, device, vehicle) evoke the image of the model as a tool, not a truth-generating theory, and thus prompt the insight of judging the trustworthiness of the model on such a basis, of whether the model is ill or well designed for its purpose (Beck and Chen, 2000).

Evaluating IAMs, therefore, is likely to be less dependent on the previous conventions of classical peer review and history matching, and more dependent upon protocols and tests yet to be developed. This is because they are distinctively defined as serving the needs of mostly lay stakeholders and, perhaps ever more frequently, of incorporating the human dimension of environmental problems. The constitution of a "peer group" can therefore be expected to be very different and more varied than just the former sub-groups of model builders and model users, steeped largely in the professional training and standards of science. While there may be scope for modest parts of the model to be evaluated by mono-disciplinary (scientific) peers, there will be few renaissance (wo)men capable of reviewing the whole and fewer still who will be able to claim no conflict of interest, as the model evolves over possibly many years in the light of successive reviews by these few. And whereas, likewise, there may be scope for parts of the model to be evaluated against part histories, it is highly unlikely that a "whole history" will be available for evaluating the whole IAM, considered as a single computational complex (even if it were technically possible to match a model to such a whole history, which arguably is no longer feasible; Beck and Chen, 2000).

In short, as apparent in Ravetz (1997), the path towards a "good practice" of evaluating IAMs and of integrated assessment itself is likely to be one the community of persons engaged in IA will have to construct itself, and uncomfortably so, precisely because of this. Insofar as it is the process, not the final product of an integrated assessment, for which a code of good practice is required, so too may it be that a protocol of validation (or evaluation) is needed more for the evolving structure and content of an IAM, as opposed to the eventually finished product. The difficulty and discomfort of this challenge notwithstanding, we should be encouraged by the fact that we are today armed with a wider pallette of metaphors and analogs (quality assurance in the design and construction of tools; quality assurance in controlling procedures in an analytical laboratory; the legal process; and, as promoted by Ravetz (1997), the discipline of historical analysis) with which to fashion a broader protocol for the conduct of evaluating IAMs.

9. Integrated Assessment

In addition to the integrated modeling described above, many researchers are engaged in the development of better sustainability indicators to provide an integrated assessment of societal performance and well-being. The dominant indicator of societal well-being is the GNP as a measure of income and our ability to purchase goods and services. The advantage of this indicator is that it is additive and new activity can simply be valued in dollar terms and added together. Many weaknesses in this measure have been identified in the ecological economics literature and more broadly (Costanza, 1991; Daly and Cobb, 1989;

Van den Bergh et al., 2000). Two types of responses have been made to develop more integrative assessment tools.

First, efforts have been directed to develop a socioeconomic measure that recognises many harms or negative impacts that are counted as additional goods by GNP measures simply because they stimulate economic activity. The genuine progress indicator (GPI) is an example of this approach. It subtracts the cost of economic activities that are not beneficial, such as traffic congestion, pollution-related health costs, prison costs, environmental clean-up costs, etc., from total economic activity. The approach was first used at the national level (Daly and Cobb, 1989) and is now being applied at the local level to assess community sustainability (Charles et al., 2001; Charles, 2001).

Second, efforts have been directed to counter the economic evaluation provided by GNP with a biophysical measure of the environmental consequences of society's activity. The ecological footprint is an example of this approach. It uses land as its basic unit of measure and converts society's demand for goods and services into the area of land required to produce those goods and services (Wackernagel and Rees, 1996). The conclusion reached from these studies is that most nations (and the aggregate global economy) have demands that exceed the ecological carrying capacity of their territory and that the natural capital of the environment is being depleted to maintain current economic output and consumption (Parker, 1998; Wackernagel et al., 1997). Clearly, the ecological consequences of economic activity need to be integrated into assessments of social well-being rather than simply rely on a single economic indicator such as the GDP.

In contrast to the single aggregate indicator (GDP, GPI, EF, etc.), many studies have developed series of sustainability indicators. These indicators typically cover several dimensions of sustainability: ecological, socioeconomic, community, institutional, and participation/communication. The performance of each category of indicator, and its sensitivity to thresholds, is as important as the overall measure because when one category (such as ecology) is not sustained, then the whole system is not sustained (Charles et al., 2001; Charles, 2001).

10. Agent-based models

The recognition of the central role of humans in environmental policy studies and contemporary IAM models leads to the revisiting of the behavioural assumptions that underlie the neo-classical economic model (Van den Bergh et al., 2000). These assumptions are important as they are often used as the basis for allocation decisions in many economic models or modules that are incorporated in IAM case studies. Agent-based models recognise a much wider range of motivators and behaviour patterns that stakeholders may adopt. For example, Hare and Pahl-Wostl (2002) model groundwater decisions by farmers by abandoning the traditional assumptions of perfect information and utility maximization. Instead, they depict farmers in a more realistic manner as agents who are non-identical, partially-informed, and pseudo-rational.

Agents are also influenced by their cultural setting. Rather than assume that the generic models can be used in any setting, models need to be modified to recognise local preferences and variations in choices and interactions. IAM processes often emphasize the link between modeling and participation. The formal specifications of agent preferences is one approach while Delphi techniques, expert meetings, and key informant/stakeholder participation have often been used for less formal integration. In all cases, communication is considered to be of utmost importance.

11. Communication

Communication was agreed to be the critical factor in the success or failure of integrated studies. To achieve integrated environmental management, integrated modeling, integrated assessment, or integrated knowledge, communication is required. Communication among members of the scientific team is just as important (and challenging) as communication between modellers and stakeholders. Communication within the scientific community is required to advance the collective pool of knowledge as well as to encourage review and dialogue.

The old idea that communication consisted of presenting the results to the public is no longer acceptable. A dialogue is required that ensures the exchange of information among parties. The traditional "prima donna" factor where scientists assumed that they had better knowledge and could simply present their conclusions is no longer accepted. New skills such as the importance of listening are required. The need for iterative dialogue is recognised to both improve the exchange of information and to build the trust relations among partners. Communication is important in all stages from initial project specification through to final exploration of results and decision support.

Different forms of communication may be required and used by different parties depending on the range of functions they are performing. For example, scientists may engage directly in the delivery of models to decisionmakers or they may have software designers who specialise in interface design undertake the design of the tool that enables decisionmakers to use the scientist's model. When many tasks are separated and undertaken by different parties, communication is essential to ensure that the final product achieves the initial objectives without compromising the integrity of the model as a whole.

Communication tools may be designed to simplify complex models to simple indices or indicators that are easily understood by the general public and decisionmakers. Graphic presentations may effectively summarise a large quantity of information. In each case, decisions need to be made that are appropriate for communication with the audience (scientific, policymaker, or general public).

12. Values in models

The different values represented by multiple stakeholder groups need to be recognised and explicitly included in models. Previous models often made allocations based on implicit value choices. In the future, the selection of values should be open and transparent. In this way, the effect of changing values can be seen in the allocation scenarios.

One point of continuing disagreement is cultural relativism. Although most voices in the working group called for an open and clear identification of the values incorporated in models, others argued that models are objective. The two views are articulated below.

- "All models have values throughout. The human dimension permeates everything. There is no pure science free of subjective opinions, we make choices about what we include and exclude".
- "The product of modeling is as culturally bound as other social products".

The counter argument was also voiced during the working group session.

- "No! Cultural relativism is limited, the physical word has constants".
- "We build rational (objective) models to be viewed and evaluated by peers (scientists) and stakeholders (public, etc.)".

In addition to concern over the challenge about values systems, concerns were also expressed in the working group about the conflict and debate over the numerical values attributed to particular variables.

- "Now everything is challenged, even atmospheric dispersion coefficients are challenged".

To preserve the integrity of science, the focus should be on transparency to allow for criticisms of models, their selected variables, and the values used. The accounting for environmental services and other features important to people should be made explicit. In this way, honest endpoints can be generated which are key policy endpoints where sensitivities can be examined to explore scenarios and inform decisionmakers. In this way decisionmakers will have confidence in the models knowing which parts are built on solid knowledge and what is dependent on particular values.

13. Future of IAM

The future of integrated assessment and modeling builds on the best elements from its past and adopts inclusive partnerships to address new environmental problems. The consensus among participating scientists is that models and the whole modeling process needs to be *open, honest, and transparent.* Communication is an integral feature of the process at all stages. It is accepted that other people view the world differently. The development of a model starts with a storyline or preliminary model. Qualitative as well as quantitative elements should be included as an iterative approach is taken to enable the model to evolve with improved understanding of the system under investigation. Failure is recognised as an important part of the learning process that helps refine the model and move research in an agreed (the right) direction. In particular, it was agreed that IAM requires a validation of processes and outcomes, rather than strictly of model outputs. It was seen that the first step in achieving this is the development of guidelines on methods and standards representing best practice in IAM.

New tools are needed to achieve integration, but the components required may change from one environmental problem to the next. Integrated models often include biophysical and socioeconomic components and some have decision support functions. However, the biophysical component may itself integrate modules, e.g., physical hydrology modules with biological modules, while the socioeconomic component integrates economic modules with psychological or sociological modules. The particular combination depends on the environmental problem being investigated such as water allocations within a catchment, fisheries, forest functions, wetlands, nuclear waste storage, etc.

An example of a challenging component that is receiving increased attention is agent-based modeling. While agents and their decisions are widely recognised as important to the outcome of models, the move away from simple optimization based on monetary values creates significant challenges. Agent-based modeling integrates social and psychological insights to generate more realistic models of the behaviours observed in society. Motivations are recognised as being more complex than was traditionally assumed and new allocation routines are designed and the outcomes studied.

A second set of integrated assessment techniques are designed to produce sustainability indicators that replace the single economic measure GNP (gross national product) with an indicator that integrates ecological, socioeconomic, community and institutional sustainability. The genuine progress index (GPI) is being developed as one measure to incorporate these elements. Biophysical measures such as the EF (ecological footprint) or TMR (total material requirements) have been developed to assess the total demands placed on the environment by human consumption. Recent work seeks to integrate ecological footprint analysis as part of future GPI measures. In this way, rather than having an indicator (GDP) that simply rises whenever production or consumption occurs, a new indicator differentiates between environmental and social goods vs. bads and reflects the decline in welfare created when pollution creates health costs or natural stocks are depleted. The challenges facing these integrated assessment approaches are similar to those of integrated modeling as processes may be non-linear and sensitivities to thresholds need to be identified even when the system has no historical record to provide data points for reference. Once again, the scenario generation approach can be used to extrapolate beyond the historical record to explore likely outcomes and to evaluate alternate decision paths.

14. Links to other groups

The integrated assessment and modeling group recognizes the overlapping interests they share with the other EcoSummit groups. Indeed some members chose to move between groups because of these overlapping subjects and the desire to contribute to more than one discussion. It was agreed that all the groups produced some kind of models and that other groups were also problem driven. Therefore, as we approach the problems being modelled, we must incorporate developments from other groups. Future models will need to incorporate new science. Uncertainty was expressed over whether this was the end of the era of mechanical models. Equally others questioned the new models by the Santa Fe group. Regardless of which new models became accepted, there was agreement on the need for communication and exchange among groups.

The Complex Adaptive Hierarchical Systems group has substantial overlaps as they work to develop new models and methodologies for the same environmental problems that IAM researchers examine. The established exchange of models to deal with complexity is expected to continue as scientists from both groups provide new models for use by the research community. New challenges that need better representation include the issue of criticality and how to identify and represent thresholds where systems may flip from one state to an entirely different, discontinuous state. The transfer of new models from physical systems to biological systems poses particular challenges. There are, however, many new developments in modeling, which show the potential for improved understanding, prediction, control, and forecasting of complex systems. These include catastrophe theory, chaos theory, artificial neural networks, and areas of artificial intelligence generally. One development worthy of mention is that of structurally dynamic models (Jørgensen, 1999a,b). These use goal functions to reflect changes in the properties of biological components of models, changes which are due to the adaptability of the biological components to prevailing conditions. By continuing the joint investigation of these and related issues, the research agenda of both the IAM working group and the Complex Adaptive Hierarchical systems working group will move forward.

The Ecosystem Services group focused on studies in ecological engineering that incorporate technological changes that enable ecosystem services to be provided by either natural capital or built capital. Debates over the substitution of different types of capital are central to the debate over "strong" sustainability goals and affect the processes of transformation that are represented in integrated models.

The Science and Decisionmaking group focussed on a communication issue that the IAM group identified as being of paramount importance. Scientific insight and the results of models are only effective if they can be communicated and put to use. The importance of stakeholder participation to understand both the model and the results was highlighted as central to current IAM processes and a weakness in earlier efforts.

The Ecosystem Health and Human Health group highlighted the complex interactions between ecosystems and humans. The IAM group shares this perspective and hopes to assist research in this area by assisting in the development of integrated assessment techniques and models of some of the processes of interest to both groups.

The Quality of Life group highlighted the value system that determines the problems selected for study by IAM researchers. The role of values in IAM was subject to different viewpoints by various participants. However, shared insights may be gained by further examination. Strong overlaps were found in the area of integrated assessment and sustainability indicators. Each group recognised the need to go beyond conventional measures such as GDP and presented progress toward better measures for future use. Overall, the hope is that by working together the groups can achieve better understanding and improved environmental decisions for this generation and those that follow.

15. Conclusion

Integration is essential to address future environmental problems. Integration goes beyond IAM to include integrating the sciences, knowledge, and our understanding of the future. The old descriptive systems models have been replaced with integrated models that are designed to address particular management issues. The role of humans, both as decisionmakers in management roles and as agents causing environmental change, has become an integral part of new models. The next challenge is to improve on the existing suite of modules by relating each one to broader ecological sustainability. The environmental problems of the 21st century cannot be considered in isolation, but need to be set within the broader sustainability context.

Optimistic observers expect that IAM will be used to address environmental challenges before crises are reached. Pessimistic observers expect that environmental disasters will occur before resources are made available to pursue IAM more broadly to achieve the goal of ecological sustainability. IAM needs to focus on moving forward beyond single issues to improve broader ecological sustainability, to improve decisionmaking, to integrate insights from natural and social sciences, to seek validation of IAM process, and to maintain integrity and rigor through openness, transparency, and honesty in the processes used.

IAM researchers share the objectives of many other groups and hope to continue the exchange of ideas and insights. Researchers studying complex adaptive hierarchical systems, ecosystem services, science and decisionmaking, ecosystem health and human health, and quality of life share similar challenges. Insights gained in one area are important to inform and help guide work in related areas. Overall, there is a strong desire to work together as part of a team to address the environmental challenges ahead. An integrated approach, a participatory process, an interdisciplinary team, and a visible research program that is conducted with integrity and trust, are key ingredients that need to be combined in a framework where communication is continuous.

Integrating the sciences starts with an integrated problem. Environmental problems of the 21st century are public and non-exclusive. They affect the Earth's ecosystems and all inhabitants. Separating mediums, inputs, or vectors is inadequate. Integrated science also means generating integrated knowledge. We invest time and ability to create a shared pool of knowledge and understanding. This result can only be achieved through integrated scientific efforts. To achieve an integrated future, instead of a fragmented disciplinary future, there is a call for integrated environmental science as its base. Rather than simply debate over the components of the curriculum, it is time to take action. The goal cannot be achieved unless we take integrated action. IAM is not limited to a particular model, but argues the importance of process. We seek to bring dispersed data and stakeholders together to gain improved understanding and enhanced environmental quality for future generations.

References

Beck, M.B. and Chen, J., 2000, Assuring the quality of models designed for predictive tasks. In: A. Saltelli, K. Chan and E.M. Scott (Editors), Mathematical and Statistical Methods for Sensitivity Analysis (Wiley, Chichester) pp. 401–420.

Beck, M.B., Ravetz, J.R., Mulkey, L.A. and Barnwell, T.O., 1997, On the problem of model validation for predictive exposure assessments. Stoch. Hydrol. Hydraul. 11(3):229–254.

Born, S.M. and Sonzogni, W.C., 1995, Integrated environmental management: strengthening the conceptualization. Environ. Manag. 19(2):167–181.

Charles, A.T., 2001, Sustainable Fishery Systems (Blackwell Science, Oxford) 384 pp.

Charles, A.T., Lavers, A., Benjamin, C. and Boyd, H., 2001, Fisheries and Marine Environment Account: A Preliminary Set of Ecological, Socioeconomic and Institutional Indicators for Nova Scotia's Fisheries and Marine Environment (GPI Atlantic, Tantallon, Canada).

Consortium for International Earth Science Information Network (CIESIN), 1995, Thematic Guide to Integrated Assessment Modeling of Climate Change (University Center, Mich). http://sedac.ciesin.org/mva/iamcc.tg/TGHP.html.

Costanza, R. (Editor), 1991, Ecological Economics: The Science and Management of Sustainability (Columbia University Press, New York) 525 pp.

Daly, H.E. and Cobb, J.B., 1989, For the Common Good: Redirecting the Economy Toward Community, the Environment, and a Sustainable Future (Beacon Press, Boston, MA) 482 pp.

Downs, P.W., Gregory, K.J. and Brookes, A., 1991, How integrated is river basin management? Environ. Manag. 15(3):299–309.

Grayson, R.B., Doolan, J.M. and Blake, T., 1994, Application of AEAM (Adaptive Environmental Assessment and Management) to water quality in the Latrobe River catchment. J. Environ. Manag. 41:245–258.

Hare, M. and Pahl-Wostl, C., 2002, Stakeholders' stakeholder categorisations: an empirical study. Integr. Assess., submitted.

Janssen, W. and Goldworthy, P., 1996, Multidisciplinary research for natural resource management: conceptual and practical implications. Agric. Syst. 51:259–279.

Jørgensen, S.E., 1999a, State-of-the art of ecological modelling with emphasis on development of structural dynamic models. Ecol. Model. 120:75–96.
Jørgensen, S.E., 1999b, Recent trends in environmental and ecological modelling. An. Acad. Bras Ci. 71:1017–1035.
Margerum, R.D. and Born, S.M., 1995, Integrated environmental management: moving from theory to practice. J. Environ. Plan. Manag. 38(3):371–391.
Morgan, M.G. and Dowlatabadi, H., 1996, Learning from integrated assessment of climate change. Clim. Chang. 34:337–368.
Oreskes, N., 1998, Evaluation (not validation) of quantitative models for assessing the effects of environmental lead exposure. Environ. Health Perspect. 106(Suppl. 6):1453–1460.
Palmer, D., 1992, Methods for analysing development and conservation issues: the Resource Assessment Commission's experience, Research Paper Number 7 (Resource Assessment Commission, December).
Park, J. and Seaton, R.A.F., 1996, Integrative research and sustainable agriculture. Agric. Syst. 50:81–100.
Parker, P., 1998, An environmental measure of Japan's economic development: The ecological footprint. Geograph. Z. 86:106–119.
Ravetz, J.R., 1997, Integrated Environmental Assessment Forum: Developing Guidelines for 'Good Practice', Working Paper WP-97-1, ULYSSES Programme (Darmstadt University of Technology, Darmstadt, Germany).
Risbey, J., Kandlikar, M. and Patwardhan, A., 1996, Assessing integrated assessments. Clim. Chang. 34:369–395.
Rotmans, J. and Van Asselt, M., 1996, Integrated assessment: growing child on its way to maturity. An editorial essay. Clim. Chang. 34:327–336.
Scoccimarro, M., Walker, A., Dietrich, C., Schreider, S., Jakeman, A. and Ross, H., 1999, A framework for integrated catchment assessment in northern Thailand. Environ. Model. Softw. 18:567–577.
Syme, G.J., Butterworth, J.E. and Nancarrow, B.E., 1994, National whole catchment management: a review and analysis of processes, Occasional Paper 1/94 (LWRRDC, Canberra).
Van den Bergh, J., Ferrer-i-Carbonell, A. and Munda, G., 2000, Alternative models of individual behaviour and implications for environmental policy. Ecol. Econ. 32:43–61.
Van Waveren, R., 1999, Application of models in water management in the Netherlands: past, present and future. Water Sci. Technol. 39:13–20.
Voinov, A., Costanza, R., Wainger, L., Boumans, R., Villa, F., Maxwell, T. and Voinov, H., 1999, Patuxent landscape model: integrated ecological economic modeling of a watershed. J. Ecosyst. Model. Softw. 14:473–491.
Wackernagel, M. and Rees, W., 1996, Our Ecological Footprint: Reducing Human Impact on Earth (New Society Publishers, Gabriola Island, B.C., Canada) 160 pp.
Wackernagel, M., Onisto, L., Linares, A., Falfán, I., García, J., Guerrero, A. and Guerrero, M., 1997, Ecological Footprints of Nations: How much do they use? – How much nature do they have? "Rio+5 Forum" study, Earth Council – San Jose, Costa Rica. http://www.ecouncil.ac.cr/rio/focus/report/english/footprint/.
Waldrop, M.M., 1992, Complexity: the Emerging Science at the Edge of Order and Chaos (Touchstone, New York) 380 pp.

Chapter 3

Complex Adaptive Hierarchical Systems

B.C. Patten, B.D. Fath, J.S. Choi, S. Bastianoni, S.R. Borrett, S. Brandt-Williams, M. Debeljak, J. Fonseca, W.E. Grant, D. Karnawati, J.C. Marques, A. Moser, F. Müller, C. Pahl-Wostl, R. Seppelt, W.H. Steinborn and Y.M. Svirezhev

Abstract

As the human influence upon its environment accelerates, it is becoming urgently apparent that this influence must be modulated in some fashion for the benefit of both *our* environment and *our* selves. But just how *should* we modulate and regulate this influence, especially when we are increasingly aware of the vast complexity of our environment and our selves. Indeed, we are trying to make rational decisions in the face of overwhelming information overload. We are trying collectively to adapt to this new status quo, as we grapple with the concepts of sustainability and ecosystem health.

A potentially powerful guide to such adjustments is found under the guise of CAHSystems analysis (Complex Adaptive Hierarchical Systems). This chapter offers a variety of perspectives on the pluralistic emergence of complex systems knowledge and their environmental applications. In particular, a critical need in such an analysis is the useful measurement of the organizational complexity of CAHSystems. In what follows, some of these approaches and their implications are illustrated. What is both promising and exciting is that there are premonitions of the integration of these approaches into a more unified body of a CAHSystems Theory. A second need that is also treated in this chapter is the development of a more rigorous and coordinated modeling approach to serve as a medium of comprehension and application. A final need is the direct integration of human ethical–social–political–economic systems into the decisionmaking process in the form of social learning and polycentric integrated assessments, rational allocation of resources for remediation attempts and the development of more global 'eco-ethics' that can go beyond the purely reductionistic modulation and regulation of *our* environment and *our* selves.

Understanding and Solving Environmental Problems in the 21st Century
Edited by R. Costanza and S.E. Jørgensen
© 2002 Elsevier Science Ltd.   All rights reserved

1. The new confrontation – biocomplex wholeness

One of the most powerful images of our time, an image that has changed the way we think of ourselves and the way we think about our relationship to our environment, is the image of Earth viewed from the surface of the moon. As we view "Spaceship Earth" we sense an organized whole in which the complexities of nature and those of human societies are intimately intertwined within a single ecosphere. We also sense that despite the inherent intractability, science is ready for the new confrontation – understanding and managing biocomplex wholes. This is the compelling scientific challenge of our times.

Certainly there are particular subsystems that are of special interest at particular times in particular places. We are interested in our environment, our economy, our culture, our families, and our selves. Sometimes our interest is focused on long-term, global environmental trends, sometimes on short-term, local socio-economic problems. But regardless of the subsystem or the spatial-temporal scale, we are also commonly concerned about the viability and sustainability of the whole – the collection of living systems in their entirety that populate the planetary ecosphere. The intuition is that in wholeness there is something deeply different from partness, and that the long-term survival and well-being of humanity hinge on knowing this difference.

On the applied front, the need for appropriate indices for monitoring vitality and sustainability has evolved as the concept of "sustainable development" which has become prevalent in environmental policy dialogue (Harger and Meyer, 1996; Deville and Turpin, 1996; Schultink, 1992; Mitchell, 1996; Rees, 1996; Suter, 1993; Gilbert, 1996; Munasinghe and Shearer, 1995; Wynne, 1992). Sustainable development, an oxymoron (e.g., see Choi and Patten, 2002), represents an uneasy wedding of interests in economic development, measured in such terms as GNP, to interests in biological conservation, measured in such terms as biodiversity. The goal of environmental indices is broadening from monitoring potential environmental contaminants based primarily on toxicological measurements to broader focus on environmental "health". In general, indices of a system's sustainability are not reducible to sets of individual environmental indicators (Gallopin, 1997), nor to individual economic or social indicators. Nor are indices derived from after-the-fact mathematical manipulations of sets of individual indicators likely to be particularly useful. True sustainability indicators are inherently relational, monitoring the symmetries of exchange among components that maintain system integrity.

General Systems Theory provides insights into the behavior and other characteristics of whole systems that can help address questions concerning indicators of sustainability. The "Theory of Complex, Adaptive, and Hierarchical Systems (CAHSystems)" represents a special branch of this theory that can address holistically and reductionistically, as appropriate, the crucial problems of man-

and-environment in a crowded, complicated but still unitary ecosphere. Systems theory is based on the premise that all systems possess four properties: (1) wholeness and order (the systemic or state property, "S"), (2) intra- and intersystemic hierarchies (the holon property, "H"; Laszlo, 1972), (3) adaptive self-stabilization (system cybernetics I), and (4) adaptive self-organization (system cybernetics II – the technical basis of adaptation, "A"). Or, stated more simply, natural systems are (1) wholes with irreducible properties, (2) coordinating interfaces in nature's holarchy (definition in chapter glossary), that (3) maintain and (4) adaptively create new patterns of organization in the face of an ever-changing and self-organizing context (Laszlo, 1996). These are the systems whose deep nature must be understood as the basis for identifying reliable indicators of natural and human vitality, their coexistence and mutual sustainability, and environmental health.

How do such complex systems emerge? What are the characteristic behaviors of such systems and how are they related to their structure and organization? How can such complex systems be efficiently and usefully described? What does sustainability mean in such embedded systems? And, how should we as humans, embedded in such complex systems, interact and behave to promote the sustainability of ourselves and that of the biosphere? These are the questions that arise naturally in the study of CAHSystems. However, analyzing complexity is problematic. There exist problems of (1) description; (2) measurement; and (3) comprehension due in the main to the non-trivial relationships that exist between such systems (large numbers of elements, indirect causalities, nonlinear functional relationships, distributed effects and cross-scale interactions and control issues).

In what follows, we outline our current understanding of complexity and its relationship to the hierarchical (or holarchical) organization of systems and their adaptive capacities. Highlighted are approaches to the measurement of CAHSystem complexity (sections 1.1 and 2); their extension to the study and understanding of evolutionary/adaptive facets of CAHSystems (sections 1.2 and 2.6); and new methodologies and approaches in dealing with the human element (section 3). Wherever possible, examples and case studies are provided.

1.1. Complexity

The first property of CAHSystems is complexity and in many ways this is the defining feature of CAHSystems. Complexity is also a slippery notion. As described by Casti (1986), we have an intuitive understanding of complexity that eludes formalization. This elusiveness is in part due to two related reasons: (1) the presence of a subjective or interpretative element in the observation of complexity; and (2) the resultant plethora of methods and approaches to the description of complexity.

These two subjects are briefly detailed in this section to provide the context of CAHSystems and an awareness of how current approaches are trying to go beyond these limitations to arrive at more useful and integrated measures of the organizational complexity of CAHSystems. The concept of complexity, especially as it relates to CAHSystems, must necessarily remain vague, until more objective measures can be developed (e.g., Bosserman, 1982) and brought into wide acceptance.

1.1.1. Subjective and objective aspects of complexity

What makes a system complex? Flood (1987, p. 177) begins his analysis by examining an intuitive proposition: "In general, we seem to associate complexity with anything we find difficult to understand." From this, he concludes that complexity must be associated with both people and things. The association with people comes from the word "we", and the association with things comes from the term "anything". Thus, there are two sources of complexity to address in CAHSystems. One is subjective and observer dependent. The other derives from intrinsic system characteristics.

The subjective nature of system complexity is related to how an observer interacts with a system (Rosen, 1977; Klir, 1985, 1991). The more such ways, the more complex the system. This means that the knowledge, skills, beliefs, and interests of the observer will affect how simple or complex a system is (in relation to that observer). A good example of this kind of complexity is provided by Ashby (1973) as quoted by Klir (1991):

> To the neurophysiologist the brain, as a feltwork of fibers and a soup of enzymes, is certainly complex; and equally the transmission of a detailed description of it would require much time. To a butcher, the brain is simple, for he has to distinguish it from only about thirty other 'meats' ...

Rosen (1977) goes so far in the subjective direction as to say "complexity is not an intrinsic property of systems". This extreme position is not accepted here because, though some aspects of a system's complexity are clearly related to an observer's interaction with it, there is still a real system beneath the observer's lens. If Rosen's proposition is accepted, advancement would become paralyzed by a "Post-Modern" perspective that suggests everything is relative to the observer; there is either no real system or it is inconsequential because it is absolutely unavailable to the observer.

In this report, we do not adopt such an extreme stance but rather reason that if the underlying real system is intrinsically simple there will be fewer ways for an observer to interact with it; whereas if a system is intrinsically more complex, then the possible ways to interact will increase proportionately. Therefore, there must also exist some intrinsic characteristics of systems that contribute to their complexity.

Ch. 3: Complex Adaptive Hierarchical Systems

Table 1.1
Some characteristics of complex systems

Structural characteristics	Behavioral characteristics
Large number of components	Nonlinear
Large number of connections	Chaotic
High diversity of components and connections	Catastrophic
Asymmetry	Self organized
Strong interactions	Multiple steady states
Hierarchic and holarchic organization	Adaptive

1.1.2. Parameters of complexity

A large number of characteristics can cause a system to be classified as complex. These divide into two related categories: structural and behavioral/functional (sometimes referred to, respectively, as static and dynamic complexity; Casti, 1979). A few of these characteristics are listed in table 1.1 (adapted from Flood and Carson, 1988).

With respect to structural characteristics, the traditional way of examining system complexity is to enumerate components. A system with more component parts is generally more complex. The same kind of measurement can be made for the number of direct connections. These are of two kinds, conservative energy–matter *transactions* and nonconservative (or distributive) informational *relations* (Fath and Patten, 1998). The number of possible direct connections increases as the square (n^2) of the number of components, n. A system is more complex if there is a greater diversity of types of components or relationships (Casti, 1979; Holland, 1995). For example, a system with 10 identical components is less complex than one with 10 different kinds of components. Structural asymmetry also tends to increase complexity. This is clear if we look at the development of an embryo. Early cell division creates symmetric cells; however, to differentiate and develop into an adult organism the embryo requires asymmetric cell growth. Strength of the interactions between system components also affects the system complexity (Casti, 1979). An asymmetric pattern of interaction strengths would create a more complex system. Finally, if a system has hierarchic or holarchic organization, then it tends to be more complex (Simon, 1962; Casti, 1986; Kay et al., 1999).

A system may also be complex in its behavior. Several categories of behavior are notoriously complex, namely nonlinear, chaotic, and catastrophic behaviors. Nonlinear behavior fails to follow input–output superposition (Patten, 1997a), meaning a linear combination of inputs generates the same combination of outputs. Patten (1997a) provides an example. If solar radiation is the only input to a green leaf and this is doubled, then for a linear leaf we would observe a doubling of the photosynthetic output. If this were not observed, then the system would be nonlinear and to that extent complex. If there are multiple inputs (such as light, water, nutrients, etc.) required for photosynthetic production, then each of these must be doubled to observe a doubling of photosynthesis. If only some inputs are doubled, and the leaf is known to be linear from previous observations, then a nonlinear response will be observed. Patten (1983) calls this "pseudononlinearity" and holds it to be more common in nature than true nonlinearity. This can never be determined, however, because complexity obscures the true input–output relationships. Chaos is erratic behavior of nonlinear (but never pseudononlinear) systems. Catastrophic behavior includes dynamics such as bifurcations, flips, and shifting steady-state mosaic patterns (Kay et al., 1999). Bifurcations are points where system dynamics can diverge in one of two directions. Flips are points where there are behavioral discontinuities. An example of the shifting state mosaic pattern of behavior is Holling's (1986) "lazy-8" developmental cycle of growth, maturity, senescence, and regeneration. System complexity also increases to the extent its trajectories can reach multiple steady states. Kay et al. (1999, p. 727) describe this condition as a system in which "There is *not necessarily* a unique preferred system state in a given situation." Other characteristics of systems that tend to make them more complex include the ability to self-organize and also adapt to new conditions.

There are many intrinsic variables that can make a system complex. The list in table 1.1 based on system structure and behavior is not exhaustive. Complex systems will have some of these characteristics, and possibly others; in general, just one of them by itself may not be sufficient to classify a system as complex.

1.2. Adaptation and hierarchy

The second property of the CAHSystems concept is adaptation and the third property is hierarchy. Adaptation and hierarchy are implicitly coupled characteristics. When an adaptive response is caused by some stimuli, the causes and effects of the stimuli exist across a number of spatial, temporal and organizational scales. Take for example the simple adaptive response of insects to pesticides. The stimuli in the form of chemical pesticides induce physiological, behavioral, genetic, population-dynamic, and morphological alterations in the affected organisms, which in turn have feedback effects upon humans (populations, industry, economics, disease, behavior), other organisms, and the environment in general.

Adaptation is readily accepted as some genotypic/phenotypic response to environmental stimuli. Implicit in the concept of adaptation is the notion of a goal. At the organismal level, this goal is generally equated to survival to reproduce in the Darwinian sense (or more generally, survival to replicate). To extend the concept of adaptation to all CAHSystems requires an expansion of this notion of adaptation, as the mechanisms of information transfer are not as explicit as is the case of genetic material for organisms. This expansion is usually attempted by reference to thermodynamic principles as they alone of all known scientific laws have a direction of time and the generalization of information transfer in the form of signal processing of matter and energy flows.

One of the first steps required to accomplish this expansion of the concept of adaptation of CAHSystems is the development of theoretically sound and empirically useful indicator(s) of the organizational complexity of the CAHSystem (section 2). The use of such indicators is essential as it is impossible to obtain any unambiguous understanding of such highly multidimensional systems [e.g., due to pseudononlinear, sensu Patten (1983) and truly nonlinear phenomena]. In fact, it is in many ways the "information overload" experienced by all sectors of the human system in trying to deal and adapt to the complexities of reality that is now driving us to a CAHSystems approach. Although such attempts have been criticized and ridiculed in the past (e.g., Månsson and McGlade, 1993), the CAHSystem approach represents a direction that now seems to be providing a coherent means of approaching and dealing with the intrinsically complicated nature of complex systems.

In what follows, some of these recent and quite encouraging attempts and results are detailed to allow the dissemination of this approach to the more general scientific, engineering, policy, and management communities. In particular, these indices in the form of various guises that stem from divergent historical origins include informational exergy, energetic exergy, emergy, empower, transformity, dissipation, degradation, informational entropy, physical entropy, cycling, throughflow, and many more measures to describe what is becoming increasingly apparent as complementary descriptions of the spatial, temporal, and organizational complexity of CAHSystems. This awareness is due in part to the network formalism of Patten, Fath, and colleagues, which is highlighted in section 2.5.

1.2.1. Organizational hierarchy (holarchy)

In the simplest and the truest sense, adaptation represents the *continued* maintenance of a system away from thermodynamic equilibrium. A system achieving this *continuously* in the presence of numerous interactions (stimuli, perturbations, stresses, etc.) is an adapted, or adaptive system (i.e., some thermodynamic quasi-steady state).

What must be emphasized in the above is the word *continuously*. That is, it is maintenance away from thermodynamic equilibrium and not simply unbounded growth (e.g., as in the current human/economic mode). Unbounded growth *cannot* represent an adaptive state as growth is accomplished at the cost of other forms of organization and useful free energy (see Patten's AWFUL Theorem; Patten, 1997b). This is ultimately harmful to the growing system itself due to

(1) the feedback of these effects upon the growing system, and (2) the increased dynamic and structural instabilities associated with increased deviations from thermodynamic equilibrium (Choi et al., 1999; Choi and Patten, 2002). We must therefore focus upon the quasi-steady state and try to understand what determines this "adapted", and therefore sustained or sustainable state.

This state of balance may be simplistically seen as the balance between the "assertive" tendencies (see glossary – or the "goal functions" of section 2.5 operating at a particular level of organization) of a system's internal milieu and "integrative" tendencies (sensu Koestler, 1969; see glossary – or the "goal functions" of Section 2.5 operating at all other levels of organization) that constrain and hold these assertive tendencies in check (Choi and Patten, 2002). Thermodynamically, this state of balance may be represented as the quasi-steady-state realized by the "*Generalized thermodynamic feedback model*" (see glossary).

Associated with such a most adapted state are "normal" ranges in the intensities of metabolic activities (Choi et al., 1999). Deviations from these "normal" intensities represent a destabilized or maladapted state; Koestler (1969) referred to such imbalance as a "pathological" state. When these intensities are more elevated than "normal", they indicate dominance of the external milieu (i.e., local disorder – perturbations or disturbances), whereas intensities lower than "normal" are indicative of the dominance of the internal milieu (i.e., local order – growth). Let us emphasize that we have here a case where a simple indicator can directly measure the health of the holon (or ecosystem) by simply monitoring the intensities of metabolic activities of the system, much as doctors monitor the temperature of a homeotherm to determine if there are any health abnormalities (Choi and Patten, 2002).

What is quite fascinating, beyond the immediate applications of the above, is that associated with the relative dominance of the assertive tendencies of a particular holon, are implied increases in the size and number of degradative steps internal to the holon (i.e., the relative efficiency of free-energy utilization before final degradation). In fact, one finds that the closer one approaches a thermodynamic equilibrium the greater are the local reversibility and efficiency of energetic interactions. Very complex biochemical systems seem to operate at very elevated efficiencies with a great number of intermediary steps, ensuring that each degradative step is as small and thus thermodynamically efficient (reversible) as possible (e.g., Spanner, 1964). Similarly, for ecological systems, the elaboration of "aggradative" structure in the form of a greater number and diversity of organisms also increases the number of degrading (feeding) steps and as such an associated increase in the thermodynamic efficiency of energy transfers. When mechanisms of co-ordination exist, such active control (e.g., in enzyme systems where environmental constraints are often rigidly controlled) results in greater efficiency at the cost of power output.

Odum and Pinkerton (1955), in characteristic foresight, remarked that useful power output has a simple relationship with the efficiency of processes and that there is a tendency for many open systems to operate at not the most efficient configuration but rather a configuration producing maximum power output (somewhere near an efficiency of 50%). Perhaps we may re-interpret this empirically observed operating point as not being a goal "chosen", but rather one that is negotiated via adaptive processes. That is, CAHSystems are "trying" to be as adaptable as possible because CAHSystems do not and can not have complete or active "control" over their environment (i.e., the Red Queen's Hypothesis of van Valen, 1976).

One may therefore speculate that the minimum size of energy degradation steps is constrained by the structure and mechanics of a holon. That is, how efficient can a system be with the constraint that it maintains itself and survives periodic catastrophes (within-and-across-scale negotiated balances). The maximum size of an exergy-degrading step would be that which would keep things together without structural and mechanical parts melting away or burning up via strong nonlinearities creating overshoots and undamped autocatalysis. Inversely correlated with the size of degradation steps is the number of such steps. The greater the number, the more efficient the processes – however, the constraints pertaining to predicting and controlling the environment are also greater. To increase efficiency of the steps would require increased reversibility of processes that can only be possible when these are really tightly integrated. After a perturbation, the continual "refinements" or "fine-tuning" (adaptations) or "negotiations" to some new balance involve an ever-diminishing approach to the steady state.

Can this most adaptive state be the same as that illusive concept of a most sustainable/healthy state? This question is quite intriguing and merits greater study and deliberation in our efforts to better understand sustainable development and ecological and environmental health (Choi and Patten, 2002; Choi, 2002).

1.2.2. Spatial hierarchy

Spatial patterns in nature are evident at many different scales, changing with succession and disturbance. These patterns are evidence of adaptive hierarchies (Cousins, 1993; Milne, 1991; Turner et al., 1991), with forcing functions (generative forces) low in the hierarchy working in opposition to control functions (constraints) higher in the hierarchy creating the pattern (Milne, 1991). This juxtaposition of low- and high-level processes creates a discernible pattern at discrete hierarchical scales (Turner et al., 1991).

Several spatial hierarchy structures are immediately obvious in natural systems (Cousins, 1993; Merriam et al., 1991; Noss, 1990). Individuals or populations, adapting to diurnal and microtopographic resources, drive the processes at the ecosystem or community level. Communities create patterns of connectivity and patchiness that generate patterns of convergence at the larger landscape scale. Landscapes and global processes, created over longer geological time frames, act as constraints for the lower levels. This three-tiered hierarchy is evident in all environmental systems.

A more complex hierarchy emerges when human activities are imposed on landscapes (Brandt-Williams, 1999; Odum, 1994; Cousins, 1993). Community-scale processes, the scale at which landscape patterns become most evident, organize into recognizable ecosystems. Atmospheric and geologic weathering processes create distinguishable watershed patterns, where runoff converges into larger streams of increasing order. These two hierarchical levels (ecosystem and watershed) shape the next hierarchical level, human resource energy (trade and information), where convergence of services, information, and materials within the landscape concentrate into cities. Trade alone creates spatial patterns. Large central cities are surrounded by a few smaller cities, a larger number of smaller towns and even more clusters of villages (Christaller, 1966), interspersed with the agricultural and natural systems providing support and environmental services. The smaller towns ship goods to the city and the city in turn exerts control over them by returning information, services, and dictates for production. Cities create repeatable patterns where areas of higher energy at the two lower hierarchy levels converge, deep bays in fertile river valleys, for example. Evidence of human control over ecosystem and watershed, lower in the hierarchy, are well known. The global processes, such as deep earth heat and oceanic circulation, then become a fourth (or higher) level, creating structure through plate tectonics and currents.

These spatial hierarchies are determined by the scaling of energy hierarchies. Energy hierarchies arise when energy flows of many small processes converge and are transformed to make larger-scale processes (Odum, 1994). The energy hierarchy determines the spatial and temporal scales at which the system controls are exercised (Odum, 1994; Salthe, 1985; Allen and Starr, 1982). Larger entities occur as a result of more energy transformations and concentration, and can exert influence over smaller items having more diffuse energy storage. Items with larger territories and storages control smaller-scale functions with faster turnover times. Consequently, just as hierarchies of convergence are evident in flows of materials and energy through food webs, pathways of the energy hierarchy also form converging patterns in space on a landscape scale (Brandt-Williams, 1999; Lambert, 1999; Huang, 1998; Odum, 1994; Cousins, 1993). Larger energy flow builds greater spatial structure (Odum, 1994), evident, for example, in urban centers and mountains. Built structure becomes dependent upon inflows from the surrounding landscape, and the more structure built, the larger the support area required.

Within each hierarchical level, there are adaptive spatial patterns that result from energy delivered to the landscape in different forms. The three energy "shapes" are point sources, line sources, and planar sources (Odum, 1994). Most landscapes receive energy from several different sources, but the highest, most concentrated energy will determine the spatial shape taken (Odum, 1994). Point sources, such as desert springs or high concentrations of rare elements, create spatial patterns reflecting the energy driving the system. Nearly circular growth of lush vegetative communities occurs in decreasing concentric density around the spring, while clearings and settlements surround the immediate location of extractable resources, with no other settlements or patches in the region. Line sources, highways and coastlines for instance, create linear patterns of both vegetative zones and development. These energy sources often intersect, creating the light pattern recognizable from space. Extractable resources often occur as line sources, and patterns at the human hierarchy level reflect the linearity of the trade good. Planar sources create less distinct landscape features, because by definition these sources are distributed more evenly over large regions. Solar energy and its carrying media, rain and wind, are examples of planar energies. Expanses of grassy plains and deserts are spatial examples of planar energy shapes.

There are, of course, landscapes where no adaptive pattern or hierarchy is discernible. Several properties likely contribute to this. The area may lie somewhere between the distinct hierarchical levels evident at differing scales (Turner et al., 1991). However, receiving equal energy from different source "shapes" also leads to overlapping landscape patterns (Odum, 1994). Last, as spatial hierarchies become more complex, with energy convergence at the ecosystem and economic levels creating overlapping hierarchical levels, "noise" might obscure the signatures evident at lower levels of energy or higher scale (lower resolution) of reference, just as highly differing temporal scales create noise in simulations.

Land features create the constraints to which all terrestrial processes on earth adapt. As we recognize the connectedness between changing landscape processes created by humans and the backward feeding control of global processes, both on the landscape and economics, understanding the theory of complex adaptive spatial hierarchies will become a valuable management tool.

1.2.3. Temporal hierarchy – mathematical heterogeneity of CAHSystems and the Petri-Net approach

An important form of hierarchical/holarchical complexity derives from the different "natural times" in which the various sectors of CAHSystems operate. For example, environmental impact assessment of human activities necessitates a comprehensive analysis of both industrial and ecological systems. In classical ecosystem models, anthropogenic effects enter the system as environmental covariables or indirectly via model parameters (Seppelt, 1999). Models of technical systems are primarily devised for process control and optimization, and yield at most the order of magnitude of emission rates. In reality, both the industrial and ecological systems are closely interlocked and should be treated as a whole at least at higher scales. A general concept for the global scale has recently been presented by Schellnhuber (1998) based on a system-theoretic approach expressed in terms of ordinary differential equations.

A conceptual difficulty arises in connection with CAHSystems with industrial and natural components because processes of different dynamic qualities interact. The dynamics of technical systems tend to be time-discrete and their dynamics are closely related to discrete spatial structures. By comparison, many environmental processes are continuous in time and space. The mixture typical of whole CAHSystems can be characterized as structured time-discrete and time-continuous. An example is crop production where continuous biological processes such as crop growth or soil water transport are imbedded within time-discrete agrotechnical management procedures.

Mathematically, continuous systems are described by differential equations, whereas discrete systems like production lines are described by condition–event systems or matrix equations. Therefore, one is faced with the problem of mathematical heterogeneity. It is not, in general, feasible to model integrated CAHSystems in the framework of a single mathematical theory like ordinary differential equations only.

A general framework for mathematically heterogeneous CAHSystems is provided by Petri-net theory. This combines structural and behavioral features of complexity, as outlined previously (section 1.1), and is amenable to various kinds of extensions such as time-weighted and stochastic transitions, and integration of differential equations. These nets are called hybrid low-level Petri nets. A simulation tool incorporating mathematical heterogeneity has been devised which allows for graphical development of the Petri nets (fig. 1.1). Nets with such extended properties are capable of simulating the dynamics of mathematically heterogeneous systems, which are structured by net topologies. They serve as a theoretical basis for the analysis of topological properties and enable the efficient simulation of complex integrated technical–ecological systems. A major innovation is the capability of handling in a single framework event-based dynamics together with mass and information flows.

This is demonstrated by several case studies (Seppelt and Temme, 2001). The first is a meta-population model of a species at the Galapagos archipelago. The network structure, which is derived out of a geographic information system, contains time-discrete migration processes within the habitat structure as well as continuous processes of growth and extinction in habitats. A second study focuses on a life-cycle assessment of a magnesium car component. In this, a life-cycle analysis of a car door production process is integrated with an atmospheric spread model and an environmental impact assessment model within the framework of hybrid low-level Petri nets. A third case study focuses on agricultural production.

Ch. 3: Complex Adaptive Hierarchical Systems

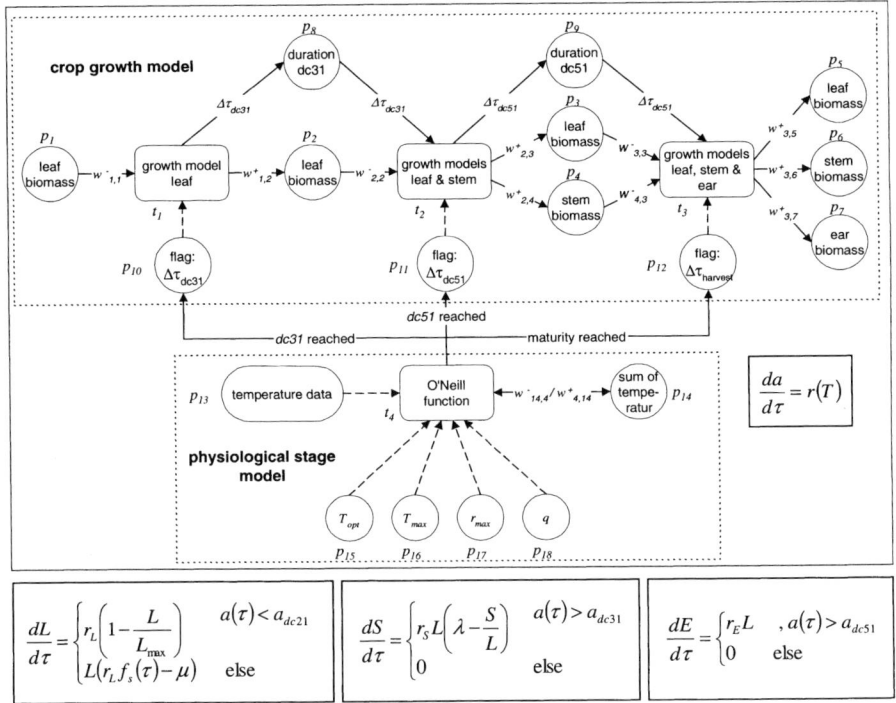

Fig. 1.1. Petri net integration of biological models.

In this, the Petri net is used to integrate biological models for crop growth (mass flows), physiological stage estimation (event-based simulation), technical models for energy consumption by plowing, seeding, etc., and optimization of fertilizer application (fig. 1.1).

2. Measuring the organizational complexity of CAHSystems

The following illustrates various approaches to the measurement of the organizational complexity of hierarchically organized systems. The approaches are many and not exhaustive. There is, however, much overlap. What differences exist are differences of emphasis mainly due to the influence of different schools of thought and different scales of application of CAHSystems.

What is in common is the repeated application of thermodynamic principles in the study of CAHSystems. This too has a historical component, however. Thermodynamic theory is appealed to aid in the study of CAHSystems for two main reasons: (1) thermodynamic theory is the only known physical law with a directionality of time, a time sense that is shared by CAHSystems; and

(2) the treatment of systems as a complex whole and not simply a number of parts. A second point of commonality is found in the convergence of ideas, especially as it relates to the mutual relationship of the various schools of thought (sections 2.4 and 2.5).

There exist terminological difficulties, especially with the term "exergy", which is used in a number of different manners ranging from its physical interpretation as useful free energy to the other extreme of informational content. Regarding these ambiguities, the readers' patience is appealed to in the hopes that they will see the broader commonalities in these as yet incomplete mosaics. (Use of the glossary is strongly recommended.)

2.1. Measuring the complexity of genomes and organisms – biocomplexity

2.1.1. Genomes and organism complexity

Although system complexity can be interpreted as the amount of *information* necessary to describe the state of the system or its behavior, the measurement of complexity must be translated into more formal (mathematical) terms in order to allow for comparisons (Badii and Politi, 1997; Vilela Mendes, 1998). It is in this translation where both fundamental and practical difficulties arise.

Regarding the complexity of an organism, one may assume that the genome (defined as the amount of haploid DNA in a genomic set) contains most of the information needed to specify physiological design and function. Therefore, the upper bound on the amount of information contained in DNA sequences could be an acceptable (though very approximate) estimate of organismal complexity. This could be connected to increasing length of the genomic representation, although it is only when longer DNA is used for some purpose that organisms become more "complex" (Bar-Yam, 1997).

Determination of the information content of genomes is not free of difficulties. Although eukaryotic genomes are much more complex than prokaryotic ones, the genomes of higher organisms seem to contain a large excess of DNA. It is hard to draw a clear systematic trend of increasing genome length that correlates to the complexity of organisms (Li, 1997; Futuyma, 1998). At the organismal level, considering prokaryotes, genome length is an acceptable measure of internal effective complexity, but for eukaryotes emergent effective complexity (for instance, cultural complexity in humans) may no longer be adequately measured by effective genome complexity (length). Taking total genome length, small genetic alterations permit complex adaptive organisms to create new levels of effective complexity at certain periods in their evolution, increasing their potential complexity for that time period (Gell-Mann, 1994). At the genome level, a DNA sequence is the source for different types and levels of information. A molecule codes for amino-acid sequences, for proteins, for RNA sequences for tRNA, rRNA, mRNA, telomeraseRNA, and other RNA's, and for protein binding sites. In addition, DNA also codes for architectural and structural information such as intrinsic DNA curvature, nucleosome positioning, transcription initiation, origins of replication, and mutational "hot-spots" (Lewin, 1994). All of this serves to increase the difficulties of estimating the information content of genomes and from this the complexity of CAH organisms.

2.1.2. Exergy as a measure of biocomplexity

Ecosystems, like other CAHSystems, living and nonliving, show different levels of order emerging from both order and disorder (Schrödinger, 1944). At a certain moment, the complexity of these systems is not a direct measure for levels of complexity that may be reached in the future. Ecosystems are composed of many complex elements, each evolving patterns to describe and preview behavior from one another, which makes improbable a final stationary state (Gell-Mann, 1994). Nevertheless, during development, thermodynamic laws will determine allowable directions for the many processes (Schneider and Kay, 1994a) taking place with energy as the driving factor (Jørgensen, 1992a). Therefore, the energy–matter balance and structure of ecosystems are expected to evolve to a state of optimal thermodynamic balance, far from equilibrium, with the system being characterized by an optimized storage of the available energy and an increased dissipation to maintain the levels of biomass and (higher) complexity (Jørgensen, 1992a,b; Schneider and Kay, 1994a,b, 1995; Marques and Nielsen, 1998; Jørgensen et al., 2000).

Exergy can be used in ecology to measure complexity. Its mathematical definition contains both conservative (energy and matter) and nonconservative (informational) terms. Complexity in ecosystems is expected to be associated with more complex organisms, which in principle correspond to higher information content and greater distance from thermodynamic equilibrium (Marques and Nielsen, 1998), as well as to increased utilization (implying energy dissipation) of food and other resources to compensate for storing and processing higher information levels in the form of DNA, RNA, and protein sequences. The exergy content of biomass ("ecological exergy") can be estimated using "weighting factors" ($\exists$). For each system compartment, a corresponding weighting factor is multiplied by the biomass content in order to discriminate the information embedded in the biomass as a function of its structural complexity.

To estimate weighting factors for biological compartments, it has been proposed to assess the information content of genomes in terms of (1) probabilities associated with "genome dimensions" (number of genes) (Jørgensen et al., 1995), or (2) "overall coding capacity" expressed as total genome length (estimated from corresponding "c-values") as a topmost limit (Marques et al., 1997; Fonseca et al., 2000). Proposal (1) is limited by scarcity of data concerning the numbers of genes for different species. For (2), estimating the information content from total DNA content will suffer strong bias due to the "c-value paradox" and should be used as a preliminary approach only (Fonseca et al., 2000). The "c-value" stands for the total amount of DNA per cell nucleus, which is a characteristic value for each species. The "c-value" corresponds to the amount of a haploid cell nucleus, while the "2c-value" corresponds to a diploid one. The paradox associated with the total amount of DNA per nucleus arises from the fact that, on one hand, similar organisms (in 'complexity') may have significantly different nuclear DNA contents, and, on the other hand, some organisms which exhibit less morphological 'complexity' than mammals may possess larger "c-values" (Futuyma, 1998). Also, in some phyla, the range of c-values can be either narrow or very wide (Lewin, 1994). These findings are essentially a consequence of repetitive DNA sequences in eukaryotic genomes (Li, 1997), and reinforce the concept of the "c-value paradox" (Cavalier-Smith, 1978; Futuyma, 1998).

Studies regarding reassociation kinetics of DNA have identified three kinetic classes of genomic DNA. These are fast, intermediate, and slow components, corresponding to highly repetitive DNA, middle-repetitive-DNA, and single-copy DNA, respectively. Hybridization experiments indicate that most genes coding for mRNA's fall into the single-copy class. Assuming complexity to correspond with the total length of different sequences, slow or single-copy genome components better estimate genome size and therefore complexity (Lewin, 1994). Methodologies should be developed that use data for the single-copy portion of the genome to determine the ($\exists$) weighting factors for exergy measure.

2.1.3. Case study: Mondego Estuary

The concept of "ecological exergy" as discussed above has been applied to analyze data from benthic macrofaunal communities periodically sampled along an estuarine eutrophication gradient in the south arm of the Mondego estuary (Western Portugal). Increasing nutrient drainage and level of confinement contribute to eutrophication along a spatial gradient. This gradient extends from areas less affected (non-eutrophied) near the mouth of the estuary, where a macrophyte community is found, to a highly affected (eutrophied) area where green-macroalgal blooms periodically occur. Along this gradient, the macrophytic community is progressively replaced by free- (or partially-) floating, fast-growing species. This shift in primary producers has consequences for the structure of communities found in these areas, reflected in the specific composition and productivity of the communities (Pardal, 1998; Marques et al., 1997; Lopes et al., 2000; Pardal et al., 2000).

Exergy estimates (ecological exergy and specific exergy) were calculated from the biomass of organisms, using weighting factors determined as described above according to Fonseca et al. (2000) for the specific exergy content of biomass.

Analyses were done at two different structural levels: (1) total biomass in each area, and (2) biomass for organisms grouped into trophic guilds (herbivores, filter feeders, detritus feeders, carnivores, and omnivores). Changes in ecological exergy reflected most closely changes in the biomass of corresponding areas, both presenting identical patterns of seasonal variation. On the other hand, estimates of specific exergy gave a different picture. Lower average values for the structural exergy index were found in the area less affected by eutrophication [non-eutrophic: 451 ± 29.8 exergy/(g biomass); intermediately-eutrophic: 559 ± 75.2 exergy/(g biomass); and eutrophic: 496 ± 100.3 exergy/(g biomass). Note that structural exergy units are expressed in detritus equivalents per unit biomass]. In order to perform a deeper assessment we estimated contributions of secondary consumers for the index of structural exergy considering organisms grouped into trophic guilds. The specific exergy guild in the intermediate-eutrophic area presented a pattern of variation oscillating between values determined for non-eutrophic and eutrophic areas. This sampling area probably corresponds to an intermediate state between two different directions of development, or two opposite levels of relative stability along a spatial gradient of environmental impact: the less eutrophic area and the area profoundly affected.

Although these results are preliminary, the exergy index seemed capable of providing a clear indication about alteration of communities when considering macrofaunal biomass grouped into trophic guilds. For instance, higher contributions to specific exergy were obtained from the groups of filter-feeders and detritus-feeders in more eutrophic areas.

2.2. Exergy-based orientors in a natural (virgin) forest

2.2.1. Background

The combination of general systems theory (von Bertalanffy, 1968), thermodynamics of irreversible processes (Prigogine, 1967; Prigogine and Stengers, 1984), and systems ecology (Odum, 1983) have resulted in holistic approaches to the study of ecosystems. It has been found that ecosystems are able to meet changes in external factors with various regulatory processes on individual and whole-system levels. These characteristics result from systemic complexity, which allows selection of developmental directions and self-organization of structure to meet external changes. Evolution as CAHSystems reflects adaptations of species and the ecosystems as wholes to move as far as possible away from thermodynamic equilibrium.

Among virtually infinite alternatives for development, the one uniquely selected can be inferred to best meet external changes and optimize evolutionary directions. This direction can be described by *orientors* (Müller and Leupelt, 1998). Bossel (1998) defined these as aspects, notions, properties, or dimensions of systems that can be used as criteria to describe and evaluate a system's developmental stage.

The central idea of the orientor approach refers to self-organizing processes that are able to build up gradients and macroscopic structures from the microscopic "disorder" of non-structured, homogeneous element distributions in open systems, without receiving directing regulations from the outside. In such dissipative structures the self-organizing process sequences in principle generate comparable series of constellations that can be observed by certain emergent or collective features. Thus, similar changes of certain attributes can be observed in different environments. Utilizing these attributes, the development of the systems seems to be oriented toward specific points or areas in the state space. The state or other variables used to elucidate these dynamics are the orientors. Their technical counterparts in modeling are called goal functions (Müller and Fath, 1998, p. 15).

A pluralistic approach to study ecosystem complexity has resulted in the appearance of a number of different orientors — maximum ascendancy, biomass, persistent organic matter, emergy and empower, stored exergy, dissipated exergy, and indirect effects, as well as least specific energy dissipation, and information-based concepts such as the minimax principle developed in section 2.3 below. Because orientors all focus on ecosystem development, Jørgensen (1994) and Patten (1995) looked for relations between different orientors and found a high correlation between them. This triggered even stronger attempts to unify orientors with the purpose of formulating a new ecosystem theory. In accordance with these efforts, exergy storage and dissipation are here explored as orientors in the growth and development of the natural (virgin) forest Rajhenavski Rog, Slovenia.

Jørgensen and Mejer (1977, 1979) proposed exergy storage as an orientor. They hypothesized that ecosystems change in a direction to attain highest exergy storage levels under given environmental circumstances with the available genetic pool. Stored exergy expresses distance from thermodynamic equilibrium and reflects the size of the organized structure and its content of thermodynamic information (Jørgensen, 2000).

Schneider and Kay (1995) hold a competing hypothesis about exergy dissipation. This states that ecosystems develop in a way to systematically increase their ability to degrade incoming solar exergy (Kay, 1984; Schneider and Kay, 1994a). This means that ecosystems develop in such a way as to degrade as much exergy as possible in the shortest possible time. The exergy drop across an ecosystem is related to the difference in black-body temperature between captured solar energy and reflected or heat energy reradiated by the ecosystem. If a group of ecosystems is bathed by the same insolation, Schneider and Kay expect that the most mature system would reradiate its energy at the lowest exergy level. Having degraded the most incoming exergy, such ecosystems would have the coldest black-body temperature.

The storage and dissipation hypotheses are at first sight contradictory. They can, however, be merged into a single new hypothesis (suggested by Patten in Jørgensen et al., 2000), stating that a system receiving an exergy input will use all means available to degrade this as fully and quickly as possible (Schneider–Kay) but, recursively through the work performed and structure created, maximize its storage (Jørgensen–Mejer) faster than the said dissipation such that its storage/dissipation ratio (dissipation-specific storage, or turnover time) will tend to a maximum. Inversely, storage-specific dissipation or turnover rate will tend to a minimum (Choi et al., 1999; Fath et al., 2001).

2.2.2. Experimental approach

The above-described hypotheses have been evaluated on five developmental stages of the natural (virgin) forest Rajhenavski Rog in Slovenia. The stages were successional, from earlier to later development. Distinguished were gap, new growth, mature stand, regenerated stand, and stands with a mixture of all previous four stages on one plot. Sixty plots (35 by 35 m) were sampled, 12 for each developmental stage. Each was selected randomly and described by the following attributes:
- quantity of dead wood above ground (metric tons dry matter per hectare);
- quantity of living trees (tons dry matter per hectare);
- vascular plant composition and percent of plot surface covered by each species; and
- surface temperature at the summer solstice, 1998 at midday in cloudless and still weather.

The study also included one plot located in a pasture near the virgin forest. This served as an example of a man-made system artificially maintained far from its natural condition by human intervention. It is a simple system compared to the other forest stages, and thus serves as an extreme example.

Exergy storage was calculated according to Bendoricchio and Jørgensen (1997) and Fonseca et al. (2000). Because genome size is required (see section 2.1), this was measured for all species appearing in the sample plots by flow cytometry (Dolezal et al., 1998).

Exergy dissipation can be directly determined from surface temperatures of study plots. Unfortunately, canopy temperatures could not be obtained due to technical limitations of the available thermal camera. Instead, temperatures of the forest surface were obtained in artificial units (1 to 255) at a time when all study plots received the same insolation. Differences in reradiated energy (surface temperatures) were related to the different developmental stages.

2.2.3. Results and conclusions

Results of the exergy storage calculations are summarized in fig. 2.1. The graph shows the same pattern observed in earlier studies that examined relations between

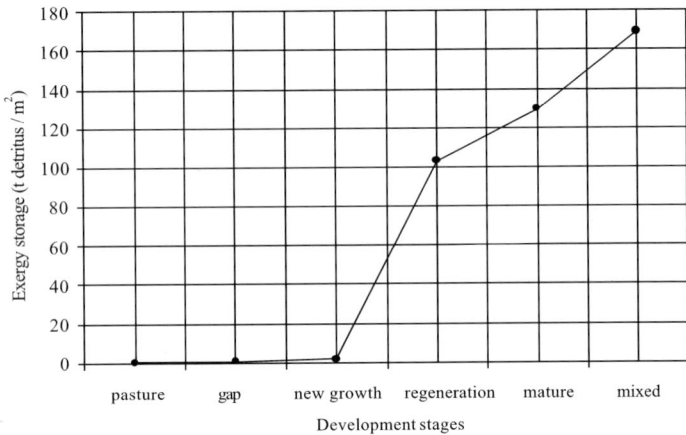

Fig. 2.1. Exergy storage in different development stages.

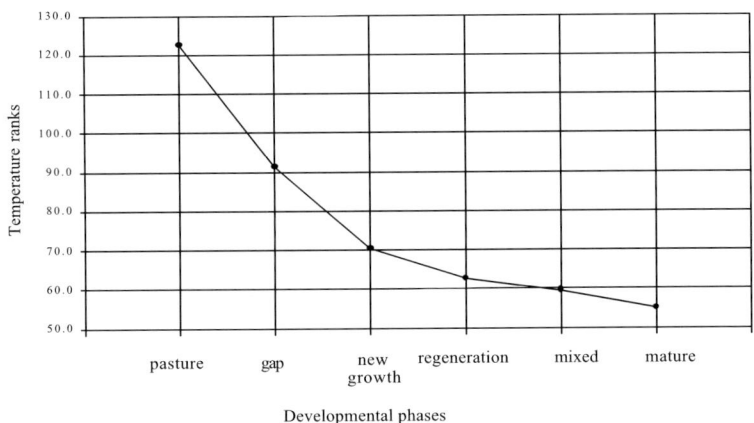

Fig. 2.2. Average surface temperature ranks of development stages.

exergy storage and ecosystem development (Jørgensen, 1986; Herendeen, 1989; Jørgensen, 1997):
(1) *Developmental stages with highest concentrations of organic mass also have highest exergy storage.*

These are the mature and mixed-development stages, which are the most dominant and long-lived stages in the study forest. They conform to the maximum storage hypothesis.

Average temperature ranks of the different successional stages are shown in fig. 2.2, with pasture taken as reference. The descending curve indicates cooling temperatures.

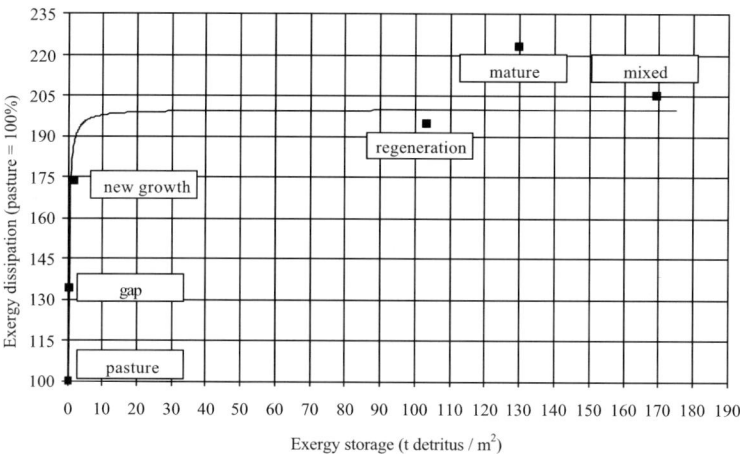

Fig. 2.3. Exergy dissipation versus exergy storage.

With other variables constant, Luvall and Holbo (1989, 1991) found that more developed ecosystems had cooler surface temperatures, reflecting greater degradation of their reradiated energy. Figure 2.2 confirms this result. Kay (2000) cited authors who conducted similar research and came to similar conclusions:

(2) *More developed ecosystems are cooler, meaning that compared to less developed ones they transform more incoming radiation in nonradiative processes.*

This is consistent with the maximum dissipation hypothesis.

The third hypothesis, maximum dissipation-specific storage, combines the previous two. If exergy storage is plotted as the abscissa on a graph with exergy dissipation the ordinate (fig. 2.3), then the plot shows the same characteristics discussed in Jørgensen et al. (2000).

These results accord with the third hypothesis. Thus:

(3) *Exergy dissipation increases very quickly in succession to its maximum, indicating this orientor applies to younger stages with simple structure and low biomass. Exergy storage, on the other hand, appears appropriate for the more mature successional stages with complex structure and high biomass concentrations.*

The combination maximizes dissipation-specific storage throughout the sere in successional development.

The data presented are consistent with all three of the indicated hypotheses about ecosystem growth and development. They are also consistent with informational development as taken up in the next section. It is concluded that orientors guide patterns of growth and development in natural forest ecosystems.

2.3. Exergy and information of solar radiation

2.3.1. Introduction

A radiation balance for different types of vegetation cover gives information about the transformation of incoming solar radiation by that vegetation. It is possible to calculate the dynamics of radiation balance using the First Law of Thermodynamics (energy conservation). However, for a complete description of vegetation as a "thermodynamic machine", in addition to the energy conservation law, the principles connected with the entropy concept must also be used. These are the Second Law of Thermodynamics, Prigogine's Theorem, and the exergy concept (Jørgensen, 1992b). Here, we propose a relationship between the information content, specifically Kullback Information, and exergy of the system.

Kullback (1959) Information measures the increment of information as a result of the transition from a "reference state" to a current one and can be shown to have a connection with the exergy measure. Also, because of the relationship between the exergy and entropy concepts, one can formulate the exergy concept for a natural process such as the interaction between solar radiation and a surface (e.g., soil, ocean, vegetation, etc.) where the increment of information expressed in energy units is the exergy. In this manner, it is possible to apply Jørgensen's exergy concept to the process of interaction between solar radiation and a reflecting, transforming, and absorbing surface.

2.3.2. Exergy and information calculations

Using the Kullback measure for the increment of information, we can rewrite exergy as:

$$Ex = E^{out}K + E^{out}\ln\frac{E^{out}}{E^{in}} + R,$$

where E^{out} is the outflowing energy, K is the information, E^{in} is the inflowing energy, and R is the fraction of incoming solar energy that is absorbed by a surface. When we speak about the energy of radiation (incoming or outgoing), we imply a flux of energy. Therefore, the exergy of radiation implies a flux of exergy. Since $E^{out} = E^{in} - R$, the expression for exergy can be considered as a function of two *independent* variables, K and R, and one external parameter, E^{in}. The function $Ex = Ex(R,K)$ for fixed E^{in} increases monotonically with an increase of K.

Now we introduce two new definitions. The ratio $\eta_R = R/E_{in}$ is called the *radiation efficiency coefficient*. This value describes a fraction of total energy absorbed by the surface. Analogously, the ratio $\eta_{Ex} = Ex/E_{in}$ is called *exergy efficiency coefficient*. Since exergy is a measure of "useful" work a system is able to perform, we can say that η_{Ex} is an efficiency coefficient of some "radiative" machine, which is our active surface. The working process of this machine is an interaction of incoming radiation with the active surface.

This allows us to view the surface as two components, one that operates as a classic thermodynamic machine, performing mainly mechanical or chemical work, and another that operates as an information machine, producing mainly information. If radiation (energy) balance were a measure of mechanical work of the thermodynamic machine, then the value of Kullback's measure, K, could be a measure of "information work". At one extreme, the system works only like a classic thermodynamic machine and at the other it has the "non-classic" information branch working only as a purely informational machine producing an infinite amount of information. Substantively, the active surface is a composition of these two ideal machines, and it produces both mechanical (or chemical) work and information.

As the system approaches an ideal classic thermodynamic machine the exergy efficiency coefficient increases with decrease of the radiation efficiency coefficient, i.e., with the decrease of energy absorbed by the active surface. In the limit, when the absorbed energy tends to zero, the exergy efficiency coefficient tends to a non-zero value, being equal to K. At first view, this is a paradoxical result. But it is not surprising, if it is remembered that the energy content of any information is very low (Volkenstein, 1988), especially in comparison with classic thermodynamic (for instance, thermal) processes. On the contrary, the value of the exergy efficiency coefficient does not depend on K since the main performance of a classical machine is to produce no information but only mechanical work.

2.3.3. Seasonal dynamics of exergy and the energy balance

Let us consider seasonal dynamics of the following observed and calculated parameters: energy balance, exergy, exergy and radiation efficiency coefficients, and the Kullback measure for a particular case study. In principle, all parameters follow an annual radiation balance dynamic. But there are differences, which can be used to characterize the specific behavior of different sites, as was done with exergy in section 2.2. For example, during seasonal dynamics, exergy is typically higher than energy balance, but during the summer, when productivity of plants is highest, this difference becomes smaller.

For the case studies investigated, crop field values of all parameters are a bit lower than for the forest. Looking at exergy and energy balance, it can be seen that the difference between the two parameters is bigger in the forest. This means that the forest not only absorbs more energy from solar radiation but it also has a much higher flow of exergy. This trend especially becomes obvious during the summer. Lower values of the crop field efficiency may be due to a higher emission of long-wave radiation, which is the result of a higher temperature. A thin layer of vegetation usually evaporates less and therefore has a lower cooling ability (Herbst, 1997). Incoming energy cannot be dissipated in the crop field to the same extent as in the forest. Another reason for the lower exergy on the field is that albedo of crops is higher, which results in a lower amount of available energy. Therefore, the forest uses incoming solar radiation both more efficiently and more effectively; per unit of absorbed energy, more exergy can be dissipated, whereas the amount of absorbed energy is also higher. Exergy efficiency is higher than energy efficiency, but during the summer their values almost coincide. For the crop field, the values of these parameters are slightly lower. All these results allow us to formulate the following teleological hypotheses:

(1) Vegetation works as an information machine in the course of almost the whole year.
(2) Exergy (Ex), energy balance (R), and the increment of information (K) achieve maxima when the productivity of vegetation is also maximal (usually in June–July). The same statement is also true for the exergy and radiation efficiency coefficients.

Before formulating a third hypothesis, it is necessary to remember that the exergy efficiency coefficient is a function of two independent variables, energy efficiency coefficient and information content. When η_{Ex} increases in the process of growing productivity, this means that the increase is the sum of two independent processes. The first is an increase of η_{Ex} with increase of K, the second is a *decrease* of η_{Ex} with increase of η_R. Using the terminology of game theory we say that in order to maximize productivity (under corresponding constraints) the vegetation uses a *minimax* strategy.

Finally, a third hypothesis can be formulated:

(3) In the process of productivity maximization, vegetation follows a *minimax* protocol, minimizing the exergy efficiency coefficient with respect to the radiation efficiency coefficient and maximizing it with respect to the increment of information.

This hypothesis is consistent with the results of the previous section and can be considered a generalization of Jørgensen's (1992b) exergy-maximization principle, although applied to one concrete case, i.e., to the interaction between solar radiation and vegetation. If the third hypothesis is a *minimax-exergy principle*, then we get a relationship between such originally unconnected values as the Kullback measure K and the radiation coefficient, R. It is natural to assume that these values are independent if our "active surface" is passive in relation to incoming radiation. However, if the "active surface" is really "active", i.e., it actively interacts with incoming radiation partly using it and transforming its spectrum (like a vegetation cover), then we can expect these values will be connected.

2.4. Emergy and exergy

2.4.1. Preliminary observations

As a measure of network organization, *emergy* (see definition in glossary) can be used to evaluate both environmental and economic processes. It considers a system with larger boundaries, encompassing all inputs contributing to form a product, including environmental services usually regarded as "free" in energy analysis. Emergy accounts for nature's "labor" required to generate a given product or flow. Ratios of renewable emergy to non-renewable or imported emergy reflected in a process can also serve as useful sustainability ratios (Brown and Ulgiati, 1997; Ulgiati et al., 1994, 1996).

Spatial areas with convergence of emergy, such as cities, will have a higher concentration of emergy and empower than areas with less emergy, such as forests (Odum, 1996). Emergy density measures the emergy per area of land or water surface. Emergy per unit area is similar to measures of development density used by city planners and to the concept of ecological footprint as developed by Wackernagel et al. (1999). This areal *empower density* (see definition in glossary) is useful in identifying the centers of energy hierarchy and quantifying the extent of environmental support.

Emergy (sej) and empower (sej/t) are related to usable energy (exergy, J_{ex}) and exergy flow (J_{ex}/t) by the fact that they measure the history of energy content (J) and flux (J/t) in a system whereas exergy (J_{ex}; see glossary) and exergy flow (J_{ex}/t) specify usable portions at points to which boundary-input energy has passed. To express emergy analysis in exergy terms, it is necessary to measure usable rather than total energy storages and flows in a system. The two measures would also, obviously, be numerically different. Calculations are often made of empower, usually on an annual basis (sej/y) to average out seasonal oscillations. When a CAHSystem follows a process of selection and organization, the ratio of exergy to empower, $J_{ex}/(sej/t)$ with time dimensions (T), expresses energy-use "efficiency" in terms of time required (Bastianoni and Marchettini, 1997), and the ratio of exergy flow to empower (J_{ex}/t)/(sej/t), gives a dimensionless measure of the information and organization already in place to transform new usable inputs. The higher these ratios, the more efficient is the system in transforming available energy into structure and organization.

Might time-dimensioned exergy/empower ratios, $J_{ex}/(sej/t)$, be taken as orientors with the objective of minimizing the time required to produce a product? In general, assuming both emergy and exergy as orientors, it is suggested that CAHSystems *first* tend to maximize empower (sej/t) and *then*, as a consequence, their internal organization will be reflected in higher exergy/energy ratios (J_{ex}/J). The two orientors are consistent, and the optimization process can proceed in a negative feedback loop. Therefore, exergy/empower and exergy/energy ratios would be expected to be lower early in any developmental process and increase as the system adapts and "learns" efficient use of resources to build up (as stored exergy) more and more organization.

2.4.2. Application: watershed organization and management implications

Watersheds are naturally engineered, multi-step, cascading treatment processes for materials draining downhill in a basin. A natural energy hierarchy is formed when material and energy flows from a landscape scale converge on a watershed's focal point. The watershed also has a hierarchy of human settlements with an economy and concentration of information. Developed areas recycle materials to the watershed in the form of runoff and its constituents, and the intervening natural terrestrial systems sequester nutrients on the way to the watershed focal point, a lake for instance. If these treatment stages are decreased or eliminated, the larger-scale watershed process is short-circuited, creating pulses and increased convergence of materials and their energy within the lake.

As human developments increase in watersheds, impervious surface areas increase. Consequently, water retention capacity decreases, but it is not just stormwater runoff that increases over time. Patterns of delivery change, developing more complex networks that differ with parent material. Many points of convergence develop in material and energy flow patterns, but emergy transformities (ratio of boundary input to each internal flow, see definition in glossary) higher than normal for a particular material only occur at certain junctions in the watershed (Brandt-Williams, 1999). These junctures are appropriate points for intervention and simulations indicate their effectiveness at reducing lake loads to predevelopment levels.

The network that develops because of human influence develops at a much faster rate than that of the geologic processes dominant in the original organization of the watershed, and also faster than terrestrial processes and structures useful in intercepting runoff nutrients. Eutrophication of surface waters results. However, it is likely that in the longer term organization of the watershed, the concentration junctions will adapt to utilize the additional energy available, building structure capable of retaining nutrients and maximizing the total energy throughflow for the entire watershed, not just the focal point.

Fox (1985), in discussing energy-driven systems states, "Energy flow is a necessary but not a sufficient condition for the living state of matter ... the living state is as much a consequence of special substances and their emergent properties as it is a consequence of energy flow". Phosphorus, as both material and energy transducer in production of evolutionary building blocks, represents a special substance with emergent properties sufficient for evolution (Fox, 1985). Phosphorus is usually the material limiting productivity in freshwater (Wetzel, 1983), and water moving downhill provides the kinetic energy necessary to bring phosphorus into a lake. However, mass quantities alone often disguise the real power of the material in relation to other materials in the system. Phosphorus, usually a limiting nutrient, is measured in g, while water is best dimensioned with cubic meters and sediment with metric tons, making it difficult to use simple temporal models for in-lake functions.

Emergy better defines the real power in phosphorus and provides a higher ranking by including the previous energy used in moving the element into a concentration usable by primary producers (Brandt-Williams, 1999). As runoff moves downwards through the watershed, the actual concentration of runoff constituents may not change significantly. However, the spatial concentration of water and phosphorus increases and carries with it the emergy of all the runoff and constituents used in moving to the point of concentration. This includes the geopotential emergy inherent in the watershed slope.

Changing patterns of emergy hierarchy in the watershed can be used to develop adaptive intervention strategies (Brandt-Williams, 1999). Mapping phosphorus transformities results in a hierarchy of small and mid-sized areas of higher concentration that can be useful as retention areas on land, similar to the natural hierarchy of wetlands formed in many Florida watersheds (Brown and Sullivan, 1987). This is undoubtedly a modified network from a pre-development era, and it illustrates a way to incorporate ecological principles of adaptive spatial hierarchy organization into watershed management.

2.4.3. Applications: exergy/empower ratios

The exergy to empower ratio has been applied to various aquatic ecosystems of different types:
- Two water bodies used for comparison are in North Carolina, USA, and are part of a group of similar systems planned and constructed to purify urban wastewaters. Of six ponds near Morehead City, three are "controls" that receive a mixture of estuarine waters and purified waters from the local sewage treatment plant and three are "waste" ponds that receive estuarine waters mixed with more polluted, or nutrient-rich, wastewaters. Plants and animals were introduced into the ponds and surrounding land to create new ecosystems by natural selection. The different conditions have produced quite different ecosystems in the two types of ponds, with phytoplankton and crustaceans prevalent in the waste ponds and a great abundance of aquatic macrophytes in the control ponds.
- The third water body used for comparison is the lake of Caprolace in Latium, at the edge of Italy's Circeo National Park. This is an ancient natural formation fed mostly by rainwater, plus an input rich in nitrogen, phosphorus and potassium that percolates from nearby agricultural land. Human impact is low. A sustainable quantity of fish is taken each year, which does not decrease the populations.

Table 2.1
Exergy/empower ratios for five example systems

Quantity[a]	Control pond	Waste pond	Lake Caprolace	Lake Trasimeno	Figheri basin
Empower density (sej/y)/ℓ	20.1×10^8	1.6×10^8	0.9×10^8	0.3×10^8	12.2×10^8
Exergy density (J_{ex}/ℓ)	1.6×10^9	0.6×10^9	4.1×10^9	1.0×10^9	71.2×10^9
Exergy/empower ($J_{ex}/(\text{sej}/y)$)	0.8	0.2	44.3	30.6	58.5

[a] Units: y, years; ℓ, liters.

- The fourth ecosystem is Lake Trasimeno in Umbria, among the biggest lakes in Italy. It is not included in a National Park but it is quite well protected since the surrounding area is not heavily developed.
- The fifth system analyzed is an aquaculture basin in the southern part of the Lagoon of Venice. Fish-farming basins consist of peripheral areas of lagoon surrounded by banks in which local species of fish and crustaceans are raised. Salt water from the sea and freshwater from canals and rivers are regulated by locks and drains. Control of water levels, salt content, and drainage towards the sea are part of an ancient tradition that is an economic and cultural heritage. Man learned how to exploit the instinct of certain species of fish that enter the lagoons, delta, and coastal ponds attracted by available food and calm waters. In spring, young specimens and adult fish come from the sea. Traditionally, they were herded into the basin where they grew quickly in the shallow nutrient-rich waters. The same principles are exploited today, however the basins are stocked with artificially raised juveniles or fry. In autumn, the fish would normally return to the sea to reproduce, attracted by warmer water. Now, they are directed into special structures that act like traps by regulating the flow of water. There they are selected on the basis of size and type.

Table 2.1 shows empower and exergy density values and their ratios. Densities were used to enable comparison between ecosystems of different dimension. The Figheri basin is an artificial ecosystem, but has many characteristics typical of natural systems. This depends partly on the long tradition of fish-farming basins in the Lagoon of Venice, which has "selected" the best management practices.

Table 2.1 shows that the natural lake (Caprolace) has a higher exergy/emergy ratio than the control and waste ponds, due to higher exergy density and a lower empower density. These observations are confirmed by the study of Lake Trasimeno, carried out by a research group at Perugia University. In natural systems where selection has acted undisturbed for a long time, the exergy/emergy ratio is higher and decreases with the introduction of artificial stress factors. The human contribution at Figheri basin is manifested as a higher empower density (of the same order as that of artificial systems) than in natural systems. However, there is a striking difference in exergy density, with much higher values than in any of the other systems. The fact that Figheri can be regarded as a stable ecosystem makes this result even more interesting and significant.

2.5. Integration of orientors

2.5.1. Introduction

One of the defining characteristics of CAHSystems is that they are adaptive. In section 1.2.1, adaptation was referred to as the property in which a system maintains deviation from thermodynamic equilibrium. However, the question was also raised regarding adaptation to what ends (in the terminology of section 1.2.1, what are the "assertive tendencies"?). The very nature of adaptation implies orientation toward or at least away from something, that there is some direction, force, or organizing principle that moves a system toward or away from a particular state. The question becomes the following: As a system adapts, to what is it orienting? Is there a general pattern or organizing principle that can explain, model, or even anticipate this orientation? In bioevolutionary theory one universal orientor is used, *fitness*. In ecology above the organism level, the search for organizing principles that apply to adaptation has produced a variety of energy "orientors" (Müller and Leupelt, 1998), as previously discussed (sections 2.2 and 2.4).

As a precursor to the orientor debate, Odum (1969) hypothesized about the trends to be expected in ecosystem development. In this seminal work, Odum proposed several metrics to indicate the direction of ecological succession. Many of these, such as increasing biomass, cycling, internal organization, residence time, and information have been further expanded and investigated as potential ecosystem orientors. More recently, this orientation has been described in ecological systems using thermodynamic principles such as those discussed in the previous sections. In particular, Bendoricchio and Jørgensen (1997) made the case that the primary ecosystem goal function is exergy storage and Schneider and Kay (1994a) suggested exergy degradation. Bastianoni (1998) and Bastianoni and Marchettini (1997) suggested minimum empower to exergy ratio as the primary ecosystem goal function. Johnson (1981, 1994), Choi et al. (1999) and Jørgensen et al. (2000) suggested specific dissipation as the primary pattern observed in growth phenomena.

Each goal function was developed for a particular purpose and gives useful information about the system to which it was applied. However, their abundance has led to some confusion, especially among those outside the immediate community that engages in this research. Others interested in and looking for guidance on what metric to use to measure the organizing principles of an ecosystem saw largely a confusing, competitive promotion of various goal functions for use with various systems. Rather than focusing on differences, highlighting similarities, usefulness, and applicability of goal functions would be more useful to others. Recently, the similarities between different approaches have been a focus of study.

2.5.2. Integration

There is a consensus building in ecological theory that many suggested goal functions are consistent, each giving slightly different, but complementary information about tendencies in ecosystem development. Several authors have investigated these similarities. Jørgensen (1992b, 1994) found a strong correlation between several of them and suggested that perhaps their integration could lead to consideration of only one of them. Patten (1995) showed that many goal functions have a common basis in the path structure and associated microscopic dynamics of systems. In section 2.2.1, experimental data from Slovenian forests illustrated how exergy storage and exergy dissipation can be merged into a single new hypothesis: systems degrade exergy input as quickly and as fully as possible and while doing so maximize storage. These results (of Debeljak) help resolve the divergence between proponents of maximizing dissipation (Schneider–Kay) and maximizing storage (Jørgensen–Mejer). Fath et al. (2001) used a network analysis method to develop a theory that gives a common basis for both ideas because it allows, against dissipation which must be bounded by prior energy acquisition and utilization efficiency, the indefinite development of total system storage through increased organization. This theory is consistent with Debeljak's empirical work.

Fath et al. (2001) use network analysis to demonstrate the consistency of ten goal functions. These include maximum power (Lotka, 1922; Odum and Pinkerton, 1955), maximum storage (Jørgensen and Mejer, 1979, 1981), maximum empower and emergy (Odum, 1988), maximum ascendancy (Ulanowicz, 1986, 1997), maximum exergy degradation (Schneider and Kay, 1994a,b, 1995, 1996), maximum cycling (Morowitz, 1968), maximum retention time (Cheslak and Lamarra, 1981), minimum specific dissipation (Onsager, 1931a,b; Prigogine, 1955), and minimum empower to exergy ratio (Bastianoni and Marchettini, 1997). Two fundamental system properties, total system throughflow (power) and total system storage (exergy), were taken as reference conditions upon which the others build. Such results show that these seemingly disparate extrema are all mutually consistent, suggesting a common pattern for ecosystem development. The point here is not to reprint the details of the integration, but rather to highlight its significance in context to CAHSystems.

The ten energy-organizing principles discussed above are all based on increasing boundary flow and three primary internal properties: first-passage flow, cycling, and retention time. Boundary flow, along with first-passage and cyclic flows, all contribute to increasing total system throughflow. Boundary flow follows from exogenous inputs and first-passage flow from endogenous transfers. Cycling is a function of system connectivity, organization, and efficiency. Retention time depends on cycling and system structure because these mechanisms retain and store flow, thus increasing the retention time. Cycling at one scale can appear as structural storage at another. The reliance on basic network configurations adds to the congruence of thermodynamic properties. Not only are all ten energy-organizing principles consistent with these basic properties: input, first-passage flow, cycling, and retention time, but they are interdependent for fulfillment. For example, maximizing dissipation and maximizing cycling both contribute to maximizing total system throughflow and maximizing throughflow contributes to maximizing total system storage, subject to turnover considerations. The use of a plurality of goal functions is warranted because each organizing principle reflects a slightly different aspect of overall system function. A next step is to see how these orientors might be preferentially prioritized during different stages of ecosystem development (Jørgensen et al., 2002).

2.5.3. Conclusion

A goal of this workshop was to build bridges to other disciplines. Whereas before the goal-function message was muddled because there were conflicting messages among us, now there is growing agreement, if not consensus, that many of the different thermodynamic orientors that ecosystem theory has produced are in fact consistent with each other and complementary. All the details are still not worked out and there will always remain minor discrepancies, favorites, rivalries, or priorities. As was stated in the introduction to this chapter, it may be well into the present century before CAHSystems are understood well enough to have an unambiguous statement regarding their behavior and the organizing principles that guide them. But, the integration of ecological goal functions provides a common basis and strong foundation to build a testable and applicable theory of ecosystem development. It now may be possible to take this message, however incomplete it is, that ecosystem development does tend to follow a natural, thermodynamically based pattern and that there is a suite of metrics that can be used to measure the direction and degree to which these systems have developed. Future research will include finding better empirical methods to arrive at these values, better and more consistent ways to model such systems, and better ways to include anthropospheric (and other anticipatory) systems into the theory. Such a program will continue to advance and test orientor theories. A main reason to have such a theory of ecosystem development is the hope that this may improve our understanding of how combined natural and anthropospheric systems behave. Anthropospheric systems might not follow exactly the same orientors or patterns of development, but ultimately they too are bound by underlying thermodynamic constraints. In particular, we wish to be able to assess whether the direction and degree to which these systems are developing is sustainable. The orientor approach provides a potentially extremely useful tool for that purpose.

2.6. Adaptation and hierarchy, again

Most CAHSystems have not completely "oriented" in accordance with the organizing principles mentioned above. There are likely many reasons for this but two classes of explanations in particular are worth highlighting. The first is historical, dealing with the structural and functional constraints and inertia in the organization of CAHSystems. For example, there exist evolutionary constraints and inertia in the nature of biochemical, genetic, and morphological structures and interactions that preclude the complete optimization of biological CAHSystems. The second class of explanations is ahistorical, dealing with the action and interaction of the organizing principles themselves. All CAHSystems are subjected to a myriad of forces pulling and tugging in different directions. As such, there must exist a balance or quasi-balance of forces that maintains CAHSystems poised in a nonequilibrial position (i.e., some quasi-steady state, much as Bak's (1996) sandpiles; or Choi et al.'s (1999) size–abundance relationships). The goal-function integration described above shows a consistency among the goal functions. However, this consistency does not rule out an interplay among them or variations in the relative strength or intensity of their manifestations over different temporal, spatial, and organizational scales (Choi and Patten, 2002).

These historical and ahistorical explanations merge when they are re-interpreted within a framework analogous to that of Sewall Wright's (1988) "adaptive fitness landscapes". In such a framework, the CAHSystems' organizing principles represent different dimensions or facets of the well-known fitness function of evolutionary biology. The form of this adaptive landscape reflects the interplay between the CAHSystem with both its internal and external milieu (i.e., the environ or holon). Historical constraints or inertia allow only a limited set of dynamics in the landscape by dictating the form of the landscape. The physical and contextual landscape in which these systems operate changes with the activity of the system, altering the landscape in which it functions. Any attempt to orient is slowed and hindered by the ever-changing landscape in which it operates. This is the Red Queen effect (see glossary) in which one runs faster just to stay in the same place. It may explain why systems haven't already oriented to a fixed position. Also, long-term evolution of the planet precludes realization of such a fixed position. The goal-function properties can be used to determine the likely direction of adaptation, but they do not imply that system must reach their final "oriented" states.

3. Systemicity

CAHSystems are "systemic" in the sense that they act, adapt, and change as constituted wholes. Wholeness is of the essence in CAHS-organization. Decomposition is necessary to achieve the appropriate level of understanding of structures, processes, and dynamics – the traditional scientific method accomplishes this very efficiently. Integration is necessary to achieve an understanding of the "wholeness" of CAHSystems. Modeling represents a powerful method of integration of theory, empirical knowledge and scientific rigor. A distinction must be made between models and modeling. The first is "product" and the second "process". The use of a modeling approach is promising as an aid in understanding and "managing" (control, predictability) whole CAHSystems.
- The pragmatic challenge is to integrate models and the modelling process into participatory assessment studies in an optimal manner.

In the following, three such approaches to dealing with embedded systems such as CAHSystems are highlighted. The first deals with the treatment of the global system as a whole, and asks the question, How can we live and interact in a coherent fashion with the enormous complexity of CAHSystems? The second subsection deals with the interaction of human social, economic and policy systems with the environment and asks the question, How can we interactively negotiate a new and hopefully more sustainable balance? The final subsection describes a methodology in integrative assessment and resource allocation at landscape levels of organization and asks the question, How is it best to approach systems that have dynamics that interact at numerous spatio-temporal scales (much as the Petri-net approach of section 1.2.3 and more generally section 1.2)?

3.1. Eco-anthropic CAHSystems

3.1.1. Complexity: eco-anthroposphere commingled

The "world as a whole" is a case of a CAHSystem requiring integration between anthropospheric and ecospheric components based on ecological principles. There are several significant parts of the anthroposphere, both objective and subjective:
- *"the true"* – science, technology, economy;
- *"the right"* – ethics, religions, theology;
- *"the beautiful"* – aesthetics, art, literature, music; and
- *"the whole"* – commons in society, the "bonum commune".

The ecosphere is another "whole" on the other side. The practical aim of CAHSystems theory is to integrate the eco- and anthropospheres in order to fully reconcile economy and ecology (Moser, 2000a,b).

An integrated eco-anthroposphere of great complexity and heterogeneity will have a number of salient characteristics: (1) a great number of variables with unidentified interdependencies resulting in interdependent model parameters;

(2) a complex network of laws required for holistic descriptions exhibiting pseudo- and eu-nonlinearities (section 1.1.2), without obvious cause-and-effect relationships; (3) experimental quantification at best uncertain and at worst impossible; (4) subjective and qualitative aspects difficult to verify with conventional models; and (5) initial and boundary conditions largely unidentified, and little quantified.

Due to the intractable complexity of the global CAHSystem, an approach to human sustainability has been developed based on a number of "macroscopic" orientors, or goal functions. These involve the eco-active surface area of soils as a kind of conservative but exhaustible new "currency", in general resulting in an "Eco-Social Product" instead of the Gross National Product as a holistic index of sustainability. The orientors include: (1) the anthroposphere/ecosphere ratio as a measure of ecological "security"; (2) money earned per unit area of anthroposphere as a measure of economic viability; and (3) the ratio of jobs per area of anthroposphere as a measure of social equity. The latter two indices refer to a region with a certain area of eco-active soil and a given number of people.

3.1.2. An integrative methodology

In the 20th Century a methodology for "the whole" was stressed by philosophers like Teilhard de Chardin and systems scientists such as L. von Bertalanffy, U. Maturana, and J.F. Varela, but nothing practical was ever designed. To adequately handle "the world as a whole" a two-phase integrative methodology is envisioned:

(1) *"Deep science" approach* – Understanding science in general as a process directed towards progress in cognition, by: (1) setting up hypotheses, often in mathematical terms; (2) looking for experiments and experiences; and (3) comparing the above to produce falsified or verified hypotheses.
(2) *"Macroscopic pattern" analysis* – Due to the impossibility of employing traditional reductionistic, mechanistic, and microscopic approaches of conventional science, the "wisdom of the whole" is elucidated by analyzing macroscopic patterns in the ecosphere, economy, society, policy, philosophy, and sustainability. Essential to understanding this macropattern approach is that, using the concept of the deep science approach, the main focus becomes the "picture of the whole", looking on this without prior bias and with no theory in mind. One example is the macropattern of nature where high diversity of components and self-organized interactions result in sustainability, evolution, and beauty. Often the structure of reality is so complex that the model of "the whole" can only be described based on analogous or even arbitrary mathematical functions, which are fitted to data. These models may not represent reality as it really is, but they must (as in quantum mechanics) conform to empirical data – "mirror the truth!" To deal

with the enormity of the unknown, "eco-principles" (translated from formal analogies) can be used as the basis for modeling, rather than speculative cause–effect relationships. The integrative basis thereby shifts from causality to plausibility.

The significant innovation here is that not only is rational thinking used, but also "right-side-of-the-brain" feelings and ideas. Spiritual, emotional, and aesthetic dimensions of human reality are taken into consideration, leading to a holistic model of external global reality.

There are four "eco-principles" that must be considered:
(1) *Sufficiency* – everything on earth is limited;
(2) *Efficiency* – bounding natural and human capitals within limits;
(3) *Embeddedness* – expressing the integrative tendency leading to everything on earth being interconnected; and
(4) *Non-invasiveness* – anti-assertiveness, necessary to achieve full embeddedness based on the recognition that humankind is bounded by carrying capacity of both the ecosphere (production, assimilation, evolution of life, etc.) and anthroposphere (health, basic needs, jobs for people, etc.).

These eco-principles would provide guidelines for eco-restructuring the anthroposphere in all its aspects: (1) eco-social economy, having ecological and social boundaries for economic value and a self-organizing pattern of behavior inside; (2) eco-social product (ESP), to replace the GNP; (3) eco-technology, limited by and adapted to natural capacities; (4) ecosophy, the "new" philosophy based on nature's wisdom; (5) eco-social ethics, holistic ethics based on non-invasiveness; (6) eco-literacy, the "new" education system based on "the whole" earth; (7) eco-theology, the "new" belief system also based on "the whole"; and finally, from the above (8) "Deep sciences", including the macro-approach; and (9) "Deep arts", based on universally shared spiritual values.

Two distinct achievements of this new approach are foreseen to be (1) a higher problem-solving capacity, which would lead to (2) a "deep sustainability" of a balanced eco-anthroposphere, featuring economic security achieved by regional eco-technologies, social security achieved through jobs created in eco-social economies, and nature respected and its healthy evolution sustained based on ESP's (above).

3.2. A polycentric approach to integrated assessment

Transitions towards sustainability will require major changes in today's socio-economic systems. Such changes cannot be brought about by conventional policy measures. We advocate a new approach of a polycentric understanding of policymaking that invokes instances of social learning at different levels of societal organization (Minsch et al., 1998; Pahl-Wostl, 2002). Such an approach is based on the idea that human–environment systems are CAHSystems.

The notion of polycentric involves the relationship of different levels of human choice and geographical domains. The spatial component involves the sequence local–regional–national–global. It involves the combination of different types of human choice at different levels of societal organization (e.g., legal regulations, taxes, subsidies, local initiatives). Trying to understand what is the impact of dealing with diverse "global change phenomena" at diverse levels of organization will be one of the central tasks of institutional theorists studying global change processes.

Social learning is defined here as the mutual shaping of expectations of the involved groups. Shaping of expectations depends on institutions. In light of the new approaches of institutional economics, an institution may be defined very broadly as shared rules of human conduct. For example, if one is driving on a road one expects other drivers to respect the red light and stop. Without such shared rules of conduct, life in a society would be impossible. Some institutions (laws) are enforced by legislation (e.g., traffic regulations). Others (customs) are shared by the members of a society and evolve and change in a social setting (e.g., shake hands for welcome). Rules enable individuals to form expectations concerning the actions of others. One obstacle to change is to overcome lock-in situations where mutually dependent expectations stabilize each other. An important research question is the development of concepts and models of institutional change that focus on the mutual relationship between processes of social and individual learning, between individuality and the embeddedness in social networks.

Models of collective choice processes should link theoretical and applied research and approaches focusing on agents (complex individuals) and on system behavior (interaction in social networks). Starting from the focus of complex individual agents and acknowledging cognition as a source for uncertainty and even indeterminacy is rather new. Cognitive psychology itself offers here important insights. Subjective probabilities in the micro-economic sense of bounded rationality depend only on the state of information. Two actors with the same state of knowledge should per definition have the same subjective probabilities. In a more advanced perspective, one has to acknowledge that the processing of information is inherently subjective. Personal values, previous experience, the embedding in a social network define what one may call a cognitive filter for the acquisition and processing of information. Here it is more appropriate to use the concept of dynamic belief networks.

To foster processes of social learning modeling approaches should be closely linked to participatory processes as outlined in fig. 3.1.

Figure 3.1 suggests relationships between different activities of research and different types of models to derive a new research agenda for improving the understanding of human–environment systems and for approaches to joint problem solving in participatory settings. Agent-based models may be particularly

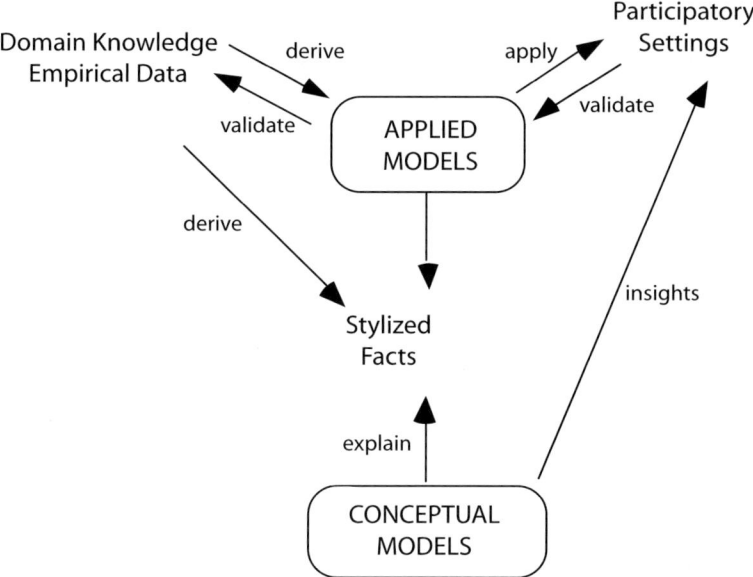

Fig. 3.1. Linking social learning modeling and participatory processes.

suited for applications in participatory settings to foster processes of social learning and to provide assistance in specific decisionmaking processes.

The suggested tight link between modeling and participatory approaches implies to derive an improved understanding of how one can build models in a participatory fashion and how to apply them in participatory settings. Applied agent-based models are based on factual knowledge derived from domain experts and empirical data. Validation in the participatory setting is based on their potential, how they structure the debate about a problem by integrating factual and local knowledge.

3.3. Eco-geological assessment towards sustainable coastal development in Yogyakarta, Indonesia – scale adjustment to observe and analyze CAHSystems

3.3.1. Introduction

Eco-geological systems are natural systems in which geologic phenomena significantly control all processes (physical, biological, social, and economic) and significantly shape their interactions. As discussed by UNESCO (1998), eco-geological systems are complex and unstable. Change is the rule, not the exception, and surprise is common. These systems rarely remain long in equilibrium as they adjust continually to new conditions of, for example, climate, hydrology, and sea level. Some changes are sudden, catastrophic, and newsworthy. However, there is also a background of continuous small-scale change whose cumulative impact over time may be of even greater significance. In geo-ecological systems, not all environmental change is anthropogenic.

There is a need to discover ways to observe and analyze such volatile systems. Observation and analysis are commonly focused on fixed system characteristics, on changes, and on human impacts, as well as on interactions between these. To enable such observation and analysis, simplification by adjustment of space and time scales is crucial. Some changes may reflect global trends, but it is important not to generalize all local and regional changes to the global scale. For example, not all local increases in sea level are necessarily indicative of global sea-level rise.

3.3.2. Subject system and methodology

The south coast of Yogyakarta, Indonesia is a landscape-scale CAHSystem. It has two distinctive landforms, the Coastal Highland and Coastal Lowland. The coast is dynamic due to unique geological phenomena. These involve interactions of volcanic, fluvial, aeolian, and marine processes, as well as limestone solution in a karsted area. These interactions control the development of landforms and also impact regional human populations.

Human and natural complexity must be taken into account in setting up a sustainable coastal land-use management program aimed at minimizing and mitigating conflicts between societies and environment as well as those within and among societies, and also degradation of the natural system. The approach advocated and taken involves adjustments in both space and time scales.

To observe the eco-geological phenomena of a region, observations and analyses must first be carried out on a regional scale where, as hierarchical organization establishes, only a relatively few selective components are significant. Based on preliminary indications, some local areas and distinctive processes that impact human populations can later be singled out for further discrimination at smaller scales. Thus, the first stage of observation and analysis is aimed at overviewing the system, and more refined investigations follow from this.

Regional landscape complexity is observed over a large spatial scale (say 1:50 000) encompassing a variety of geological elements:
- *geomorphology*, with sub-components relief, landform conditions, and exogenic processes such as erosion and sedimentation;
- *geology*, with sub-components lithologic types and their distributions;
- *geodynamics*, with sub-components including volcano-fluviatile, marine, and aeolian processes;

- *hydrogeology*, with sub-components aquifer characteristics and their distribution and hydrogeological dynamics; and
- *environmental geology*, with sub-components geological resources such as mining deposits, soil resources, water resources, and geological hazards such as landslides, coastal erosion, and flooding.

In accordance with the above, a regional overview of geomorphologic, geologic, geodynamic, hydrogeologic, and environmental–geologic conditions pertaining to sustainable coastal management was obtained.

Next, zoning or clustering systems along lines of natural decomposability ("H" property) as shown in fig. 3.2, aids management by simplifying the process of identifying the limiting resources and constraints for each eco-geological unit. As such, zoning aids in minimizing and localizing conflicting processes and needs, as well as ultimately the balanced and sustainable (controlled and planned) utilization of resources within the system and across its subsystems. Two such zones in the most developed parts of the coastal system were selected for investigation at more detailed spatial scales, which allowed the identification of the key processes driving the eco-geological systems: cliff destabilization for Parangtritis Beach and coastal abrasion for Baron Beach.

The geomorphology, geology, geodynamics, hydrogeology, and environmental geology of Parangtritis Beach were observed at a spatial scale of 1:25 000. Clustering or zoning was also conducted to decompose this area by similarities and differences observed at the smaller scales (fig. 3.3). The most important conflict observed at these smaller scales was between human activities and the structural stability of the environment, as the development of human structures greatly destabilized the cliffs adjacent to the beach. Numerical modeling (by finite-difference analyses) of cliff stability was conducted at spatial scales of 1:400 which allowed the evaluation of how cliffs can be stabilized with existing structures, and also the prediction of the impact of potential future development of the area. Thus, time scale also entered into the evaluation. The conclusions were that a protective buffer of 50 m to 300 m from the cliff edge, depending upon steepness, could effectively minimize further damage. No further structures could be permitted in the buffer area and the maximum allowable load of structures would be 1 kg/cm^2 beyond this zone (Karnawati et al., 2000).

Another interesting zone was Baron Beach where the distinctive process observed was coastal abrasion. Again, detailed investigations in terms of time and space were conducted to assess cliff resistance to abrasion and predict the rate of erosion. Multitemporal topographic maps and aerial photographs were utilized in analysis and prediction. The rate of abrasion was found to be about 13 cm per year, and from this the zone required for cliff protection was established at about 50 m (Karnawati et al., 2000).

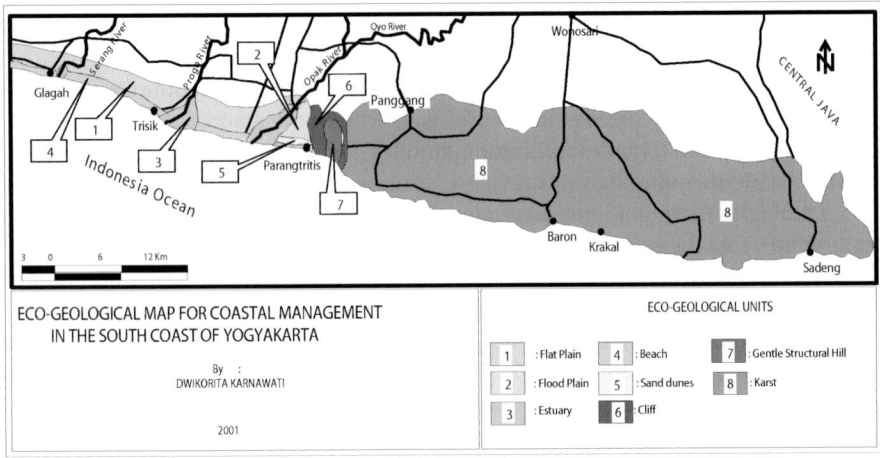

Symbol	Units	Characteristics	Potential Resources	Potential Limitations/Constrains	Present Landuse	Proposed Landused Management
1	Flat plain	Flat topography; silty sand & fairly fertile soil	Shallow groundwater-table (4 to 5 m depth)	Flood	Agriculture (seasonal): Paddy fields (rice, chilly, onion, soya)	No permanent settlements; maintain the development of agriculture
2	Flood plain	Flat topography; silty sand with high permeability	Shallow fresh groundwater table (3 to 4 m depth)	Flood (in particular in early rainy season)	Agriculture (seasonal): chilly, onion, soya	No development of structures; periodically provide through-flow channel to drain water through sandpits into the estuary/ocean; maintain development of agriculture
3	Estuary	Sand (point bar) deposits; fluctuating stream-flow; river shifting towards the west	Magnificent view for tourism & good potential fishery	Flood	No landuse (unutilized)	Eco-tourism; no permanent settlements; fishery, agriculture & sand mining
4	Beach	Fine to coarse sand; flat to moderately steep slope (2° to 15°) extending along the beach with 4 to 40 m width	Magnificent view for tourism	Coastal abrasion, tsunami, backwash of sea waves, flood around the mouth of main rivers, low quality of groundwater	Tourism	Coastal zone protection from permanent structures & sand mining; Eco-tourism
5	Sand dunes	Barchans & transversal dunes (unvegetated); fine grained sand	Shallow, fresh, groundwater table (2 m to 3 m depth)	Dry & un-fertile soil; migration of dunes covers some structures & vegetation	No landuse (unutilized), very few settlements	Natural laboratory of sand dunes; Eco-tourism
6	Cliff	Steep slope (20°-35°) & cliff formed by jointed (fractured) andesite & andesitic breccia	Relatively fertile soil	Unstable slope (susceptible for rockfalls & landslides); intensive erosion; deep water table (more than 15 m)	Dry agriculture	Eco-tourism (tracking area); Conservatory forest; dry agriculture; cliff protection zone should be established
7	Gentle structural hills	Gentle slopes (5°-15°) formed by limestone	Relatively fertile soil	Unstable slope (susceptible for landslides & rockfalls); deep water table (more than 10 m)	Villas, settlements	Minimize the development of structures & settlements; Conservatory forest; Eco-tourism
8	Karst	Limestone with conical hills, tower karst, dolina, & sinkholes	Un-fertile soil; limestone mining; magnificent view for tourism	Deep water table (more than 20 m)	Mining; non-irrigated agriculture fields; settlements	Natural laboratory of Karst; Conservatory forest; isolate & minimize the mining of limestone; Eco-tourism

Fig. 3.2. Eco-Geological Map for Coastal Management in the South Coast of Yogyakarta, Indonesia (Karnawati et al., 2001), based upon regional observations and analyses (scale 1:50 000).

3.3.3. Conclusion

This case study demonstrates that landscape complexity is not only inherent in the characteristics and behavior of natural components and their interactions, but

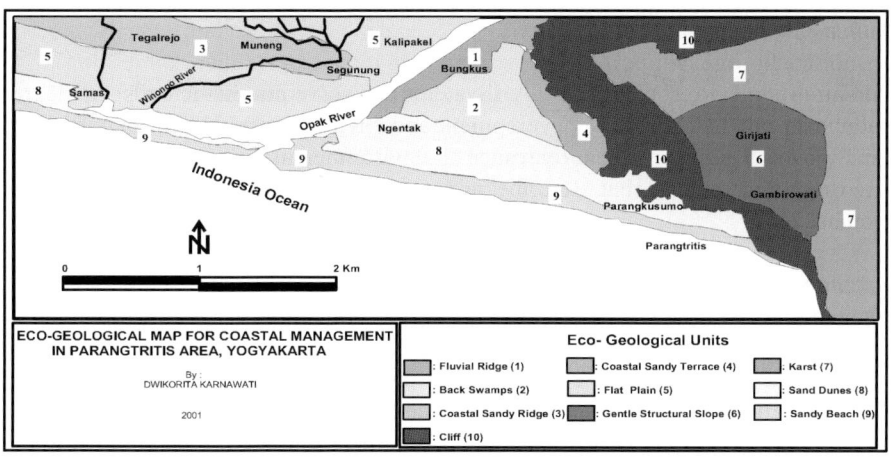

Symbol	Units	Characteristics	Potential Resources	Potential Limitations/ Constrains	Present Landuse	Proposed Landuse Management
1	Fluvial Ridge	Clayey sand; good permeability	Shallow, fresh groundwater table (3 to 4 m)	Flood	Settlements	Development of flood protection system (vegetated system/no concrete) structure; establishment of river protection zone.
2	Back swamps	Flat topography; black silty soil; low soil permeability	Shallow groundwater table (4 to 5 m); soil with high fertility	Flood	Agriculture (seasonal) : paddy field & vegetables (chilly, onion, soya beans)	No permanent settlements; maintain the development of agriculture; development of drainage system to overcome flood.
3	Coastal sandy ridge	Fine to medium sand; vegetated; stable (unmigrated)	Shallow, fresh, groundwater table (2 to 3 m depth)	Soil with low fertility.	Settlements	Agriculture (plantation of vegetation for cattle's food)
4	Coastal sandy terrace	Flat to gentle topography; gravelly sand	Relatively fertile soil	Poor drainage for agriculture	Un-utilized (undeveloped land)	Agriculture (with drainage improvement)
5	Flat plain	Flat topography; silty sand; high permeability	Shallow, fresh groundwater table (3 m to 4 m depth)	Flood (in particular in early rainy season)	Agriculture (seasonal): chilly, onion, soya beans	No development of structures; periodically provide through-flow channel to drain water through sand pits into the estuary/ocean; maintain the development of agriculture
6	Gentle structural slope	Gentle slope (5°-15°) with cliff at the southern part formed by limestone	Relatively fertile soil	Unstable slope (landslides & rockfalls); deep water table (> 10 m)	Villas; settlements	Minimize the development of structures & settlements; Conservatory forest; establishment of slope protection zone; Eco-tourism
7	Karst	Limestone with conical hills, tower karst, dolina, & sinkholes	Un-fertile soil; limestone mining; magnificent views for tourism	Deep water table (> 20 m)	Mining; non-irrigated agriculture fields; settlement	Natural laboratory of Karst; Conservatory forest; isolate & minimize the mining of limestone; Eco-tourism
8	Sand dunes	Barchan & transversal dunes (un-vegetated); fine grained sand	Shallow, fresh groundwater table (2 to 3 m)	Dry & un-fertile soil. Migration of dunes covers some structures & vegetation	No landuse (unutilized) & few settlement	Natural laboratory of sand dunes; Eco-tourism
9	Sandy beach	Fine to coarse sand; flat to gentle slope (2° to 15°); 4 m to 40 m width	Magnificent view for tourism	Coastal abrasion, tsunami	Tourism	Coastal zone protection from permanent structure; Eco-tourism
10	Cliff	Steep slope (20°-35°); formed by jointed or fractured andesite & andesitic breccia	Relatively fertile soil	Unstable slope (rockfalls & landslides); erosion; deep water table (> 15 m)	Dry agriculture	Eco-tourism (tracking area); Conservatory forest; establishment of cliff protection zone; dry agriculture

Fig. 3.3. Eco-Geological Map for Coastal Management in Parangtritis Area, Yogyakarta, Indonesia (Karnawati et al., 2001), developed based on local observation and analysis (scale 1:25 000).

also occurs in aspects of human impacts and management. It is apparent that to observe and analyze landscapes seen as CAHSystems, paying special attention to spatial and time scales is very important.

4. Concluding thoughts

Pursuit of the understanding of CAHSystems can serve pragmatic ends – a clearer understanding of sustainable development and environmental health. In particular, the following questions have relevance to environmental concerns:
- How can indicators that describe and evaluate CAHSystems be derived and quantified?
- What are the basic requirements that must be met to realize sustainable development and environmental health strategies for economic, ecological, and social (sub)systems and how can these parts be integrated across the hierarchy?
- What is the specific role of institutions in sustainable management, and how should they deal with the coevolution of the socio-ecologic complex?
- What are the roles of model products and the modeling process in organizing institutions as agents of sustainable development and environmental health?

The state-of-the-art of CAHSystems knowledge is not such that these questions and issues can be definitively addressed, but some points of consensus are beginning to emerge, and these are considered in the chapter that follows.

Glossary

Adaptation – The process of becoming, or the state of being made suitable for a new use, a new need, or a new situation (Hornby, 1974); change in the excitability of a sense organ of an organism as a result of continuous stimulation (Begon et al., 1990); change in the form or behavior of an organism during its life as a response to environmental stimuli (Begon et al., 1990); the set of organismal characteristics evolved over an evolutionary past through natural selection, resulting in a close match to features of the environment (Begon et al., 1990); and a response of a biological system to its environment so as to accommodate to its constraints and take greater advantage of the environmental circumstances (Allen and Starr, 1982).

Assertive tendencies – Empirically observed tendencies of holons to self-assert, compete, separate, dominate, individuate, etc. and so increase their autonomous, modular, independent natures (Koestler, 1969). This is analogous to the local-order, or least-specific-dissipation, principle (Choi and Patten, 2002).

Asymmetric interactions – Signals that can be characterized and classified according to their temporal and spatial scales.

Canon – The permissible set of rules that govern a holon's activity. These rules are distinguished from the set of "strategies" or protocols by which holons, in interacting with other holons and their associated canons, exhibit or "choose" appropriate rules and so negotiate a balance between the canons (Koestler, 1969; Regier and Kay, 1996).

Complexity – That which is complicated, difficult to understand and hard to explain (Hornby, 1974) which is generally the consequence of multiple inter-relationships between system elements. This complexity increases with the number of interacting units and their interaction intensity (Nicolis, 1986). In information theory, it is represented as a property of an entity that makes long messages necessary to describe the system, and that takes long time periods to develop (Salthe, 1993), or alternately, the minimal amount of information needed to describe a system's structure (Kolmogorov, 1965) or the observer's ignorance of a system (Salthe, 1993). It has also been suggested to be the logical depth of an object (Bennett, 1988); or that, which cannot be further simplified (Chaitin, 1975). Complex systems range across large hierarchical scales: from genomes, to cells and their organelles, organisms, to populations, communities, geosystems and biogeocoenoses, ecosystems and their complexes, and landscapes, human economic and social entities and interactive systems of man-and-nature.

Constraints – Signals that cannot be filtered and thus define the degree of freedom for an entity.

Emergy – The available energy [in joules, J] of one kind previously used up directly and indirectly to make a service or product. Its unit is the emjoule [ej] (Odum, 1996 – the 'em" being an allusion to '*em*bodied'), and its physical dimensions in the mass (M)–length(L)–time(T) dimensional system are those of energy, $[ML^2T^{-2}]$. Emergy analysis puts all energy forms on a common basis, and the reference form usually selected is solar as the ultimate energy source. Solar emergy is expressed in solar emjoules (sej), and related to other forms of energy by dimensionless stoichiometric coefficients, solar transformity, expressed in solar emjoules/joule (sej/J).

Emergy analysis – Analysis of flow- and storage-based emergy, empower, and transformities. An energy-oriented approach to quantifying and understanding the historical *em*bodiment and organizational status of systems. Emergy and empower can be viewed as benchmarks – points of reference against which the complexities of energy forms and configurations can be measured. They represent external sources of different energy forms embodied in the internal products $[ML^2T^{-2}]$ and processes $[ML^2T^{-3}]$ within systems motivated and organized by exchanging and transforming energy. While emergy and empower have energy $[ML^2T^{-2}]$ and power $[ML^2T^{-3}]$ dimensions, respectively, this does not mean they are stored, flowed, or can be "used up" in the usual senses applicable to energy and matter. They are measures of system organization, and may be a truer measure of the capacity to do work than heat, which only measures molecular motion. When we consider, for instance, whole CAHSystems (of much greater complexity than steam engines) the multitude of energies that drive them, each of a different form, cannot be reduced solely to their heat content and result in meaningful expression of system organization.

Emergy and empower measure the "value" of energy storages and flows within systems in terms of a common basis. The concepts are donor-referenced rather than receiver-referenced, measuring the convergence of source (donor) energies and flows at system boundaries into products [ML^2T^{-2}] and processes [ML^2T^{-3}] within system interiors. This is sometimes referred to as "energy memory" (Odum, 1996) – the embodiment or enfolding in products or processes of energy from distal sources.

Empower – A concept corresponding to emergy, based on energy flows [ML^2T^{-3}] rather than stored quantities [ML^2T^{-2}].

Environ – Term introduced by Patten (1978, 1985) to describe, looking upward–outward, the incoming and outgoing environments of holons (see below) circumscribed by the boundaries of open systems in which the said holons are component parts. Input environs are afferent transactive networks extending in past time from the system input boundary to a component-level holon in the present, and output environs are efferent networks extending through future time from the component-level holon in the present to the system output boundary. Environ theory thus defines two system-bounded environments associated with each holon within the system, and is mathematically an ecological extension of economic input–output analysis. Environs are nested networks across levels of hierarchic and holarchic organization.

Exergy – The portion of energy of a given kind usable to do work by a given process. In chemical thermodynamic terms it is *free energy* and, being energy, it has energy units and dimensions. Unlike energy, however, exergy is not conserved because any unused energy fraction at any point in a system may always be used by a more efficient or qualitatively different "technology" (process).

Generalised thermodynamic feedback model – General phenomenological feedback mechanism for maintenance of a thermodynamic steady state. Any thermodynamically open system fluctuating about some local quasi-steady state experiences the following cycle of states:

1. (+) gradient → (+) exergy degradation rate
2. (+) exergy degradation rate → (−) gradient
3. (−) gradient → (−) exergy degradation rate
4. (−) exergy degradation rate → (+) gradient → step (1)

where: → signifies "causes" or "leads to", in the probabilistic, phenomenological sense; (+) signifies "increases of", and (−) signifies "decreases of".

Step (1) indicates the Order through fluctuation (see below) scenario. The autocatalytic steps (1) and (2) detail the formation of "dissipative structures". Step (3) illustrates the "relaxation" phase, prior to recurrence of step (1). This feedback mechanism was recognized by von Bertalanffy (1950), Ulanowicz and Hannon (1987), Schneider and Kay (1994a) and many others as the LeChatelier–Braun principle. Such systems may be expected to approach some local, thermodynamic quasi-steady state. That is, there will always exist some region of applicability of linear approximations to the thermodynamic "equations of motions" (Denbigh, 1951; Spanner, 1964; Katchalsky and Curran, 1967; Choi et al., 1999), and therefore the least-specific-dissipation principle, that may be only empirically verified for appropriateness or inappropriateness (Choi and Patten, 2002).

Hierarchy – Term pertaining to levels of organization, usually defined as an epistemic property to simplify CAHSystemic complexity. The question of whether hierarchies are "real" or only mental constructs is usually sidestepped and generally remains unresolved. Hierarchies probably have an ontic basis in near-discrete breaks in energy levels associated with bond energies that couple systems (and holons) together (Simon, 1973). Hierarchy models are well known in ecological literature (e.g., Allen and Starr, 1982; O'Neill et al., 1986) as the basis for such concepts as "scale" and "grain" expressed in terms of both space and time. Hierarchies of "scale" are nested (supersets, sets, subsets, ...) whereas hierarchies of "control" (as in the military – generals, colonels, majors, lieutenants, ...) are not.

Holarchy – Term pertaining to the nested structure of holons, as in scale hierarchies. Analogous to the term "hierarchic", but created to remove from this the connotation of a lineal, chain-like order and thereby emphasize a more anarchistic or polyarchistic nature of across-scale organization (Regier and Kay, 1996).

Holon – Term introduced by Koestler (1969) which is the juxtaposition of two concepts: *holos* (Greek for "whole") and *on* (suffix indicating a particle or part – e.g., neutron, proton). A holon is a "Janus-faced" (two-way) open system, one that is simultaneously a whole (to its sub-systems, looking hierarchically downward–inward) and a part (in relation to its super-systems, looking upward–outward).

Information-input overload – Concept attributable to Ralph W. Gerard (Mental Health Research Institute, University of Michigan, Ann Arbor; Patten, personal communication) and more recently in an evolutionary context to David B. Brown (Department of Zoology, University of Toronto, Canada; R.I.C. Hansell, personal communication). It describes the phenomenon of a system that becomes too saturated with information (e.g., irregular, erratic, or mixed environmental signals or biotic interactions) to process it coherently. It responds incoherently, or breaks down and does not respond at all. Information overload forces an adaptive response: alternate or novel attractors (e.g., phenotypes/genotypes, behaviors, species) are turned to as current ones become inviable.

Integrative tendencies – Empirically observed tendencies of holons to integrate, diminish, or restrain assertive tendencies in favor of the assertive tendencies of larger wholes within which the holons in question serve as parts. This is analogous to the local-disorder principle (i.e., interactive uncertainty or perturbational influences; Choi and Patten, 2002).

Least specific dissipation (LSD) – The intensity of entropy production decreases with time for systems that are near some local quasi-steady state (Onsager, 1931a,b; Prigogine, 1955). This quantity is usefully approximated by the ratio of respiration rate to biomass (R/B ratio; Choi et al., 1999) and more generally as the ratio of system boundary output to system storage (Fath et al., 2001).

Most adaptive state – The presence of a quasi-equilibrium or balance (i.e., a local steady state) between assertive and integrative tendencies was suggested by Koestler to be indicative of a stable (or "adapted") holon. Any local imbalance (i.e., over-exaggeration of the integrative or the assertive tendencies) represents a "pathological" state. As a system deviates further and further from such a local balance, the degree of nonlinear characteristics in its dynamics increases, and as a consequence so also its likelihood to adaptively find (or "negotiate") a new balance. This represents a meta-mechanism of adaptation (Choi, 2002).

Order through fluctuation (OTF) – In the presence of a free-energy exergy gradient, statistical asymmetries/inhomogeneities of free-energy densities become coherent flows of energy inside the system, giving rise to organized structures (Glansdorff and Prigogine, 1971; Nicolis and Prigogine, 1977). As such, internal energy flows are never completely efficient (i.e., the Second Law); they act as gradient dissipating structures.

Red Queen's Hypothesis – In *The Adventures of Alice in Wonderland,* the Red Queen said one must keep running just to remain in place. The "hypothesis" from evolutionary biology refers to the necessity of continuously adapting in the face of ever-changing context to stay in the game of life (van Valen, 1976; Choi et al., 1999; Choi, 2002). In the face of ever-changing sets of interactions (i.e., the Red Queen's), the probability of continued existence is a function of trade-offs between refined information-handling capacity (specialization) to avoid disaster at one extreme, and robustness and flexibility (generalization) at the other to be opportunistic and capitalize on new signals or information and so cope with drastic changes in information content.

Soho systems – An acronym for <u>S</u>elf-regulating <u>O</u>pen <u>H</u>olarchic/<u>H</u>ierarchic <u>O</u>rder systems (Koestler, 1969; Regier and Kay, 1996). The concept extends the holon concept to systems explicitly open to matter/energy/information exchange (though this is implicit in the two-way-looking faces of holons). The concept is referred by von Bertalanffy's (1950, 1968) General Systems Theory and incorporates the empirical observation of self-organizational tendencies in such systems. The self-organizational property is generally attributed to the negative-entropy concept of Schrödinger (1944), which has recently been shown to have canonical origins in transactive coupling (Choi and Patten, 2002).

Transformity – A dimensionless quotient of two energies or energy flows that serves as a stoichiometric coefficient relating the quantities involved. For instance, the solar transformity is the ratio of the solar emergy that is required to generate a product or service to the actual energy in that product of service. Transformities have the units of emergy/energy (sej/J). A transformity for a product is calculated by summing all the emergy inflows to the process and dividing by the energy of the product. Transformities are used to convert resources of different types to emergy of the same type. As such, it is also a measure of the "value" with the assumption that systems operating under the constraints of the maximum-emergy principle generate products that stimulate productive process at least as much as they cost (Odum, 1996). Transformities are system-specific because systems vary both in number and kinds of external energy inputs and also in the connectance pattern and transformation efficiencies of their internal transfer pathways. [However, like all other intensity factors (including atomic mass number for instance), a mean or median value is often used in practice. For instance the transformity for electricity can vary from 9.0×10^4 to 2.0×10^5 sej/J depending on the rules and process that generate it, however a mean value of 1.5×10^5 is often used.] Because energy storages and flows tend, sometimes and always, respectively, to decrease along individual transfer chains, transformities tend to increase down-chain causing emergy and empower to also increase. This progressive increase with increasing displacement from energy sources is taken to reflect increase in system organization. Using material and energy pathway diagrams, it is possible to visualize the emergy at different points in an energy-flow sequence increasing, for example, from a boundary source to a producer to a consumer and as a general rule transformity increases as energy dissipates along food chains. Contrary to what may have appeared in the emergy literature, this does not mean emergy "flows" along transfer pathways in energy networks, nor is it "used up", "dissipated", or "lost in transfer". Emergy, though having energy dimensions, is not a conservative quantity. Measured by transformity, it reflects the intrinsic qualitative contribution a system component makes to overall system organization.

References

Allen, T.F.H. and Starr, T.B., 1982, Hierarchy: Perspectives for Ecological Complexity (University of Chicago Press, Chicago, IL) 310 pp.
Ashby, W.R., 1973, Some peculiarities of complex systems. Cybern. Med. 9(2):1–6.
Badii, R. and Politi, A., 1997, Complexity: Hierarchical Structures and Scaling in Physics (Cambridge University Press, Cambridge) 318 pp.

Bak, P., 1996, How Nature Works: The Science of Self-Organized Criticality (Copernicus Books, New York) 223 pp.

Bar-Yam, Y., 1997, Dynamics of complex systems. In: R.L. Devaney (Editor), Studies in Nonlinearity (Addison-Wesley Longman, Reading, MA) 848 pp.

Bastianoni, S., 1998, A definition of 'pollution' based on thermodynamic goal functions. Ecol. Model. 113:163–166.

Bastianoni, S. and Marchettini, N., 1997, Emergy/exergy ratio as a measure of the level of organization of systems. Ecol. Model. 99:33–40.

Begon, M., Harper, J.L. and Townsend, C.R., 1990, Ecology: Individuals, Populations, and Communities (Blackwell Scientific Publications, Brookline, MA) 945 pp.

Bendoricchio, G. and Jørgensen, S.E., 1997, Exergy as goal function of ecosystem dynamics. Ecol. Model. 102:5–15.

Bennett, C.H., 1988, Dissipation, information, computational complexity and the definition of organization. In: D. Pines (Editor), Emerging Synthesis in Science, Proc. Founding Workshops of the Santa Fe Institute, Santa Fe, NM (Addison-Wesley, Redwood City, CA) 237 pp.

Bossel, H., 1998, Ecological orientors: emergence of basic orientors in evolutionary self organization. In: F. Müller and M. Leupelt (Editors), Eco Targets, Goal Functions, and Orientors (Springer, Berlin) pp. 19–33.

Bosserman, R.W., 1982, Complexity measures for evaluation of ecosystem networks. ISEM (Int. Soc. Ecol. Model.) J. 4(1,2):37–59.

Brandt-Williams, S., 1999, Evaluation of watershed control of two Central Florida Lakes: Newnans Lake and Lake Weir, Ph.D. Dissertation, Environmental Engineering Sciences (University of Florida, Gainesville, FL) 257 pp.

Brown, M.T. and Sullivan, M.F., 1987, The value of wetlands in low relief landscapes. In: D.D. Hook (Editor), The Ecology & Management of Wetlands (Croom Helm, Beckenham, England) pp. 133–145.

Brown, M.T. and Ulgiati, S., 1997, Emergy based indices and ratios to evaluate sustainability: monitoring technology and economies toward environmentally sound innovation. Ecol. Eng. 9:51–69.

Casti, J.L., 1979, Connectivity, Complexity, and Catastrophe in Large-Scale Systems (Wiley, Chichester) 203 pp.

Casti, J.L., 1986, On system complexity: identification, measurement, and management. In: J.L. Casti and A. Karlqvist (Editors), Complexity, Language, and Life: Mathematical Approaches (Springer, Berlin) pp. 146–173.

Cavalier-Smith, T., 1978, Nuclear volume control by nucleoskeletal DNA, selection for cell volume and cell growth rate, and the solution for the DNA c-value paradox. J. Cell Sci. 34:247–278.

Chaitin, C.J., 1975, Randomness and mathematical proof. Sci. Am. 232:47–52.

Cheslak, E.F. and Lamarra, V.A., 1981, The residence time of energy as a measure of ecological organization. In: W.J. Mitsch, R.W. Bossermann and J.M. Klopatek (Editors), Energy and Ecological Modelling (Elsevier, Amsterdam) pp. 591–600.

Choi, J.S., 2002, Dealing with uncertainty in a complicated thermodynamic world, Ph.D. Thesis (Université de Montréal, Canada) 250 pp.

Choi, J.S. and Patten, B.C., 2002, Sustainable development: lessons from the paradox of enrichment. J. Ecosyst. Health. In press.

Choi, J.S., Mazumder, A. and Hansell, R.I.C., 1999, Measuring perturbation in a complicated, thermodynamic world. Ecol. Model. 117:143–158.

Christaller, W., 1966, Central Places in Southern Germany. Translated from *Die zentralen Orte in Süddeutschland* by C.W. Baskin (Prentice-Hall, Englewood Cliffs, NJ) 230 pp.

Cousins, S., 1993, Hierarchy in ecology: its relevance to landscape ecology and geographic information systems. In: R. Haines-Young, D. Green and S. Cousins (Editors), Landscape Ecology and Geographic Information Systems (Taylor and Francis, London) pp. 75–86.

Denbigh, K.G., 1951, The Thermodynamics of the Steady State (Wiley, New York) 103 pp.

Deville, A. and Turpin, T., 1996, Indicators of research relevance to ecologically sustainable development and their integration with other R&D indicators in the Asia-Pacific region. Chemosphere 33(9):1777–1800.

Dolezal, J., Greiluber, J., Lucretti, S., Maister, A., Lysak, M.A., Nardi, L. and Obermayer, R., 1998, Plant genome size estimation by flow cytometry: inter-laboratory comparison. Ann. Bot. 82:17–26.

Fath, B.D. and Patten, B.C., 1998, Network synergism: emergence of positive relations in ecological systems. Ecol. Model. 107:127–143.

Fath, B.D., Patten, B.C. and Choi, J.S., 2001, Complementarity of ecological goal functions. J. Theor. Biol. 208(4):493–506.

Flood, R.L., 1987, Complexity: A definition by construction of a conceptual framework. Syst. Res. 4:177–185.

Flood, R.L. and Carson, E.R., 1988, Dealing with Complexity: An Introduction to the Theory and Application of Systems Science (Plenum Press, New York) 289 pp.

Fonseca, J.C., Marques, J.C., Paiva, A.A., Freitas, A.M., Madeira, V.M.C. and Jørgensen, S.E., 2000, Nuclear DNA in the determination of weighting factors to estimate exergy from organisms biomass. Ecol. Model. 126:179–189.

Fox, R.F., 1985, Energy and the Evolution of Life (Freeman, San Francisco, CA) 182pp.

Futuyma, D.J., 1998, Evolutionary Biology (Sinauer Associates, Sunderland, MA) 830 pp.

Gallopin, G.C., 1997, Indicators and their use: information for decision-making. Part one – introduction. In: B. Moldan and S. Billharz (Editors), SCOPE 58 Sustainability Indicators: Report of the Project on Indicators for Sustainable Development (Wiley, Chichester) 440 pp.

Gell-Mann, M., 1994, The Quark and the Jaguar – Adventures in the Simple and the Complex (Freeman, New York) 400 pp.

Gilbert, A., 1996, Criteria for sustainability in the development of indicators for sustainable development. Chemosphere 33(9):1739–1784.

Glansdorff, P. and Prigogine, I., 1971, Thermodynamic theory of structure, stability and fluctuations (Wiley, London) 306 pp.

Harger, J.R.E. and Meyer, F.M., 1996, Definition of indicators for environmentally sustainable development. Chemosphere 33(9):1749–1775.

Herbst, M., 1997, Die Bedeutung der Vegetation für den Wasserhaushalt ausgewählter Ökosysteme (University of Kiel) 119 pp.

Herendeen, R., 1989, Energy intensity, residence time, exergy, and ascendency in dynamic ecosystems. Ecol. Model. 48:19–44.

Holland, J.H., 1995, Hidden Order: How Adaptation Builds Complexity (Perseus Books, Reading, MA) 185 pp.

Holling, C.S., 1986, The resilience of terrestrial ecosystems: local surprise and global change. In: W.C. Clark and R.E. Munn (Editors), Sustainable Development of the Biosphere (Cambridge University Press, Cambridge) pp. 292–317.

Hornby, A.S., 1974, Oxford Advanced Learner's Dictionary of Current English (Oxford University Press, London) 1055 pp.

Huang, S., 1998, Spatial hierarchy of urban energetic system. In: S. Ulgiati (Editor), Proc. Int. Workshop Advances in Energy Studies: Energy Flows in Ecology and Economy (MUSIS, Rome).

Johnson, L., 1981, The thermodynamic origin of ecosystems. Can. J. Fish. Aquat. Sci. 38:571–590.

Johnson, L., 1994, Pattern and process in ecological systems: a step in the development of a general ecological theory. Can. J. Fish. Aquat. Sci. 51:226–246.

Jørgensen, S.E., 1986, Structural dynamic model. Ecol. Model. 31:1–9.
Jørgensen, S.E., 1992a, Development of models able to account for changes in species composition. Ecol. Model. 62:195–209.
Jørgensen, S.E., 1992b, Integration of Ecosystem Theories: A Pattern (Kluwer, Dordrecht) 383 pp.
Jørgensen, S.E., 1994, Review and comparison of goal functions in systems ecology. Vie Milieu 44:11–20.
Jørgensen, S.E., 1997, Integration of Ecosystem Theories: A Pattern, 2nd edition (Kluwer, Dordrecht) 388 pp.
Jørgensen, S.E., 2000, The tentative fourth law of thermodynamics. In: S.E. Jørgensen and F. Müller (Editors), Handbook of Ecosystem Theories and Management (Lewis Publishers, Orlando, FL) pp. 161–175.
Jørgensen, S.E. and Mejer, H., 1977, Ecological buffer capacity. Ecol. Model. 3:39–61.
Jørgensen, S.E. and Mejer, H., 1979, A holistic approach to ecological modelling. Ecol. Model. 7:169–189.
Jørgensen, S.E. and Mejer, H., 1981, Application of exergy in ecological models. In: D. Dubois (Editor), Progress in Ecological Modelling (Editions CEBEDOC, Liège, Belgium) pp. 311–347.
Jørgensen, S.E., Nielsen, S.N. and Mejer, H., 1995, Emergy, environs, exergy and ecological modelling. Ecol. Model. 77:99–109.
Jørgensen, S.E., Patten, B.C. and Straskraba, M., 2000, Ecosystems emerging: 4. Growth. Ecol. Model. 126:249–284.
Jørgensen, S.E., Fath, B.D., Patten, B.C. and Straskraba, M., 2002, Ecosystem growth and development. Submitted.
Karnawati, D., Subagyo, P., Sukandarrumidi and Hendrayana, H., 2000, Towards sustainable coastal development in the south of Central Java: A geological assessment for coastal management, Report on Graduate Team Research Grant Batch IV, University Research for Graduate Education (URGE) Project (Geological Engineering Department, Gadjah Mada University, Indonesia). Unpublished.
Karnawati, D., Subagyo, P., Sukandarrumidi and Hendrayana, H., 2001, Eco-geological assessment towards sustainable coastal development; a case study at Yogyakarta South Coast, Indonesia. In: E. Prasetyo Utomo, H.Z. Anwar and D. Murdohardono (Editors), Proc. Third Asian Symp. on Engineering Geology and the Environment, Yogyakarta, Indonesia, September 3–6, 2001 (Research Center for Geotechnology, Indonesia Institute of Science, Bandung) pp. 47–61. [ISBN: 979-9477-33-6.]
Katchalsky, A. and Curran, P.F., 1967, Nonequilibrium Thermodynamics in Biophysics (Harvard University Press, Cambridge, MA) 248 pp.
Kay, J.J., 1984, Self-organisation in living systems, Ph.D. Thesis, Systems Design Engineering (University of Waterloo, Ontario, Canada) 458 pp.
Kay, J.J., 2000, Ecosystems as self-organising holarchic open systems: narratives and the second law of thermodynamics. In: S.E. Jørgensen and F. Müller (Editors), Handbook of Ecosystem Theories and Management (Lewis Publishers, Orlando, FL) pp. 135–159.
Kay, J.J., Regier, H.A., Boyle, M. and Francis, G., 1999, An ecosystem approach for sustainability: addressing the challenge of complexity. Futures 31:721–742.
Klir, G.J., 1985, Complexity: Some general observations. Syst. Res. 2(2):131–140.
Klir, G.J., 1991, Facets of Systems Science (Plenum Press, New York) 664 pp.
Koestler, A., 1969, Beyond atomism and holism – the concept of the holon. In: A. Koestler and J.R. Smythies (Editors), Beyond Reductionism – New Perspectives in the Life Sciences, The Alpbach Symposium 1968 (Hutchinson & Co., New York) pp. 192–232.
Kolmogorov, A.N., 1965, Three approaches to the quantitative definition of information. Probl. Inf. Transm. 1:1–17.
Kullback, S., 1959, Information Theory and Statistics (Wiley, New York) 395 pp.

Lambert, J.D., 1999, A spatial emergy model for Alachua County, Florida, Ph.D. Dissertation, Urban and Regional Planning (University of Florida, Gainesville, FL) 569 pp.
Laszlo, E., 1972, The World System: Models, Norms, Variations (Braziller, New York) 215 pp.
Laszlo, E., 1996, A Systems View of the World: A Holistic Vision for Our Time (Hampton Press, Cresskill, NJ) 103 pp.
Lewin, B., 1994, Genes V (Oxford University Press, Oxford) 1274 pp.
Li, W.-H., 1997, Molecular Evolution (Sinauer Associates, Sunderland, MA) 487 pp.
Lopes, R.J., Pardal, M.A. and Marques, J.C., 2000, Impact of macroalgal blooms and wader predation on intertidal macroinvertebrates: experimental evidence from the Mondego estuary (Portugal). J. Exp. Mar. Biol. Ecol. 249:165–179.
Lotka, A.J., 1922, Contribution to the energetics of evolution. Proc. Natl. Acad. Sci. USA 8: 147–151.
Luvall, J.C. and Holbo, H.R., 1989, Measurements of short-term thermal responses of coniferous forest canopies using thermal scanner data. Remote Sens. Environ. 27:1–10.
Luvall, J.C. and Holbo, H.R., 1991, Thermal remote sensing methods in landscape ecology. In: M. Turner and R.H. Gardner (Editors), Quantitative Methods in Landscape Ecology (Springer, New York) pp. 127–152.
Månsson, B.Å. and McGlade, J.M., 1993, Ecology, thermodynamics and H.T. Odum's conjectures. Oecologia 93:582–596.
Marques, J.C. and Nielsen, S.N., 1998, Applying thermodynamic orientors: the use of exergy as an indicator in environmental management. In: F. Müller and M. Leupelt (Editors), Ecotargets, Goal Functions, and Orientors (Springer, Berlin) pp. 481–491.
Marques, J.C., Pardal, M.A., Nielsen, S.N. and Jørgensen, S.E., 1997, Analysis of the properties of exergy and biodiversity along an estuarine gradient of eutrophication. Ecol. Model. 102: 155–167.
Merriam, G., Henein, K. and Stuart-Smith, K., 1991, Landscape dynamics models. In: M.G. Turner and R.H. Gardner (Editors), Quantitative Methods in Landscape Ecology (Springer, New York) 536 pp.
Milne, B., 1991, Lessons from applying fractal models to landscape patterns. In: M.G. Turner and R.H. Gardner (Editors), Quantitative Methods in Landscape Ecology (Springer, New York) 536 pp.
Minsch, J., Feindt, P.H., Meister, H.P., Schneidewind, U. and Schulz, T., 1998, Institutionelle Reformen für eine Politik der Nachhaltigkeit (Springer, Berlin) 445 pp.
Mitchell, G., 1996, Problems and fundamentals of sustainable development indicators. Sustain. Dev. 4:1–11.
Morowitz, H.J., 1968, Energy Flow in Biology; Biological Organization as a Problem in Thermal Physics (Academic Press, New York) 179 pp.
Moser, A., 2000a, Macroscopic pattern analysis. Acta Biotechnol. 20(3–4):235–274.
Moser, A., 2000b, The wisdom of nature in integrating science, ethics and the arts. Sci. Eng. Ethics 6:365–382.
Müller, F. and Fath, B.D., 1998, Introduction: the physical basis of ecological goal functions. In: F. Müller and M. Leupelt (Editors), Eco Targets, Goal Functions, and Orientors (Springer, Berlin) pp. 15–18.
Müller, F. and Leupelt, M. (Editors), 1998, Eco Targets, Goal Functions, and Orientors (Springer, Berlin) 619 pp.
Munasinghe, M. and Shearer, W., 1995, Defining and Measuring Sustainability: The Biogeophysical Foundation (The World Bank, Washington, DC) 440 pp.
Nicolis, G. and Prigogine, I., 1977, Self-Organization in Nonequilibrium Systems (Wiley, Toronto) 491 pp.

Nicolis, J.S., 1986, Dynamics of Hierarchical Systems: An Evolutionary Approach (Springer, New York) 397 pp.
Noss, R.F., 1990, Indicators for monitoring biodiversity: a hierarchical approach. Conserv. Biol. 4:355–364.
Odum, E.P., 1969, The strategy of ecosystem development. Science 164:262–270.
Odum, H.T., 1983, Systems Ecology: an Introduction (Wiley, New York) 644 pp.
Odum, H.T., 1988, Self-organization, transformity, and information. Science 242:1132–1139.
Odum, H.T., 1994, Ecological and General Systems, an Introduction to Systems Ecology (University Press of Colorado, Niwot, CO) 644 pp. Revised edition of Systems Ecology (Wiley, 1983) 644 pp.
Odum, H.T., 1996, Environmental Accounting: Emergy and Environmental Decision Making (Wiley, New York) 370 pp.
Odum, H.T. and Pinkerton, R.C., 1955, Time's speed regulator: the optimum efficiency for maximum power output in physical and biological systems, Am. Sci. 43:321–343.
O'Neill, R.V., DeAngelis, D.L., Waide, J.B. and Allen, T.F.H., 1986, A Hierarchical Concept of Ecosystems (Princeton University Press, Princeton, NJ) 253 pp.
Onsager, L., 1931a, Reciprocal relations in irreversible processes I. Phys. Rev. 37:405–426.
Onsager, L., 1931b, Reciprocal relations in irreversible processes II. Phys. Rev. 38:2265–2279.
Pahl-Wostl, C., 2002, Polycentric integrated assessment. In: J. Rotmans (Editor), Issues in Integrated Assessment (Kluwer, Dordrecht). In press.
Pardal, M.A., 1998, Impacto da eutrofização nas comunidades macrobentónicas do braço sul do estuário do Mondego (Portugal), Ph.D. Thesis (Faculty of Sciences and Technology, University of Coimbra, Portugal) 315 pp.
Pardal, M.A., Marques, J.C., Metelo, I., Lillebø, A.I. and Flindt, M.R., 2000, Impact of eutrophication on life cycle, population dynamics and production of *Amphitoe valida* (Amphipoda) along an estuarine spatial gradient (Mondego estuary, Portugal). Mar. Ecol. Prog. Ser. 196:207–219.
Patten, B.C., 1978, Systems approach to the concept of environment. Ohio J. Sci. 78:206–222.
Patten, B.C., 1983, Linearity enigmas in ecology. Ecol. Model. 18:155–170.
Patten, B.C., 1985, Energy cycling in the ecosystem. Ecol. Model. 28:1–71.
Patten, B.C., 1995, Network integration of ecological extremal principles: exergy, emergy, power, ascendency, and indirect effects. Ecol. Model. 79:75–84.
Patten, B.C., 1997a, Synthesis of chaos and sustainability in a nonstationary linear dynamic model of the American black bear (*Ursus americanus* Pallas) in the Adirondack Mountains of New York. Ecol. Model. 100(1–3):11–42.
Patten, B.C., 1997b, Ecology's AWFUL Theorem: sustaining sustainability. Ecol. Model. 108: 97–105.
Prigogine, I., 1955, Thermodynamics of Irreversible Processes (Wiley, New York) 115 pp.
Prigogine, I., 1967, Introduction to Thermodynamics of Irreversible Processes, 3rd edition (Wiley Interscience, New York) 147 pp.
Prigogine, I. and Stengers, I., 1984, Order Out of Chaos: Man's New Dialogue with Nature (Bantam, New York) 349 pp.
Rees, W.E., 1996, Revisiting carrying capacity: Area-based indicators of sustainability. Popul. Environ. 17(3):195–215.
Regier, H.A. and Kay, J.J., 1996, An heuristic model of transformations of the aquatic ecosystems of the Great Lakes – St. Lawrence River Basin. J. Aquat. Ecosyst. Health 5:3–21.
Rosen, R., 1977, Complexity as a system property. Int. J. Gen. Syst. 3:227–232.
Salthe, S.N., 1985, Evolving Hierarchical Systems: Their Structure and Representation (Columbia University Press, New York) 343 pp.
Salthe, S.N., 1993, Development and Evolution. Complexity and Change in Biology (MIT Press, Cambridge, MA) 357 pp.

Schellnhuber, H.J., 1998, Earth system analysis: the concept. In: H.J. Schellnhuber and V. Wenzel (Editors), Earth System Analysis: Integration Science for Sustainability (Springer, Berlin) pp. 3–195.
Schneider, E.D. and Kay, J.J., 1994a, Life as a manifestation of the Second Law of Thermodynamics. Math. Comput. Model. 19:25–48.
Schneider, E.D. and Kay, J.J., 1994b, Complexity and thermodynamics. Towards a new ecology. Futures 26:626–647.
Schneider, E.D. and Kay, J.J., 1995, Order from disorder: the thermodynamics of complexity in biology. In: M.P. Murphy and A.J. Lukem (Editors), What is Life: The Next Fifty Years. Speculations on the Future of Biology (Cambridge University Press, Cambridge) pp. 161–174.
Schneider, E.D. and Kay, J.J., 1996, Energy degradation, thermodynamics, and the development of ecosystems. In: J. Szargut, Z. Kolenda, G. Tsatsaronis and A. Ziebik (Editors), Proc. ENSEC 93, Int. Conf. on Energy Systems and Ecology, Cracow, Poland, July 5–9, 1993 (American Society of Mechanical Engineers, New York) pp. 33–42.
Schrödinger, E., 1944, What is Life? The Physical Aspect of the Living Cell (Cambridge University Press, Cambridge) 92 pp.
Schultink, G., 1992, Evaluation of sustainable development alternatives: Relevant concepts, resource assessment approaches and comparative spatial indicators. Int. J. Environ. Stud. 41:203–224.
Seppelt, R., 1999, Applications of optimum control theory to agroecosystem modelling. Ecol. Model. 121(2–3):161–183.
Seppelt, R. and Temme, M.M., 2001, Hybrid low level Petri Nets in environmental modelling – development platform and case studies. In: M. Matthies, H. Malchow and J. Kritz (Editors), Integrative Systems Approaches to Natural and Social Sciences (Springer, Berlin) 20 pp.
Simon, H.A., 1962, The architecture of complexity. Proc. Am. Philos. Soc. 106:467–482.
Simon, H.A., 1973, The organization of complex systems. In: H.H. Pattee (Editor), Hierarchy Theory, The Challenge of Complex Systems (Braziller, New York) pp. 1–27.
Spanner, D.C., 1964, Introduction to Thermodynamics (Academic Press, London) 279 pp.
Suter, G.W., 1993, A critique of ecosystem health concepts and indexes. Environ. Toxicol. Chem. 12:1533–1539.
Turner, S., O'Neill, R.V., Conley, W., Conley, M. and Hunphries, H., 1991, Pattern and scale: statistics for landscape ecology. In: M.G. Turner and R.H. Gardner (Editors), Quantitative Methods in Landscape Ecology (Springer, New York) 536 pp.
Ulanowicz, R.E., 1986, Growth and Development: Ecosystems Phenomenology (Springer, New York) 203 pp.
Ulanowicz, R.E., 1997, Ecology, The Ascendent Perspective (Columbia University Press, New York) 201 pp.
Ulanowicz, R.E. and Hannon, B.M., 1987, Life and the production of entropy. Proc. R. Soc. London Ser. B 232:181–192.
Ulgiati, S., Odum, H.T. and Bastianoni, S., 1994, Emergy use, environmental loading and sustainability: an emergy analysis of Italy. Ecol. Model. 73:215–268.
Ulgiati, S., Brown, M.T., Bastianoni, S. and Marchettini, N., 1996, Emergy based indices and ratios to evaluate sustainable use of resources. Ecol. Eng. 5:497–517.
UNESCO, 1998, Geology for Sustainable Development, Bulletin 11. Urban Geology (Division of Earth Science, UNESCO, New York) 153 pp.
van Valen, L., 1976, A new evolutionary law. Evol. Theory 1:1–30.
Vilela Mendes, R., 1998, Medidas de complexidade e auto-organização. Col. Ci. 22:3–14.
Volkenstein, M.V., 1988, Biophysics (Nauka, Moscow) 592 pp.
von Bertalanffy, L., 1950, The theory of open systems in physics and biology. Science 111:23–29.
von Bertalanffy, L., 1968, General Systems Theory: Foundations, Development, Applications (George Braziller, New York) 289 pp.

Wackernagel, M., Lewan, L. and Borgstroem Hansson, C., 1999, Evaluating the use of Natural Capital with the Ecological Footprint. Ambio 28:604–612.
Wetzel, R.G., 1983, Limnology (Saunders, Philadelphia, PA) 767 pp.
Wright, S., 1988, Surfaces of selective value revisited. Am. Nat. 131:115–123.
Wynne, B., 1992, Uncertainty and environmental learning: reconceiving science and policy in the preventive paradigm. Glob. Environ. Chang. June:111–127.

Consensus

Chapter 4

Complex Adaptive Hierarchical Systems

B.C. Patten, B.D. Fath, J.S. Choi, with S. Bastianoni, S.R. Borrett,
S. Brandt-Williams, M. Debeljak, J. Fonseca, W.E. Grant, D. Karnawati,
J.C. Marques, A. Moser, F. Müller, C. Pahl-Wostl, R. Seppelt, W.H. Steinborn
and Y.M. Svirezhev

Abstract
A coherent theory of CAHSystems is emerging. In approaching this, there are both intrinsic and instrumental values and ends to be considered.

The orientation of First-World societies these days is decidedly pragmatic. However, this pragmatic orientation must be tempered by the awareness of the critical importance of new-knowledge generation – knowledge that cannot always be judged practical, a priori. Without this, there can be only old knowledge to apply; with the inevitable result that human adaptability will become increasingly impoverished. The area of CAHSystems represents wide-open new ground in science as the 21st Century begins to unfold.

Pragmatically, a turn to CAHSystems is motivated by the need to integrate and make sense of the information overload we experience as we attempt to grapple with complexly interwoven environmental, political, economic, social, and ethical issues that span huge scales of space and time. The development and focus of such a theory is still in its infancy and must be nurtured, but the pressing need to discover and integrate complex systems knowledge argues for its continuing promotion and coordination. There are too few tools to grapple with systems larger and more complex than ourselves, but one of these is interdisciplinary modeling which, properly employed, represents a potentially powerful means for integrating observation, theory, and practice.

We suggest that continuing development of CAHSystems Theory and modeling are essential if humanity is ever to realize a balanced, sustainable condition within the frames of healthy, diverse, and viable ecosystems over all the globe.

1. Introduction

It is apparent from the foregoing contributions by participants who spent three days in discussions about CAHSystems in Halifax that there is, superficially, little uniformity of opinions or perspectives about this class of organized systems. A plurality of approaches exists, in itself a healthy state, and as yet there is little formalism to provide theoretical guidance. CAHSystems have in fact been theoretically addressed in several different ways in prior literature, though not explicitly under the CAHS rubric, which is new. Although the ideas and approaches explored may at first glance appear relatively inconsistent, if we focus less on details and look deeper into underlying bases, it becomes evident that a pattern of something resembling an identifiable "CAHS Theory" is taking form in diverse corners of ecology motivated by applied ends – in Ecological Modeling, Ecological Economics, Ecosystem Health, and Ecological Engineering.

As seen in the CAHS background chapter, a number of different complementary approaches exist and are needed to explain the structure, organization, and dynamics of CAHSystems. This is not surprising at all. More models of different kinds are needed and to improve their predictive capabilities it will be necessary to invent and explore novel theoretical approaches much more widely. Ecological literature from the last decade may contain a large number of observations, data sets, correlations, and some rules resulting from increasing interest in sustainable development and environmental health but it must be said that such observations and rules are comparable to lonely islands in an immense ocean and only rarely do links exist between them.

2. About theory

A critical need is to develop theoretical scaffolding capable of explaining results obtained up to now in terms of an accepted pattern of related CAHS theories. In a sense, this is the right time to aim for such a framework in an ecology (and related sciences) glutted with observations and data but deficient in ordered theory that can catalyze new understandings from old. The aim should be to make this the equivalent of physical theory; "equivalent" in the sense that the laws explaining observations derive from a very few fundamental laws. In the history of physics, a logic network of all observations was gradually built with the most fundamental laws serving as nodes. Slowly, first out of Newtonian mechanics, then special and general relativity, then quantum mechanics and now superstring theory (which still lacks a physical realization), a knowledge scaffold was built that links all physics into a coherent science applicable to testing new hypotheses – do they fit the logic network? If not, there must be very strong empirical support for an hypothesis to be accepted because then the network must also to a certain extent be changed. If yes, the network is made more robust. Indeterministic vs. deterministic interpretations of quantum mechanics are still undergoing evaluation decades after definitive fits to observations were achieved.

It may be "physics envy" to want such vertical development in a science whose very diversity defies this, but it is a goal nevertheless to be striven for, as the sciences of what is "C", "A", "H", and "S" gain, late in the 21st century or perhaps not until the next, some semblance of physics' maturity. Is "scientific verticality", things building from antecedents as seen in physics, possible with regard to CAHSystems? It is not clear at this moment that even all Halifax participants in the CAHS theme would answer affirmatively to this question. Perhaps it must remain for now only a promising direction for theoretical thought. What is not in dispute is that many ecologists must contribute to broaden, deepen, and enrich whatever development occurs in realizing a more rigorous CAHS theory and facilitating its application to "real-world" problems.

This can be achieved if environmental scientists who work on CAHSystems follow some basic guidelines derived from scientific theory. New issues or branches in CAHS theory have to fulfill "inner consistency", which means that a new theory shows no contradictions within itself, and has to fulfill "outer consistency", which means that a new theory has no contradictions to other accepted theories. Such theories must also describe a new issue, new in terms of environmental issues, issues of scale, or of hierarchy, and can be tested by experiments or applications. These are necessary and sufficient conditions for new CAHS theories. Optional criteria, such as generality, depth, precision, accuracy of prognoses, etc., may be used to evaluate new CAHS theories. These criteria may be guidelines for developing a theoretical scaffolding for Complex, Adaptive, and Hierarchical, Systems.

3. About applications

At EcoSummit 2000, a low tolerance was exhibited for "theory for its own sake" – for concepts that appeared too specialized, or could not be immediately transformed to serve pragmatic ends. Whether this was part of the general "dumbing down" of science and culture that seems so widespread in today's world is not clear, but what is clear is that one cannot apply what one does not have. In other words, if CAHS-related problems are inherently difficult, progress in solving them without commitment to developing the requisite science, however specialized and esoteric it may initially seem to managers, stakeholders, and the lay public, is not likely to be made. To "sustain sustainability" (e.g., Patten, 1997) as an ongoing scientific endeavor, it is necessary to advance beyond the "buzzword" stage or else the vision will dwindle away to be replaced by the next concept destined for the same ineffective end. Such is the pattern of soft science, and permanent commitment to hard problems is the only way to build a hard science. And of course, in matters of public interest this requires permanent societal commitment that must be earned by good feedback, sound education, and demonstrably solid progress.

The efforts of many researchers over quite a few years will be necessary to develop a CAHS-theoretical network that is up to the task of solving applied

CAHS problems. Progress will be stepwise and, considering the difficulties, slow. Therefore, it is essential to get started as soon as possible. It will be critical to find ways to integrate large sets of observations and databases, to coordinate the multidisciplinary work of many scientists, and to interpret the scientific basis, results, and predictions clearly at appropriate levels to empirical researchers, environmental managers, and lay stakeholders. Why should such groups care at all about theory? The answer is, because a theoretical frame provides the context wherein researchers and others can interpret and integrate empirical results. Without integration, without interpretation of raw observations within a consistent theoretical frame, and without communication to affected constituencies, there can be no organized or effective science. Only description is possible, not basic understanding that can be transmitted to the public from which support must ultimately come.

4. About modeling

There are basically three ways to link theory and applications:
(1) The *observational way* is to attempt to interpret as many empirical observations as possible within the existing CAHS-theoretical framework, demonstrating its value and weaknesses as an explanatory and integrative tool.
(2) The *experimental way* is to design controlled experiments to verify or falsify hypotheses. This is complicated because it implies acquiring and maintaining specialized facilities as well as constituting and holding together research teams involving the whole spectrum of relevant theoretical and empirical disciplines.
(3) The *modeling way* is more complicated still, but only in that it encompasses the previous two and in addition has its own requirements. But it also has special promise, as follows.

Ecological and environmental modeling typically emphasize the development of modeling products to serve predictive purposes. Scientific modelers involved in any modeling process quickly become aware of the tremendous integrative power of modeling to organize not only concepts, information, and data about a project, but also the various specialists and nonspecialists involved – scientists, managers, and stakeholders. It has been suggested (Patten, 1994) that a process-over-product approach to conceptual modeling should be pursued in workshop settings. Concerning CAHSystems, such an approach could serve to develop common visions in relation to particular sustainable development or ecosystem health objectives. Multidisciplinary experts come together initially and, under the discipline imposed by modeling requirements for information, completeness, and consistency, contribute the integrated concepts that later become the basis for further, more technical, steps in modeling. Everything begins with the conceptual phase, and this should be facilitated but not carried out by CAHS-theorists or modelers, per se, but by scientists and others who hold primary knowledge of the systems and issues involved.

This approach was discussed at EcoSummit 2000 among individuals experienced or interested in its potential – S.R. Borrett, B.D. Fath, W.E. Grant, J.C. Marques, C. Pahl-Wostl, B.C. Patten, and M.J. van den Belt. Van den Belt et al. (1997) refer to the process as "mediated modeling"; a book on the subject is in preparation (van den Belt, 2002). Patten (personal communication) advocates "institutionalized model-making" and proposes building workshop-oriented conceptual modeling activities directly into the core administrative structure and ongoing life of scientific institutes to organize all aspects of their responsibilities – from conduct of research, to data banking and fiscal management, to public outreach.

5. Conclusions

From the diverse thinking displayed at Halifax as already broadly underway, it is clear that CAHSystems despite their inherent complexities are beginning to come up for concerted attention as objects of scientific study. It will be many years yet before theory, applications, and modeling all come together in an organized framework capable of speaking with one voice to the thorny problems of sustainable development and environmental health, but essential pluralism is being preserved, that is clear, and early results and enthusiasm spurred by need are most encouraging.

References

Patten, B.C., 1994, Ecological systems engineering: toward integrated management of natural and human complexity in the ecosphere. Ecol. Model. 75/76:653–665.
Patten, B.C., 1997, Ecology's AWFUL Theorem: sustaining sustainability. Ecol. Model. 108:97–105.
van den Belt, M.J., 2002, Mediated Modeling: Building Capacity to Solve Complex Environmental Problems. In preparation.
van den Belt, M.J., Antunes, P. and Santos, M., 1997, Mediated modeling; a tool for stakeholder involvement in environmental decision making. In: Proc. Algarve Coastal Zone Conference, July 10–12, 1997, Algarve, Portugal.

Background

Chapter 5

Ecosystem Services, Their Use and the Role of Ecological Engineering: State of the Art

A. Dakers

Abstract

There are different perspectives from which one may view the concept of ecosystem services. It may be from an economist's perspective, that of an environmental scientist or policy analyst, or that of the designers of technologies and physical facilities for society. This chapter presents the view of the engineer and ecological engineer. Because of the nature of their professions, these two disciplines tend to work closely with problem solving and design.

The EcoSummit Conference has been about "integrating the sciences". However the process of integration of effort must include many more players in addition to the scientific community. While institutional and policy frameworks for more sustainable use of ecosystem services are important, if "frontline" obstacles arise at the implementation stage, the credibility of the goals (and the goal setters) may suffer a significant setback. It is important to support innovative engineered projects that demonstrate what is technologically and physically possible (and not possible) in achieving more sustainable interactions with ecosystem services. Such engineered projects can provide very effective feedback to both the policymakers and the scientific community to assist with policy design and research priorities respectively. The critical importance of the interdependence between ecosystems and human development has very significant implications for the criteria that engineers and others should use in designing and building physical facilities.

Several case studies are presented to illustrate how projects designed to include ecological criteria can make better use of ecosystem services. While there is increasing concern about the overuse of ecosystem services, these case studies also demonstrate that some conventional engineering practices can lead to substantial under-use of ecosystem services and the loss of the opportunities they provide for a more sustainable and balanced lifestyle.

Understanding and Solving Environmental Problems in the 21st Century
Edited by R. Costanza and S.E. Jørgensen
© 2002 Elsevier Science Ltd.   All rights reserved

1. Introduction

The theme of the EcoSummit 2000 conference in Halifax was about "integrating the sciences" to achieve a better understanding of, and solutions to, the environmental problems in the 21st century. This is a background chapter for the working group given the task of considering the state and sustainable use of ecosystem services. The reader is referred to the workshop consensus chapter 6 in this Volume (Guterstam et al.), for more details of the outcomes from this workshop process. The organization and facilitation of this particular working group has been the responsibility of the International Ecological Engineering Society (IEES). The relatively new discipline of ecological engineering[1] tends to an interdisciplinary approach, applying, where appropriate, knowledge of ecological systems to engineering projects. It will be shown in this chapter that ecological engineering is primarily about applying the sciences, with creative design, in an attempt to resolve some of the environmental problems we face today and in the future. Consequently the working group has tended to focus its efforts on the more applied practices and solutions to issues relating to ecosystem services. This is not to deny the importance of other disciplines. There clearly is a need for the different groups and disciplines to work more closely together. As will be illustrated in section 5.3, the integration of the sciences is just one small but important step towards achieving this.

There is a growing and increasingly urgent voice from commentators urging the engineering profession to integrate their projects more closely with ecosystem conditions and processes (e.g., Beder, 1998; Cortese, 1999; Mitsch, 1996; Moser, 1997; Roberts, 1991; Schulze et al., 1996; Thom, 1993; Wurth, 1996). Some of these same authors are calling for a much closer interaction between ecologists and engineers. This will demand a cultural change for both disciplines.

This chapter examines the various definitions of ecosystem services, the different perspectives on the importance of identifying and valuing these services, and the pressures human development places on such services. The selection of design criteria by engineers and other designers of the physical built world can impact on whether or not ecosystem services are used more, or less, sustainably. This discussion leads to an explanation of ecological engineering, in section 6, with a number of case studies described to illustrate how this more ecologically integrated approach to design can lead to more sustainable use of the ecosystems services. While section 4 clearly demonstrates that human development is over-using and stressing many ecosystem services, the case studies also demonstrate that there is under-utilization of some ecosystem services (section 7).

---

[1] The term *ecological engineering* was first coined by Howard Odum in the 1960s (Odum, 1962).

2. Defining ecosystem services

The term, ecosystem services, was first proposed by economists and ecological economists who recognized that nature was providing humanity with a wide range of services that were not being valued. They were consequently being excluded from economic policy and decision making and generally overlooked in formal resource management structures and procedures. Essential services such as the purification services of water and air and nature's gene bank services for crops were not being included in traditional economic modeling (Peet, 1992; p. 115). While there is general agreement among the different commentators that ecosystem services are about the traditionally unvalued services provided by nature to humankind, there are different interpretations as to which specific services are included and excluded and how best to value these.

The Australian CSIRO (2000) offer a very general definition when they define ecosystem services to "include life support activities that ecosystems provide for us, largely in an unrecognized and unpriced way".

Daily is more specific when she describes ecosystem services as "... the conditions and processes through which natural ecosystems, and the species that make them up, sustain and fulfil human life. They maintain biodiversity and the production of ecosystem goods such as seafood, forage, timber, biomass fuels, natural fiber and many pharmaceuticals, industrial products and their precursors" (Daily, 1997; p. 3).

Daily clearly distinguishes between services and goods provided by ecosystems as illustrated in fig. 1. All goods and commodities produced by humans "farming" nature's resources with the input from the ecosystem services are, in most situations, part of conventional economic models.

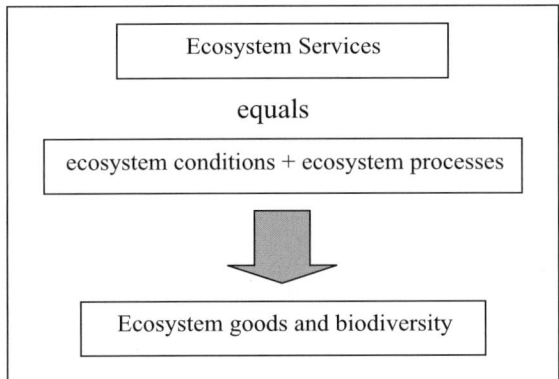

Fig. 1. The relationship between ecosystem services (i.e., conditions and processes) and ecosystem goods.

Table 1
Two lists of ecosystem services

Daily (1997)	Costanza et al. (1997)
Purification of air and water	Gas regulation, water regulation
	Water supply
Mitigation of floods and droughts	Disturbance and natural hazard regulation
	Erosion control and sediment retention
Detoxification and decomposition of wastes	Waste treatment
Generation and renewal of soil and soil fertility	Soil formation, nutrient cycling
Pollination of crop and natural vegetation	Pollination
Control of agricultural pests	Biological control
Dispersal of seeds and translocation of nutrients	
Maintenance of biodiversity	Genetic resources
Protection from UV rays	
Climate stabilization	Climate regulation
Support of human cultures	
Aesthetic beauty and intellectual stimulation	Recreation, cultural
	Refuge and shelter
	Food production
	Raw material production

Daily and others have pointed out that ecosystem services are absolutely essential to civilization but modern urban life obscures their existence.

> There are times when the most difficult decision of all is to acknowledge the obvious. It is obvious that the world's national economies are based on the goods and services derived from ecosystems; it is also obvious that human life itself depends on the continuing capacity of ecosystems to provide their multitude of benefits.
> *(United Nations Development Program et al., 2000)*

Both Daily and Costanza et al. (1997) listed specific services and these are shown in table 1.

Costanza et al. (1997) estimated a value for 17 ecosystems and 16 biomes as averaging about US$33 trillion/yr – or nearly twice the global GNP. The Costanza et al. definition seems to differ from Daily's definition as it includes goods within the list of ecosystem services.

The Australian CSIRO (2000) identified 6 specific ecosystem services of current interest. These included:

(1) Remnant ecosystems and the services they provide;
(2) Pollination services in Australia;
(3) Forest productivity, marsupial survival and truffles;
(4) Dieback in eucalypts in the heartlands;
(5) Resilience of rangeland productivity and grass biodiversity;
(6) Provision of clean water from a catchment.

The same report pointed out that ten years ago the agricultural economist Rod Gill calculated that the annual benefit of pollination services to Australian agriculture was between $600 million and $1.2 billion.

More recently the Guide to World Resources (United Nations Development Program et al., 2000) listed ecosystem services for five major classes of ecosystems, along with the ecosystem goods produced. This list is summarized in table 2.

3. Humankind's relationship with the natural environment

3.1. Anthropocentric or ecocentric valuing

There is debate and concern that present definitions and consequent valuing of the ecosystem services are too utilitarian and too anthropocentric. Cairns (1996) points out that the importance we place on ecosystem services "is a matter of societal perception because it hinges on valuation". He suggests "that all ecosystem functions could possibly be viewed as ecosystem services". Apart from the question of what the specific services are, some raise the question – service to whom or what? What about the spiritual, mystical, cultural, and aesthetic values of nature to humankind – the opportunity to view, feel, and embed oneself in "undeveloped" nature? What about the "value" of nature in providing the conditions for people to experience adventure, challenge, and fear (e.g., Antarctic/Arctic treks, mountaineering, extreme sports)? What about the value of the ecosystem to the non-human species, the many animals and plants that share with humans the services of the natural ecosystem?

3.2. Ecosystem relationships – embeddedness

Some of the concerns about, and different interpretations of, the concept of ecosystem services may be explained in terms of what level of relationship with nature the different points of view are founded on. A simple model is one in which there are two basic levels of relationship with our "neighbors." These are illustrated in fig. 2. As we individually and collectively interact with the world around us, even electronically in the case of other humans, we necessarily develop relationships with the human and non-human world, consciously or by default. In terms of our relationships with other people we may have a commercial relationship; for example with those we employ or our employer, or with those

Table 2
Primary ecosystems goods and services [a]

Ecosystem	Goods	Services
Agroecosystems	• Food crops • Fiber crops • Crop genetic resources	• Maintain limited watershed functions (infiltration, flow control, partial soil protection) • Provide habitat for birds, pollinators, and soil organisms important to agriculture • Build soil organic matter • Sequester atmospheric carbon • Provide employment
Forest ecosystems	• Timber • Fuel wood • Drinking and irrigation water • Fodder • Non-timber products (vines, bamboo, leaves, etc.) • Food (honey, mushrooms, fruit, and other edible plants; game) • Genetic resources	• Remove air pollutants, emit oxygen • Cycle nutrients • Maintain array of watershed functions (infiltration, purification, flow control, soil stabilization) • Maintain biodiversity • Sequester atmospheric carbon • Moderate weather extremes and impacts • Generate soil • Provide employment • Provide human and wildlife habitat • Contribute aesthetic beauty and provide recreation
Freshwater ecosystems	• Drinking and irrigation water • Fish • Hydroelectricity • Genetic resources	• Buffer water flow (control timing and volume) • Dilute and carry away wastes • Cycle nutrients • Maintain biodiversity • Provide aquatic habitat • Provide transportation corridor • Provide employment • Contribute aesthetic beauty and provide recreation
Grassland ecosystems	• Livestock (food, game, hides, and fiber) • Drinking and irrigation water • Genetic resources	• Maintain array of watershed functions (infiltration, purification, flow control, soil stabilization) • Cycle nutrients • Remove air pollutants, emit oxygen • Maintain biodiversity • Generate soil • Sequester atmospheric carbon • Provide employment • Provide human and wildlife habitat • Contribute aesthetic beauty and provide recreation

continued on next page

Table 2, continued

Ecosystem	Goods	Services
Coastal ecosystems	• Fish and shellfish • Fishmeal (animal feed) • Seaweed (for food and industrial use) • Salt • Genetic resources	• Moderate storm impact (mangroves; barrier islands) • Provide wildlife (marine and terrestrial) habitat • Maintain biodiversity • Dilute and carry away wastes • Provide harbors and transportation routes • Provide human and wildlife habitat • Provide employment • Contribute aesthetic beauty and provide recreation

[a] From United Nations Development Program et al. (2000).

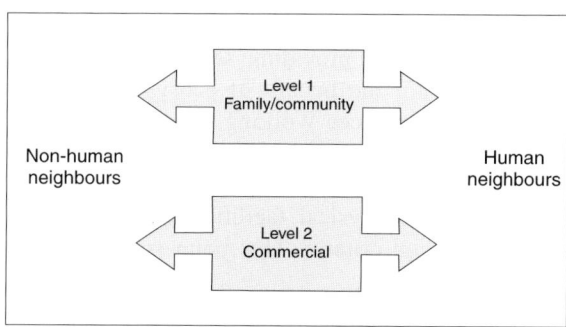

Fig. 2. Levels of relationship.

we might have joint commercial venture arrangements with. On the other hand we are likely to have a very different type of relationship with family members, friends, and loved ones. Both levels of relationship are valid in the appropriate circumstances. To achieve a healthy and stable (sustainable?) relationship with our family members, friends, and loved ones, we don't find it necessary to define and then quantify the value of the services these people provide us with. In fact, at this level of relationship, most of us would find such an attitude offensive and objectionable.

It is well accepted that humans form close bonds with animals (pets). Such a relationship is well understood by people like Aldo Leopold who has declared "It is inconceivable to me that an ethical relationship to land can exist without love, respect and admiration for land, and a high regard for its value. By value, I of course mean something far broader than mere economic value; I mean value in the philosophical sense." (Leopold, 1968; p. 223). In many societies and cultures, (and particularly the competitive, consumer, industrial/technological societies) the level 1 relationship with our non-human neighbors is ignored, unacceptable, undesired, or just not understood.

The science of ecology teaches that the human species is an integral part of ecosystems. Along with all other species humans are embedded in ecosystems. All components, including humans and their built environments, are interdependent neighbors. An ecosystem has been defined as "communities of interacting organisms and the physical environments in which they live" (United Nations Development Program et al., 2000). Townsend et al., make it clear that the abiotic environment is an essential component of an ecosystem when they define an ecosystem as the "... biological community together with the abiotic environment in which it is set" (Townsend et al., 2000; p. 390). It may then be argued that if in fact humans are integral and "community" members of an ecosystem then by definition the level 1 relationship exists between humans and the non-human world, even though we may express this differently. By definition we are of one interdependent community. Such interdependence demands mutual servicing; that is, it is not just about ecosystem providing services for human needs but also humans providing services for the benefit of ecosystems. This point is further developed in the Ecosystem Services Working Group consensus report.

The critical importance of the interdependence between ecosystems and human development has very significant implications for the criteria we should use in designing and building the physical facilities that support our societies and particularly engineering design criteria. This point is elaborated on in section 5 of this chapter.

Many cultures effectively deny that humans are an integral part of an ecosystem. Their actions often demonstrate, in terms of both levels of relationship, that humans consider themselves separate from and outside ecosystems. For example what ecosystem training and education is given to engineers, architects, technologists, tradespeople, farmers, and natural resource managers and policy analysts? The fact that ecology is not a core subject in the curriculum for the training and educational courses for such disciplines, is a clear statement that we do not always take seriously what ecological science is saying about humans, and their built environments, being an integral part of local ecosystems. We require our multi-story office block and apartment buildings, within which we work and live, to be designed by an experienced engineer with advanced training in behavior and properties of structures and their components. We might equally expect that those whose policies and designs modify and impact on the ecosystems within which we live might have advanced knowledge and understanding of ecological systems. While it is important for our buildings and other engineered structures

Ch. 5:   *Ecosystem Services: State of the Art*   109

to be safe for human occupation and use, it is essential for the continued wellbeing of humanity that our ecosystems are healthy, safe, and stable (United Nations Development Program et al., 2000). This fundamental observation will no doubt be reflected in many of the papers and workshop outcomes of this EcoSummit 2000 conference.

The popular perception is that the human world and the natural world are separate worlds, with the human world being superior and in control; a perception often reinforced by culture, language, traditional practices, technologies, values, belief systems, and religion. Furthermore, this human attitude defies logic. It denies the fact that modern humans are a very recent event (about 40 000 years old) in the story of the universe, the Earth, and its ecosystems. The universe has been evolving for about 15 billion years to the complex interactive web of life, death, and physical dynamics that now exist.

3.3. Valuing ecosystem services

In general terms, ignoring and overlooking level 2 and denying level 1 type relationships has resulted in unsustainable relationships with many ecosystems (see next section). One partial response to the question of how we can make more sustainable use of ecosystem services has been to identify and put a value on ecosystem services to enable environmental sustainability in policy and management practices. In other words, it is a worthy attempt to validate a level 2 relationship with the non-human eco-community. In the past, industrial cultures have simply taken ecosystems services for granted.

> These services are so fundamental to life that they are easy to take for granted, and so large in scale that it is hard to imagine that human activities could irreparably disrupt them.
> *(Daily et al., 1997)*

As will be discussed later in this chapter, the relatively recent discipline of ecological engineering is one genuine attempt, at a fundamentally practical level, to bring human society and its activities back into a more balanced and sustainable relationship with the non-human natural world, at both levels of relationship.

4. The use and misuse of ecosystem services

> The current rate of decline in the long-term productive capacity of ecosystems could have devastating implications for human development and the welfare of all species.
> *(United Nations Development Program et al., 2000)*

There is mounting evidence that human activity is placing considerable stress on many of the Earth's ecosystems and, as a consequence, their services. "Many signs point to the declining capacity of ecosystems" (Time, 2000). Experts working for the Worldwatch Institute (www.worldwatch.org), the United Nations

Development Programme (www.undp.org), the United Nations Environmental Programme (www.unep.ch), the World Bank (www.worldbank.org), and the World Resources Institute (www.wri.org), all agree that the overall health of each of the world's terrestrial and aquatic ecosystems is threatened. Declining ecosystem integrity translates into declining human health, declining economic activity and quality of life (www.who.org). Certainly, our health, economy, and quality of life are all dependent upon these ecosystems and their services.

Some examples of the decline of ecosystem services include
- Pollination levels reduced (e.g., Australia, South East Asia) due to loss of insect diversity which in turn is due to pesticide use. In Australia the bee population is declining by 5% per year due to an introduced European disease (CSIRO, 2000). In New Zealand, border biosecurity measures have failed and the Varroa mite (*Varrao jacobsoni*) has been recently introduced seriously impacting on the nation's pollination service.
- Climate regulation is being impacted by greenhouse gas emissions.
- Topsoil losses due to mismanagement, erosion, pollution, and degradation are threatening soil formation, regenerative processes, production and nutrient recycling.
- The logging of native forests, destruction of marine ecosystems (due to over-fishing), habitat fragmentation (for example due to roads and large dams), and drainage of wetlands are having a very significant impact on nature's ability to maintain biodiversity.
- Industrial emissions are destroying the ozone layer, causing the decline of nature's UV protection layer.
- Interference with nature's hazard mitigation processes (e.g., deforestation) is causing flooding.

5. Designing and engineering to restore a sustainable relationship with ecosystems

5.1. Making better use of ecosystem services

One of the fundamental questions we face is how we can make better use of ecosystem services. On a philosophical level, one answer may be to integrate level 1 and level 2 relationships with the non-human neighbors. As has been argued by Costanza (1996), Peet (1992), and other researchers in the field of ecological economics and policy, this will not happen until there are appropriate drivers operating within the commercial world that bring about a culture change in the way businesses and corporate bodies, and their policy decision makers, relate to the ecosystems within which they are imbedded. While institutional and policy frameworks for more sustainable use of ecosystem services are important, if "frontline" obstacles arise at the implementation stage, the credibility of the goals

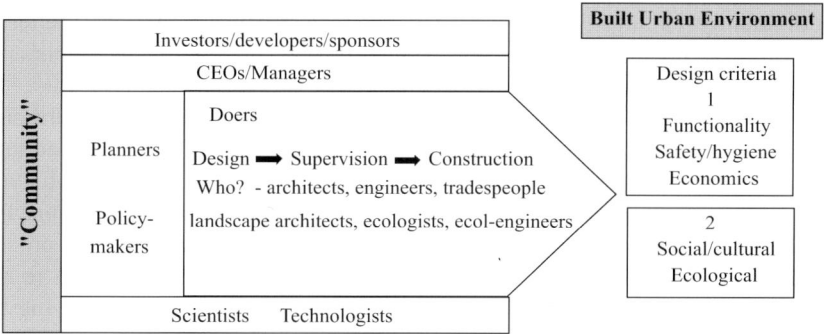

Fig. 3. Players in development of the built civil society.

(and the goal setters) may suffer a significant setback. It is important that there is support for a parallel process of development of innovative engineered projects that demonstrate what is technically and physically possible (and not possible) in achieving more sustainable interactions with ecosystem services. Such engineered projects can provide very effective feedback to both the policymakers and the scientific community to assist with policy design and research priorities respectively. Figure 3 illustrates the interrelationships between these different players.

5.2. Frontline projects

The nature of a society's impacts on ecosystems will depend on the way it designs and operates the physical components of that society such as roads, dams, industrial plants and their emission, wastewater treatment and disposal plants, solid waste processing and disposal plants, energy production and conversion plants, irrigation schemes, wetland drainage and urban stormwater infrastructures, marine ports, mining plants, military weapons, tourist resorts, and so on. Human activities that impact on the natural world are a result of excessive emissions, unsustainable resource extraction, and substantial habitat fragmentation, disruption, and destruction. Nearly all such activities can be attributed to engineered military or civil activities and projects. Each human project requires the effort of design with explicit or implicit design criteria. Setting aside military projects, the design criteria for such projects has traditionally been functionality, safety, hygiene, and economic viability. Over the last 10 to 15 years, environmental issues have highlighted the need to add ecological and social design criteria in the context of sustainability.

5.3. Players involved in achieving better use of ecosystem services

The various players in the process developing the built environment in a civil society are illustrated in fig. 3. Scientists (including the social and economic sciences) are just one of several players.

"Community" in fig. 3, is in inverted commas because it will be matter of definition, and no doubt debate, as to whether our non-human neighbors are included in its definition.

It was Plato who labeled engineers as the doers, possibly more as a criticism of their inability or unwillingness to think about the implications of their actions. Today there are some that are saying that there is too much thinking and rhetoric, about achieving sustainability and not enough doing. Figure 3 illustrates that while the EcoSummit Conference is about "integrating the sciences", the process of integration of effort must include many more players in addition to the scientific community. Designers are key players in translating and applying the knowledge and understanding discovered by scientists, the technologies developed by technologists, and the aspirations and needs of the community (reflected, hopefully, in policy). To achieve a sustainable society the various players need to work together and agree on common criteria for the design of their society. This begs the question of whether the different disciplines and players have the appropriate culture to enable them to work together as part of what the UNDP report (United Nations Development Program et al., 2000) refers to as the "ecosystems approach" – a holistic rather than a sectoral approach.

5.4. Engineer as designer

Engineering design "is a creative art, based on science, for useful purposes" (Davies and Painter, 1990). As a conscious act, it is one of the distinguishing skills that humans have that no other species have. It offers a degree of freedom that no other species of the ecosystem has. Consequently, this freedom is to be balanced with responsibility and in this case, specifically a responsibility to the ecosystem within which the designed project is embedded. Traditionally, engineers and architects have been the designers of projects such as built structures, technologies, services and infrastructures. More recently, however, landscape architects and ecological engineers have contributed to project design.

Engineering is about modifying, through conscious design, a physical environment inhabited by human beings. The President of the Institute of Engineers, Australia, noted that "virtually any engineering project modifies the environment in some way and therefore has an effect on the welfare, health and safety of the surrounding community" (Gillin, 1992). Professional Engineers have a special obligation "towards the integration of development and the environment, leading towards sustainable development" (IPENZ, 1993).

The Bruntland report (Bruntland, 1987) called for a "reorientation of technology – the key link between humans and nature". In response, David Thom, past Chairman of the World Federation of Engineering Organisations (WFEO), made it clear that "the whole ethos of engineering practice will have to be different" (Thom, 1993). The challenge to the engineering profession, if it is to remain relevant and credible into the 21st century, is to understand the nature of this difference and embrace it, particularly in the training and education of engineers. This difference is identified in the WFEO Code of Environmental Ethics for Engineers which calls for engineers to "be aware that the principles of ecosystemic interdependence, diversity maintenance, resource recovery and interrelational harmony form the bases of our continued existence and that each of those bases poses a threshold of sustainability that should not be exceeded" (Thom, 1993).

The popular response by the engineering profession and others to environmental responsibility has been a technocentric approach. That is more "efficient" technologies (clean technologies) and improved management (waste minimization and environmental management systems). Such an approach is a relatively easy political and economic adjustment for the professional engineer, politicians, and industry. It is basically business-as-usual with new or modified management practices and technologies. However, there is an increasing voice of concern saying that the technological fix, on its own, will not deliver sustainable living. While improved technologies may result in more efficient resource processing, they have also enabled, for the wealthy, higher living standards, increased expectations, and consumption. Those who cannot afford the technologies can be driven to greater poverty. A widened rich–poor gap leads to a dysfunctional society and further environmental degradation, as those trapped in poverty most often can not afford to address longer term sustainability and environmental issues.

The engineer, without a well-rounded sense of place in the social and ecological world, can become so single-mindedly focused on, if not obsessed with, the excitement and challenge of the narrow technological world, that s/he is oblivious to what is happening in the wider environment. The technological fix approach can deny an appreciation of the interconnectedness of the physical engineering and technological world with the biological, spiritual, emotional, social, and cultural worlds. This detachment can lead to a dysfunctional and unbalanced sense of responsibility towards nature.

Sustainability, by definition, requires a long-term view and a sound understanding of the ecological, economic, and social systems within which engineering projects are imposed. It requires that we rethink and reshape the nature of our relationships between the human and non-human biophysical communities as discussed earlier. It is an interdisciplinary, local and global effort to take human development beyond, but inclusive of, the more traditional goals of efficiency and technology to ecologically integrated, sustainable development.

The President of the Boston Society of Civil Engineers, in a recent address, stated that "Future engineers, scientists and business people must design technology and economic activities that sustain rather than degrade the natural environment, enhance human health and well-being and mirror and live within the limits of natural systems" (Cortese, 1999).

6. Ecological engineering

Ecological engineering is a relatively new discipline in the western world. Ecological engineering specifically recognizes that in our designs we need to "mirror and live within the limits of natural systems" (Cortese, 1999). It is a discipline that supports, in practical terms, more sustainable use of the ecosystem services.

The discipline of ecological engineering (also sometimes referred to as eco-technology) was the initiative of Howard Odum[2] who in the early 1960's recognized the need for society to respect and work with nature: "Because of the pressures humankind is now placing on the ecosystem, the management of the planet must turn more and more to a co-operative role with the planetary life support system, sometimes called *stewardship of nature*" (Odum, 1989).

Ecological engineering is about a partnership with nature. It has been defined "... as the design of human society with its natural environment for the benefit of both" (Mitsch, 1996). However, this definition was modified to "the design of sustainable ecosystems that integrate human society with its natural environment for the benefit of both" (Mitsch, 1998). Mitsch explained that the task of the ecological engineer should be the design of ecosystems and not societies.

The Chinese may claim that ecological engineering is not a new discipline and is a traditional part of their primary production practices. Yan[3] and Ma (1996) define ecological engineering as "... a specifically designed and operated system for production processes combined with the management of the natural environment, for the benefit of both".

In reviewing *Ecological Engineering* conference proceedings and journal articles, it is clear that there is a broad range of topics that can be included within the discipline of ecological engineering (table 3). There is continuing constructive discussion and debate about what should or should not be included. However, what is clear is that the essence of ecological engineering is the application of our knowledge of ecosystems[4], and our skills of technical problem solving and design, to achieve the integration of human endeavor and creativity with the natural world.

It is not a new idea that we need to integrate our engineering projects with nature. It is "in the nature of the gods to create, as it were, a second world within

---

[2] Department of Environmental Engineering Sciences and Centre for Wetlands, University of Florida, Florida.
[3] Nanjing Institute of Geography and Geology, Nanjing, China.
[4] Ecosystem includes the relevant whole of the conditions, behavior, and interactive processes of the biotic and abiotic components of the natural world, as well the social, cultural, and economic activities of the human population.

Table 3
Some design topics included in the ecological engineering discipline

• Constructed wetlands for wastewater treatment	• River restoration
• Aquaculture wastewater nutrient recycling	• Wetland restoration
• Integrated urban water and wastewater technologies and management	• Ecological engineering of coastlines
• Greywater treatment and management	• Estuary restoration and management
• Industrial and domestic water saving and recycling systems and technologies	• Minimizing ecosystem fragmentation in road design
• Solids recycling systems	• Ecotunnel, cerviduct, and ecoduct design
• Design of urban ecosystems	• Ecologically appropriate building materials
• Restoration ecology	• Modeling ecotechnology
• Habitat reconstruction	• Town planning for sustainability
• Ecosystems rehabilitation	• Biomass energy
	• Biogas technology
	• Renewable fuels
	• Sustainable agroecology

the world of nature" (Cicero, 45 BC). The following case studies demonstrate how ecological engineering has been applied.

6.1. Case Study 1: Ministry of Transport in the Netherlands (van Bohemen, 1996, 1998; and 2000, personal communication)

The Road and Hydraulics Division of the Ministry of Transport in the Netherlands have a policy that integrates transport systems with ecosystem requirements. They have a number of policy areas, one of which is to minimize the effect of habitat fragmentation by including facilities such as badger pipes and tunnels under highways, as well as overpasses for migrating red deer. These are also used by other animals such as fox, roe deer, hedgehogs, ground beetles, and amphibians.

Activities to mitigate and to compensate are fully embedded in the policy of the Dutch government. The Transport Plans include target scenarios for the short term to prevent further fragmentation of the countryside and in the long term fragmentation will be reduced. In an effort to integrate conservation and to reduce the negative effects of infrastructure and transport in the road planning and road building processes the following procedure has been adopted.

> First of all, an attempt is made to prevent habitat fragmentation by restricting infrastructural development and by integrating the infrastructure into the landscape. Secondly, the aim is to counteract fragmentation. This can be done by mitigating existing situations. Where the effects of fragmentation cannot be mitigated it is desirable to take compensatory measures (replacing lost habitats or enhancing marginal habitats through appropriate forms of environmental improvements). *(van Bohemen, personal communication)*

Van Bohemen points out that there is a growing interest in compensation, motivated by a desire to promote ecological functions and values that have been damaged or have disappeared due to the presence of infrastructure. van Bohemen (1996) also reports on the ecosystem value the Netherlands place on the land–sea transition zone. In particular the Ministry of Transport have set engineering criteria for the design of tidal inlets. The criteria recognise flood safety requirements, hydrologic effects, socio-economic implications, and ecological and scenic benefits and effects.

6.2. Case Study 2: Oxelösund Våtmark, Sweden (Personal visit, 1998)

This is a constructed surface flow wetland designed to reduce the nitrogen levels of the treated wastewater from Oxelösund township (population equivalents = 15 000) by 50% before discharging to the Baltic Sea. Following conventional mechanical/chemical treatment nitrogen removal is achieved in 22 ha of wetlands. The surface wetlands are unique in that there are two parallel series of ponds operated alternately on a 3.5 day cycle, followed by a final denitrification pond.

6.3. Case Study 3: Donaumoos – Germany (Wild, 2000; personal communication)

Three wetlands have been constructed in Donaumoos, Germany, approximately 80 km north of Munich in the valley of the river Danube. It is a percolating mire and the largest peat deposit in southern Germany. Before cultivation began 200 years ago, the peat area totaled 180 km^2. Seventy-five percent of the Donaumoos is presently used as arable land (the main crop being potatoes); the remainder is permanent grassland. Due to the long agricultural history, the decomposition of peat is quite advanced.

The total area of the 3 Donaumoos wetlands is 62 000 m^2. The input to the wetlands is heavily polluted runoff from a catchment (32 km^2) that is 95% agricultural with some urban wastewater input. The wetland plants, *Typha angustifolia* and *T. latifolia,* have been cultivated in two constructed wetlands. These plants are harvested in the winter and processed to produce insulation material. Nutrient removal in the wetlands was high for NO_3-N and PO_4-P and low for organic N. The wetlands not only treat the polluted catchment runoff but also recharge ground water, restore the natural wetland function to the ecosystems, and provide a feasible alternative for the production of a renewable raw material for production of insulation.

6.4. Case Study 4: Kaja, Ås, Norway (Etnier and Refsgaard, 1999; personal visit, 1999)

The Agricultural University of Norway, in Ås, has a student apartment building (called Kaja) with 24 flats, 54 students. Fitted within this building are 26 vacuum toilets and the system has been designed to separate the greywater and the blackwater streams. The blackwater is stored onsite in a 15 m³ belowground storage tank. Experiments were set up to demonstrate the economic feasibility of transporting this blackwater by a road tanker to a liquid composting plant (at Aremark – see Case Study 5) for sanitizing before being land disposed onto farmland. All greywater is treated onsite by a septic tank followed by an expanded clay biofilter and constructed subsurface flow wetland. A high standard of greywater treatment is achieved before it is discharged to the community stormwater system.

Researchers at the Norwegian Agricultural Economics Research Institute, Oslo, have examined the economic and environmental costs and benefits of various decentralized wastewater systems. Source separation with liquid composting of blackwater is favored (Etnier and Refsgaard, 1999).

6.5. Case Study 5: Aremark (Personal visit, 1998)

In a small town just south of Aremark in southern Norway, a farmer is paid by the local municipality to collect, sanitize, stabilize, and return domestic wastewater nutrients to local productive farmland. In this partnership between town and country, residents agree to install water saving toilets (some use vacuum toilets) and store the reduced volume (annually, 7 to 10 m³ per household) in onsite blackwater storage tanks. In Aremark, the farmer not only collects the blackwater about once a year, but also the kitchen organic waste which he uses to add to a thermophilic aerobic liquid composting plant on his farm. The greywater is treated and disposed of on site in various ways for each household.

6.6. Case Study 6: Kågeröd Recycling Project (Hasselgren, 1995; personal visit, 1998)

In Sweden there are more than 20 000 ha of *Salix* species that are grown for biomass energy fuel. This project looks at economic, energy, and environmental issues associated with biomass energy production from wastewater fed *Salix*.

The Kågeröd Recycling Project, near Lund in South Sweden, takes conventionally treated wastewater from the township of Kågeröd (1500 population plus food factory loads = 6000 pe) near Lund in South Sweden and irrigates this on to 25 ha of *Salix*. The *Salix* is managed by short rotation and harvested to produce biomass for energy production.

6.7. Case Study 7: Ruswil, Switzerland (Heeb et al., 2000; personal visit, 2000)

A large pipeline reticulates the natural gas from the North Sea to the Mediterranean countries. At various points along the pipeline are compressor stations.

At the compressor station in Ruswil, Switzerland, there is a waste heat output of about 60 MW. The pipeline company, Transitgas AG, is interested in options for utilising this waste energy.

The company has set up a pilot plant to investigate the feasibility of combining the waste energy from the compression station, with wastewater from the local community, to operate a productive greenhouse/aquaculture center. The glass house has two zones – temperate and subtropical/tropical zone. The plant also includes aquaculture tanks. A pilot phase is to run over a period of 2–3 years (from 1999–2001). The results of this phase will be the basis for a possible expansion of the project.

The pilot greenhouse, made of double-layered glass panes and a light metal structure, covers an area of approximately $1500 \, m^2$, split approximately equally between temperate and subtropical/tropical climates. The temperate climate zone includes an area for educational and demonstration purposes. In the subtropical/tropical climate zone, subtropical and tropical fruit and vegetables are produced. Rapidly growing plants, bushes, or trees with a maximum height of 4–6 m and with a large yield are required for production. Some examples include papaya, mango, kumquat, annona, passion fruit, guava, litchi, cape gooseberry, tree tomatoes, etc.

At a later stage, it is proposed that liquid manure from a neighboring farm will serve as a nutrient input for the total greenhouse system. The solid fraction will be composted and processed into fertilizer. The produced fertilizer will then be used for the plants growing in the soil of both sections of the greenhouse, along with mushrooms and earthworms cultivated on site.

The operational aquaculture system consists of shallow water basins that are suspended above fish tanks (integrated aquaculture ponds). The fish tanks have a total area of $50 \, m^2$ and a depth of 1.5 m. In these tanks, tilapias, a subtropical and tropical freshwater fish with a high market value, are raised. The water from the fish tanks is conducted to the shallow water basins and cleaned there. The nutrient rich sediment from the basins is used as fertilizer for the cultivated area in the glasshouse. In the future, the liquid fraction of the liquid manure from the neighboring farm, being rich in nutrients, will be directed to shallow water basins where floating water plants are to be grown. These floating plants will serve as food for fish in the adjacent fish tanks.

6.8. Case Study 8: Calcutta Wastewater-fed Aquaculture (Jana et al., 2000; personal visit, 1999)

Calcutta possibly has the largest wastewater-fed aquaculture system in the world. Three hundred years ago the area was dominated by mangrove forest (as currently remains in the Sundarbans region), but the original 8000 ha of wetlands have been reduced to 3600 ha. Calcutta, with over 11 million inhabitants, produces about

11 000 m³ wastewater per day. A paper presented at the 1998 IEES conference (Jana et al., 2000) by Roy stated that 2500 ha of wastewater-fed ponds produce an average of 20 T/day of fish, employing 17 000 people. In addition to fish production, it is claimed that large open ponds absorb polluting dust and act as settling ponds to contaminated storm water. Generally, the wastewater-fed fish ponds use diluted (from 1:1 up to 1:4) effluent from primary ponds.

Fish yields vary. For example:
- Mudialy ponds produce 2160–5700 kg/yr – monthly average 400 kg/ha
- Captain Beri ponds produce 3852–8770 kg/yr

The impact of wastewater-fed aquaculture ponds may not only be positive, however. Biswas and Santra (2000) from Kalyani University, Calcutta studied heavy metals and observed that the concentration of heavy metals detected in vegetables and fishes sold in different markets of Calcutta and suburbs was considerably higher than those collected from pollution free environments. Bhowmik et al. (2000) from the Central Institute of Freshwater Aquaculture, Kalyani, studied the disease risk for the fisherman and concluded that "... fisherman of sewage-fed fish farms had a high prevalence of diarrhea, cough and cold and fever although the prevalence of such diseases did not significantly differ for freshwater fish farmers".

6.9. Case Study 9: Stensund Aquaculture Centre (Guterstam, 1996; Guterstam et al., 1998; personal visit, 1998, 1999)

Stensund Folk College, just south of Stockholm, is a small rural village on the Baltic coastline of Sweden. The Folk College, owned by the Swedish Health and Sports Association, continues a centuries-old tradition of education in democracy and technology with an emphasis on sustainability and social integration.

The Stensund team recognized the need to close wastewater nutrient cycles and optimize energy use. They adapted the traditional wastewater-fed aquaculture production systems used in warm climates such as in Asia to the much colder climate of Sweden. It was necessary that their wastewater-fed aquaculture be built in a greenhouse allowing an almost year-round production cycle.

The Stensund system uses a greenhouse structure to enclose a series of tanks, each containing specifically cultured species that convert, detoxify, and recycle the components of wastewater. Up to 20 m³ per day of wastewater from Folk College have been fed through this system. At the same time, the system has been exporting, to the College, energy recovered from the wastewater and energy captured and unused by the greenhouse.

The performance of the Stensund plant has been monitored for more than 10 years now, with published results (Guterstam, 1996; Guterstam et al., 1998). Guterstam (1996) reports that the quality of the discharge from the plant almost meets bathing water standards and organic and suspended solids have been reduced by more than 95%. In addition, Guterstam reports that while the Stensund system has a higher capital and operating cost than some conventional waste treatment technologies it is stressed that the system is not only a waste treatment unit but also produces energy and food and is an education and tourist center.

In 2000, the Centre closed due to economic difficulties. It has provided services well beyond traditional wastewater treatment and disposal for the local community, which are relatively easy to analyze in terms of marginal costs. However, the additional services the Centre has provided are much more difficult to quantify in terms of traditional economics and include

- Community education and awareness raising at local, national, and international levels.
- Research and development of processes, principles of, and technologies for ecological engineering.
- Building knowledge, awareness, and experience in appropriate relationships between technology, engineering, ecology, and living sciences, economics, and social and cultural behavior.
- Interdisciplinary and transdisciplinary activities.
- International networking and technology transfer.

6.10. Case Study 10: Water Enhancement Programme, Christchurch (Christchurch City Council, 2000; personal communication)

In recent years, the governing local authority for the City of Christchurch, New Zealand, has been actively implementing a more ecologically friendly approach to the development and maintenance of its urban waterways. Factors that have contributed to this new approach are

- The enactment of the Resource Management Act. The RMA is designed to bring about an integrated approach to environmental planning and resource management. The purpose of this Act is to promote the sustainable management of the natural and physical resources.
- The enactment of the 1974 Local Government Act. This required each NZ local authority to complete an asset plan.
- The increasing cost of managing and extending drainage services.
- The community's greater awareness of environmental issues and values.

In formulating their asset plan the Christchurch City Council (CCC) realized that the traditional approach of valuing infrastructure would under-value the city's natural and social/cultural assets. Of particular concern was that the waterways not only functioned to provide a drainage service, but also to provide a natural ecological service. Christchurch City has about 90 km of rivers and almost 300 km of tributaries. Consequently, the Water Services Unit of the City Council is now implementing an asset management plan that recognizes the 6 values of landscape, ecology, recreation, culture, heritage, and drainage.

Their enhancement programme aims to provide the community with not only a functional, economic, and low risk waterways service (drainage and stormwater treatment), but also provide recreational, cultural, natural heritage, and educational opportunities. Additionally, a recent study found that adjacent property values increased as a result of willingness to pay for such enhancement (Bicknell and Gan, 1997).

6.11. Case study evaluation

The evaluation of these case studies in terms of how they aim to achieve more sustainable use of ecosystem services is presented in table 4.

Table 4
Case study evaluation

Case Study	Contribution towards more sustainable use of ecosytem services
1. Ministry of Transport, Netherlands	Enhanced biodiversity and ecosystem conservation through minimizing habitat fragmentation. Enhanced aesthetic beauty.
2. Oxelösund Våtmark, Sweden	Restoration and enhanced wetland biodiversity. Partial closing of the nutrient cycle. Enhanced aesthetic beauty.
3. Donaumoos, Germany	Restoration and enhanced wetland biodiversity. Restoration of groundwater levels. Sustainable production of insulation product, consequent energy saving and reduced greenhouse gas impact. (Contributing to the maintenance of the ecosystem service of climate regulation.)
4. Kaja, Ås, Norway	Partial closing of the nutrient cycle. Ecosystem education.
5. Aremark, Norway	Partial closing of the nutrient cycle.
6. Kågeröd Recycling Project, Sweden	Partial closing of the nutrient cycle. Renewable energy production and reduced greenhouse gas emissions. (Contributing to the maintenance of the ecosystem service of climate regulation.)
7. Ruswil, Switzerland	Partial closing of the nutrient cycle. Use of waste energy and reduced greenhouse gas impact. (Contributing to the maintenance of the ecosystem service of climate regulation.)
8. Calcutta Wastewater-fed Aquaculture, India	Enhanced wetland biodiversity. Partial closing of the nutrient cycle. Maintenance of groundwater levels.
9. Stensund Aquaculture Centre, Sweden	Closing of the nutrient cycle. Energy conservation and consequent reduced greenhouse gas impact. (Contributing to the maintenance of the ecosystem service of climate regulation.) Ecosystem education.
10. Water Enhancement Programme, Christchurch, New Zealand	Enhanced urban biodiversity. Community education and participation on ecosystem services. Enhanced aesthetic beauty.

None of the above case studies include a living machine system (Todd and Josephson, 1996; www.livingmachines.com). This particular technology is another very good example of ecological engineering. The living machine technology is a biological wastewater processing system similar to the Stensund Aquaculture Centre system described above in Case Study 9. These systems use a series of tanks, usually enclosed in a controlled environment glasshouse. Pioneered by Dr J. Todd in the USA, it is a miniature ecosystem modeled after natural systems. The innovative process is based on photosynthesis, nutrient recycling, and biodiversity. It incorporates accelerated aquatic ecological processes with the help of sunlight and a diversity of organisms including bacteria, plants, snails, and fish to break down and digest organic pollutants. The living machine design has been used at a number of sites throughout the world (e.g., USA, Canada, UK, Brazil, Australia) for treating different types of wastewater streams under different climatic conditions.

These few case studies demonstrate how the ecological engineering approach can, by adopting a more integrated approach to design, contribute to a more sustainable use of the services of ecosystems. For other case studies, the reader is referred to the website of the International Ecological Engineering Society, http://www.iees.ch/.

While this chapter has stressed the contribution of ecological engineering to more sustainable use of ecosystem services there are clearly additional measures that can be taken and these are presented in the following ecosystem services consensus chapter.

7. Under-utilization of ecosystem services

There is increasing concern about the overuse of ecosystem services. However the above case studies also demonstrate that some conventional engineering practices can lead to substantial under-use of ecosystem services and the opportunities they provide for a more sustainable and balanced lifestyle.

Most of the conventional wastewater systems throughout the world do not use the nutrient recycling services of aquatic ecosystems (as in Calcutta, Stensund and Ruswil and living machine aquaculture) or soil systems (as with Aremark). The regenerative capacity of plant/soil ecosystems in terms of the recycling of nutrients is under-utilized. There are many thousands of kilometers of transport system and urban waterways that under-utilize the ecosystem services of enhanced aesthetics and biodiversity by poor design and destruction of habitats.

Many of the integrated ecologically engineered systems may be more labor intensive in some situations but in nations where labor is low-cost and employment is needed, ecologically engineered systems such as integrated biosystems are beneficial.

8. Conclusions

This chapter has reviewed some of the literature on ecosystem services and the identification and value of these services. A wide-ranging view of ecosystem services is evident. These differences can be explained in terms of the nature and depth of one's relationship with the natural world. There are those who simply operate in terms of a simple commercial interest in the natural world and others who interact at a much deeper level with the non-human world. It also depends on how a person understands her or his relationship with ecosystems, whether one operates as if separate and outside of an ecosystem or as an integral interdependent component embedded in the ecosystem. The nature of this individual relationship will influence the type of collective relationship for a given community or organization.

Engineers and ecological engineers are interested in finding physical solutions to human development problems. The challenge is for the engineering discipline to adopt the ecosystems approach along with the additional design criteria this will require. Engineers and ecological engineers possess knowledge of the properties and behavior of materials, ecological systems, and the forces of nature. It is their professional responsibility to use this knowledge along with their creative design skills, to provide physical systems and technologies that enable society to better utilize (more sustainably) ecosystem services. Some of these ecosystem services are being overused, degraded, or destroyed, while others are being underutilized. If this is a "big-ask" for one particular discipline, then interdisciplinary project design may be an option. The author is not confident, however, that either the ecological science discipline or the professional engineering discipline is yet sufficiently capable, in general terms, of embracing the ecosystems approach to the design of engineering projects. There are specific examples (as illustrated in the case studies) where the ecosystems approach to a project design has been implemented, but such practice is not yet widely adopted by those responsible for development projects. There is ample evidence that the educational and training institutions for many of the key disciplines clearly do not consider it important to include ecological systems science in their curriculum.

The case studies presented here indicate how an ecological engineering design approach can lead to more sustainable use of ecosystem services. However, more rigorous analysis [for example by applying the sustainable process index (SPI); Krotschek and Narodoslawsky (1996)] of some of these case studies needs to be carried out to demonstrate just how sustainable the projects are.

Better use of ecosystem services is fundamental to achieving a more sustainable relationship with the natural world. This can be achieved. However, it will only be achieved by adopting an ecosystems approach to all levels of human endeavor and development.

Acknowledgements

I am grateful to all members of the EcoSummit 2000 Ecosystem Services Working Group for their insights and enthusiasm. I am particularly grateful to the following for their written comments and suggestions: Hein van Bohemen, Delft, The Netherlands; Stefan Gössling, Lund, Sweden; Björn Guterstam, Trosa, Sweden; Johannes Heeb, Wolhusen, Switzerland; Andreas Schönborn, Lucerne, Switzerland; Ralf Roggenbauer, Vienna, Austria; Alan Werker, Waterloo, Canada.

References

Beder, S., 1998, The New Engineer (Macmillan Education Australia, South Yarra) 347 pp.

Bhowmik, M.L., Chakrabarti, P.P. and Chattopadhyay, A., 2000, Microflora present in sewage-fed and possibilities of their transmission. In: B.B. Jana, R.D. Banerjee, B. Guterstam and J. Heeb (Editors), Waste Recycling and Resource Management in the Developing World (University of Kalyani, India) pp. 71–77.

Bicknell, K.B. and Gan, C., 1997, Valuing waterway enhancement activities in Christchurch; a preliminary analysis. Presented to the NZARES Conference, Blenheim, New Zealand, July.

Biswas, J.K. and Santra, S.C., 2000, Heavy metal levels in marketable vegetables and fishes in Calcutta Metropolitan Area, India. In: B.B. Jana, R.D. Banerjee, B. Guterstam and J. Heeb (Editors), Waste Recycling and Resource Management in the Developing World (University of Kalyani, India) pp. 371–376.

Bruntland, G. (Chair), 1987, Our Common Future. The World Commission on Environment and Development (Oxford University Press, Oxford) 400 pp.

Cairns, J., 1996, Determining the balance between technology and ecosystem services. In: P. Schulze (Editor), Engineering Within Ecological Constraints (National Academy Press, Washington, DC) pp. 13–30.

Cicero, M.T., 45 BC, De Natura Deorum. Book 2, Section 152.

Cortese, A.D., 1999, The role of engineers in creating an environmentally sustainable future. Presented at the annual Thomas R Camp lecture of the Boston Society of Civil Engineers.

Costanza, R., 1996, Designing sustainable ecological economic systems. In: P. Schulze (Editor), Engineering Within Ecological Constraints (National Academy Press, Washington, DC) pp. 79–95.

Costanza, R., d'Arge, R., de Groot, R., Farber, S., Grasso, M., Hannon, B.M., Limburg, K., Naeem, S., Paruelo, J., O'Neill, R.V., Raskin, R.G., Sutton, P. and van den Belt, M.J., 1997, The value of the world's ecosystem services and natural capital. Nature 387:253–260. See http://www.floriplants.com/news/article.htm.

CSIRO, 2000, Website: http://www.dwe.csiro.au/ecoservices/myerintro.htm.

Daily, G.C. (Editor), 1997, Nature's Services – Societal Dependence on Natural Ecosystems (Island Press, Washington, DC) 392 pp.

Daily, G.C., Alexander, S., Ehrlich, P.R., Goulder, L., Lubchenco, J., Matson, P.A., Mooney, H.A., Postel, S., Schneider, S.H., Tilman, D.G. and Woodwell, G.M., 1997, Ecosystem services: benefits supplied to human societies by natural ecosystems. Issues Ecol. 2(Spring):2–16.

Davies, T.R.H. and Painter, D.J., 1990, New degree in natural resources engineering. N.Z. Eng. 45(6):4–6.

Etnier, C. and Refsgaard, K., 1999, Economics of decentralised wastewater treatment; testing a model with a case study. Presented to Conference: Managing the Wastewater Resource – Ecological Engineering for Wastewater Treatment. June 7–11, Ås, Norway.

Gillin, M., 1992, Foreword. In: National Committee on Environmental Engineering, Environmental Principles for Engineers (Institution of Engineers, Australia, Canberra).

Guterstam, B., 1996, Demonstrating ecological engineering for wastewater treatment in a Nordic Climate using aquaculture principles in a greenhouse mesocosm. Ecol. Eng. 6(1–3):73–97.

Guterstam, B., Forsberg, L.E., Buczynska, A., Frelek, K., Pillaityte, R., Redzek, L. and Rucevska, I., 1998, Stensund wastewater aquaculture; studies of key factors for its optimisation. Ecol. Eng. 11(1–4):87–100.

Hasselgren, K., 1995, Wastewater irrigation of energy plantation. In: J. Staudenmann, A. Schönborn and C. Etnier (Editors), Recycling the Resources (Transtec Publication, Zuerich-Uetikon, Switzerland) pp. 183–188.

Heeb, J., Huber, F. and Wyss, P., 2000, The Greenhouse Use of Waste Heat of Transitgas AG's Gas Compression Station (GVS) at Ruswil (Switzerland). http://www.heeb-gmbh.ch.

IPENZ, 1993, Environmental Principles for Engineers, unpublished draft (IPENZ Committee On Engineering and the Environment).

Jana, B.B., Banerjee, R.D., Guterstam, B. and Heeb, J. (Editors), 2000, Waste Recycling and Resource Management in the Developing World, Proc. Int. Ecological Engineering Conference (1998), Calcutta (University of Kalyani, West Bengal).

Krotschek, C. and Narodoslawsky, M., 1996, The sustainable process index – a new dimension in ecological evaluation. J. Ecol. Eng. 6:241–258.

Leopold, A., 1968, A Sand County ALMANAC and sketches here and there (Oxford University Press, London) 226 pp.

Mitsch, W.J., 1996, Ecological engineering: the roots and rationale of a new ecological paradigm. In: C. Etnier and B. Guterstam (Editors), Ecological Engineering for Wastewater Treatment (CRC/Lewis Publishers, Boca Raton, FL) pp. 1–20.

Mitsch, W.J., 1998, Ecological engineering: the 7 year itch. Ecol. Eng. 10(2):119–139.

Moser, A., 1997, Eco-tech as a new engineering discipline. In: The Green Book of Eco-Tech, Proc. 4th Int. Ecological Engineering Conf. (Sustain, University of Technology, Graz, Austria).

Odum, H.T., 1962, Man in the ecosystem: proceedings of Lockwood Conference on the suburban forest and ecology. Bull. Conn. Agric. Stn. 652:27–75.

Odum, H.T., 1989, Ecological engineering and self organization. In: W.J. Mitsch and S.E. Jørgensen (Editors), Ecological Engineering (Wiley, New York) pp. 79–101.

Peet, J., 1992, Energy and the Ecological Economics of Sustainability (Island Press, Washington, DC) 309 pp.

Roberts, D.V., 1991, Sustainable development – a challenge for the engineering profession. Trans. IPENZ 18(1/Gen):2–8.

Schulze, P.C., Frosch, R.A. and Risser, P.G., 1996, Overview and perspectives. In: P. Schulze (Editor), Engineering Within Ecological Constraints (National Academy Press, Washington, DC) pp. 1–10.

Thom, D., 1993, Engineering to sustain the environment, Proc. IPENZ Conf., Hamilton, February 1993.

Time, 2000, A preview of the PAGE report [World Resources 2000–2001: People and Ecosystems: The Fraying Web of Life (United Nations Development Program, United Nations Environment Program, World Bank, World Resources Institute, Washington DC) 400 pp].

Todd, J. and Josephson, B., 1996, The design of living technologies for waste treatment. Ecol. Eng. 6(1–3):109–136.

Townsend, C.R., Harper, J.L. and Begon, M., 2000, Essentials of Ecology (Blackwell Science, Boston, MA) 570 pp.

United Nations Development Program, United Nations Environment Program, World Bank and World Resources Institute, 2000, World Resources 2000–2001: People and Ecosystems: The Fraying Web of Life (World Resources Institute, Washington, DC) 400 pp.

van Bohemen, H.D., 1996, Environmentally friendly coasts: dune breaches and tidal inlets in the foredunes. Landsc. Urban Plan. 34:197–213.
van Bohemen, H.D., 1998, Habitat fragmentation, infrastructure and ecological engineering. Ecol. Eng. 11(1–4):199–297.
Wurth, A.H., 1996, Why aren't all engineers ecologists? In: P. Schulze (Editor), Engineering Within Ecological Constraints (National Academy Press, Washington, DC) pp. 129–137.
Yan, J. and Ma, S., 1996, The function of ecological engineering in environmental conservation with some case studies from China. In: C. Etnier and B. Guterstam (Editors), Ecological Engineering for Wastewater Treatment (CRC/Lewis Publishers, Boca Raton, FL) pp. 21–36.

Consensus

Chapter 6

Ecosystem Services

B. Guterstam, A. Werker, with M. Adamsson, D. Barker, A. Brüll, A. Dakers, S. Gossling, J. Heeb, S. Loiselle, U. Mander, D. Melaku Canu, R. Roggenbauer, M. Roux, G.D. Santopietro, D. Stuart, M. Trudeau, H.D. van Bohemen

Abstract

The purpose of this chapter is to report on the EcoSummit 2000 working group activity and consensus building for the theme of ecosystem services. An interdisciplinary team of environmentally focused professionals spent two days during the EcoSummit to tackle semantics, present case studies, and share perspective on the importance of ecosystem services towards understanding and solving environmental problems in the 21st century. Solutions to environmental problems will be both technical and social in nature. Ecosystem services define the natural processes of the biosphere that are responsible for sustaining human life. In a world where the scale of human activity has become truly global, the significance and role of ecosystem services for assessing and managing human activity are becoming universally self evident. This chapter presents ecosystem services in terms of definitions and connotations along with discussion of the road to follow in order to explicitly build ecosystem services into the structure of future human existence. Such a road needs to be taken if humankind is to escape the current self-destructive path with the biosphere. Key outcomes of the working group were the recognition of the interplay between (1) global concerns and local initiatives, (2) thinkers, doers, and stakeholders, and (3) central authority and distributed activity. A first draft of a decision tree interrelating needs of ecosystem services with wants of human development is presented to the reader for adaptation towards evolving best practices for sustainability in the future.

1. Introduction

The ecosystem services working group gathered 15 people representing 12 nations and a wide range of experience in and perspective of ecological studies including backgrounds in agriculture, economics, erosion control, fisheries, landscape

Understanding and Solving Environmental Problems in the 21st Century
Edited by R. Costanza and S.E. Jørgensen
© *2002 Elsevier Science Ltd. All rights reserved*

planning, renewable material production, tourism, transportation, water and wastewater management, and wetlands. Representatives from the discipline of ecological engineering moderated the working group. A diversity of interests meant that discussion was well balanced and rich in the nature and range of issues and case studies that were broached.

The idea of facilitating conference participants to meet and actively take part in discussions, centered on a common theme, is in itself progress. EcoSummit 2000 took such an initiative of making a conference more efficient and meaningful. Given the large ecological footprint (Wackernagel and Rees, 1996) a gathering like this creates, it should involve more than simply the presentation of research results that could easily be communicated by other less consumptive means.

Predictably, the first day was spent reviewing what everyone already knew about the environment and ecosystems but out of that dialogue, bridges began to be built enabling communication between professionals from very disparate backgrounds and with diverse focuses on *ecosystem services*. We might describe this process as a mud-wrestling exercise that was critical for finding common ground, establishing definitions, and listing the priority issues. As expressed by the conference chairman Bob Costanza, our mud-wrestling process manifested the vision of an adaptive, self-organizing conference, and out of the amorphous mud came defined visions and ideas with working groups presenting common ground for their respective theme areas. This chapter presents the common (and not so common) ground for the ecosystem services working group.

Principal components of the mud wrestling were the semantics, the connotations, and the disparate visions for implementation of the ideas embodied in the term *ecosystem services*. All participants were interested in ecosystem services as an idea that could be implemented in order to achieve a less self-destructive relationship with the biosphere. Language is important because ideas grow out of the images that language creates. It is language that sets up the paradigm that limits our ability to see fallacy in our ways. Therefore, although not without frustration, the time spent expressing those perspectives was a critical aspect of the EcoSummit process.

The working group then embarked on the task of addressing the following key questions in order to move towards establishing common ground:
- What is the state of the art for the use of ecosystem services?
- How can we make more sustainable use of ecosystem services?
- What factors are promoting and hindering sustainable use of ecosystem services?
- What are roles of ecological engineering, ecological economics, ecological modeling, and ecosystem health with respect to ecosystem services?
- What frameworks and implementation strategies are needed for ecosystem services?
- What key problems with ecosystem services need to be solved in the future?

The first three of these questions have been considered in some detail by Dakers (chapter 5 in this Volume) and will not be covered further in the present report. The consideration of the latter three questions provided the working group with a thread for consensus building on key issues regardless of interpretation of the concept ecosystem services. Our consensus building then led us back to the concept of ecosystem services and to ways to approach and implement the concept in the future. So in a sense, we went as a group in a full circle within a very challenging conference time frame and achieved a first iterative loop for a subject that will require many more iterations and refinement within a very challenging global time frame. The purpose of this chapter is to present the efforts of the working group in hope that these efforts are iterated, refined, and taken to action if concepts of ecosystem services are to be implemented in order to achieve a less self-destructive relationship between mankind and the biosphere.

2. Ecosystem services

The book by Daily (1997), *Nature's Services – Societal Dependence on Natural Ecosystems,* formed the theoretical basis for the working group discussions with the following definition:

> Ecosystem services are the conditions and processes through which natural ecosystems, and species that make them up, sustain and fulfill human life.

Examples of ecosystem services are cited in Daily (1997) and Dakers (chapter 5 in this Volume). The pivotal question confronting the workgroup was: how can we manage ecosystems in order to maintain ecosystem services? The implicit anthropocentric perspective of this question with undercurrents of human economic valuation of nature challenged the participants as some felt the term *ecosystem services* was in itself an obstacle to curbing the current urban perception of humankind outside of nature. Some debate over the expression *ecosystem services* ensued and is illustrative of the importance of language towards encouraging society into ways of recognizing and thinking about problems. The problem is the global violation of the inescapable requirement for our co-existence *within* nature if we are to continue to develop in the future. The working group member perspectives of *ecosystem services* exhibited wide scope of possible interpretations and fell into categories of *life in general, economic outputs, give and take, addressing sustainability, public perception,* and *interconnectedness* as follows:

- *Life in general* (Roggenbauer)
 "We need to avoid definitions that reflect human-nature dualism. For example, instead of saying *ecosystem services support humans and other life* it is preferable to simply state that *ecosystem services support life*".

- *Economic outputs* (Santopietro)
  "I rather like the term ecosystem services, probably because as an economist I tend to see outputs in this sense".
- *Give and take* (Werker)
  "My comfort level for the term *ecosystem service* depends on the context with which it is understood. Generating an understanding of how the environment *services* our needs and requirements is perhaps not a bad standpoint to convey our integral part within the ecosystem, especially when one sees that extracting one form of service (perhaps a *want* such as wood fiber for paper) can impinge on another service (perhaps a *need* such as climatic stability). The concept of service could be also viewed not only anthropocentrically but also in the sense of how we in turn can or should service the ecosystem. This kind of service is analogous to the services provided by different bacterial populations in the nitrogen cycle. Different kinds of microbes extract and provide services and in so doing maintain the flow of nitrogen in the environment and their respective ecosystems. The services are *give* and *take*".
- *Addressing sustainability* (Dakers)
  "I would argue that humans are an integral part of biosphere ecosystems and if sustainable living is our goal then our development activities need to integrate with the processes, functions and conditions of these ecosystems. This requires much more than a somewhat *hit and miss* attempt at identifying and valuing (using conventional economic models) selected *services*. From both science sense and common sense points of view, my concern is that ecosystem services, because of its traditional and selective anthropocentric perspective, is insufficiently inclusive to adequately address sustainability. The concept of ecosystem services is useful, providing the limitations of the term are well understood".
- *Public perception* (Roggenbauer and Gössling)
  "We need a term that communicates to the public that ecosystems provide something valuable. Probably, we should stick with the term services. *It* creates more or less the right picture for the public at-large of what we want to communicate". (Roggenbauer)
  "I guess the term ecosystem services has been chosen in order to communicate to the public that ecosystems provide something valuable to humanity". (Gössling)
- *Interconnectedness* (Brüll) "I like the term *ecosystem services* since it reminds me of the fact that human judgment is always anthropocentric. Human recognition is limited and subjective. Therefore, it is better to admit that humans create human perspectives of nature (i.e., models, images, concepts) while at the same time show explicitly the advantages and risks of these constructs.
  The advantage of *ecosystem services* is that it generates a picture of humans as *guests on earth*. It is clear that ecosystems provide services that generate

essential living conditions that all human societies survive on. It generates awareness that nature's productivity and regenerative capacity cannot be taken for granted as classical economics would lead one to believe. It promotes the idea that ecosystem services somehow need to be incorporated into economic constructs.

One risk of the term is that it might encourage the tendency for reductive division of the *services* into elements, as is common practice by scientists. If a blinkered focus on single services is promoted in an oversimplified attempt to value each service independently of one and another, one could easily overlook the fact that ecosystem services are really the result of complex interconnected processes that have evolved over space and time.

Therefore, the perspective of ecosystem services should focus on services as a whole. One must ask the question: *which processes and interactions do ecosystems manage in order to sustain something so inextricably intertwined*?

For example, consider the development of an integral adapted landscape management strategy instead of simply planning single sites. Such an approach demands interdisciplinary cooperation and recognizes the role of unmanaged areas as well. In critical cases (such as self organized woods for the regeneration of clean water and climate stability) it is naïve to think that human management can do better than natural systems by themselves".

Consequently, other expressions such as *ecosystem functions, ecosystem goods and services,* and *ecosystem capacities* were suggested in the attempt to escape anthropocentric gravitation as follows:

- *Ecosystem functions* (Gössling)
"Ecosystem functions are less anthropocentric and still mean the same thing. It would be a more appropriate term out of a scientific perspective".
- *Ecosystem goods and services* (van Bohemen)
"Ecosystem goods and services can also be formulated in the form of ecosystem functions, in the sense of contributions to the natural environment for society: production function, carrier function, information function (i.e., education, research, and monitoring), regulation function (i.e., climate regulation, and waste assimilation). It is important to realize that in addition to the relevance of *services* or *functions,* we need a [new] system of values if we are going to be able to safeguard the planet's life-support system".
- *Ecosystem capacities* (Barker)
"Ecosystem capacities provide a suggestion of potential which is presently not fully realized (and will not be) unless we *give and take* more, to use the words of Werker".

Therefore, it can be seen that while the problem may well be recognized within and among disciplines and nationalities, semantics play a critical role in building the bridges that are required in order to bring people together to solve a common problem. It is not necessary that everyone should agree but it *is* necessary that everyone understand one and another's meaning such that discussion can move forward in a productive manner.

Ecosystem services have been defined explicitly by Daily, however the expressed individual interpretations of how this defined concept should be applied towards understanding and addressing a global problem has to be seen as instructive outcome from this working group. Each viewpoint is a fresh angle and each new angle an opportunity for discovery for routes to a solution when people with open minds from different disciplines and of different nationalities meet.

3. Key questions and common ground

Having wrestled over the concept of ecosystem services, consensus building began for the group in addressing the following three key questions:
- What are roles of ecological engineering, ecological economics, ecological modeling, and ecosystem health with respect to ecosystem services?
- What frameworks and implementation strategies are needed for ecosystem services?
- What key problems with ecosystem services need to be solved in the future?

To begin with, there exists a demand for ecological modeling to provide information about general ecosystem trends. Models of ecosystem functions are of use to engineers in seeking strategies towards finding solutions to environmental problems that require a more widespread availability of these ecosystem services. In general, each discipline has a role to play, but regular EcoSummits were seen to be very important in order to promote cooperation and cross-pollination between these linked themes. It is necessary to integrate the sciences and to improve the communication between them as evidenced by our own round of mud wrestling. This approach of integrating both engineering and social sciences will tend to induce common agreements and hopefully more balanced and appropriate decisions.

Some form of central governance for establishing priorities for implementing concepts of ecosystem services is badly needed. The necessity for sustainability of ecosystem services must be universally recognized and promoted by a central authority. However, prerequisites for successful implementation strategies were seen to be in more locally based community initiatives and, again, in more interdisciplinary projects comprising open minded specialists and well-informed citizens (stakeholders). Flexible decisionmaking and policies that would encourage such interdisciplinary processes are necessary if we are to avoid the current path of self-destructive behaviour. Therefore, the priorities for implementation at this point in time must be in educating the next generation, generating greater public awareness, encouraging direct community involvement, promoting scientific collaboration, and extending the means of communication.

Science professionals must establish open forums between the different disciplines on all plains of society from local, to regional, national, and global. There are demands for interdisciplinary and international institutions that deal with watersheds as the focal element for policy and decisionmaking by using concepts of ecosystem services within an ecological economic framework. Societal development projects must consider the longer term and natural cycling of regional economies in goods and services, the need for continuous cooperative communication and feedback, and the requirement for decentralized (stakeholder) decisionmaking.

The working group further considered a number of hurdles that must be overcome if progress with the implementation of ecosystem services is to be made. The group recognized the following mix of social, political, and economic key problems regarding ecosystem services management:

- Educational and cultural factors present an obstacle for how to improve communication among all the stakeholders. Educational and cultural differences are an obstacle but are also an opportunity since novel solutions often arise when one is forced to view the same problem from different angles.
- Financing management and implementation of ecosystem services may be problematic since ecosystem services fall under the realm of the *common good* (Daly and Cobb, 1989). Who is going to pay for the common good when each individual is lured to the prospect of greater personal profit?
- The solutions to the global problems are local. However, the relationship between local actions and global improvement is not well entrenched in the public eye. Even in smog-plagued cities during the heat of the summer, the public seeking refuge in air-conditioned buildings and automobiles do not necessarily see that their individual actions are incrementally contributing to the problem. Angry cries are made to the government to address the problem, but the solution requires local (individual) action. Achieving adequate awareness at the individual and collective societal levels in order to promote effective implementation of management strategies is a big future challenge. Especially in an age with so much rhetoric to join the bandwagon of the *global economy*, how can feelings of empowerment be nurtured at the individual and community levels?
- The implementation of existing knowledge is often hindered by uncertainty, for which management strategies need to be devised. Engineers, who do shape our urban environments, are risk averse and natural systems are filled with variability that is hard to predict. Accommodating this kind of variability requires an ecological perspective. Therefore, engineers need to build more *bridges* with ecologists. This merging of disciplinary activity may be seen with the movement towards ecological engineering as a professional designation.

Given the consensus of expressed needs and obstacles emanating from the discussions of the three key questions, the deliberations of the working group culminated, within the limited time available, in visions of ways to apply and benefit from the concept of ecosystem services in the future.

4. The role of ecosystems services tomorrow

Over-use of ecosystem services can be found in cultural consumption patterns, which exceed basic needs several times. At the same time there is an under-use of the natural recycling capacities offered by ecosystem services. Over-use includes the purification capacities and productivity of river systems in, for one example of many, Italy and the Baltic Sea (Wulff and Niemi, 1992).

Assessing the ecosystem capacities in relation to human use is essential for their future management. Today there exists a lack of systems thinking and integrated assessment of environmental issues. Institutions are needed for management policies, such as market-based incentives, that reflect the true cost of ecological impacts due to human activities. The lack of *full cost accounting* promotes *cultural traps* that lock the public into environmentally unfriendly practices and exacerbate the efforts of any fledgling positive governmental *green* initiatives. Legal and economic institutions must begin to more universally incorporate the concept of ecosystem services into their framework. Mismanagement of ecosystems is also due to our ignorance of systems and system boundaries and a lack of *thinking in interactions*. Even models that are simplistic but integrative would provide a better basis for the policymakers to go on.

As a society we must seek to use the under-used potential of ecosystem services in order to relieve the over-used ones. Governing *institutions* need to ensure that ecosystem services are managed and used properly. The role of scientists should be to provide decision support and management tools such as holistic integrative models of landscape ecosystems, in order to make clear how ecosystem services function and how they can be instituted and used sustainably.

At the same time, the *global sustainability problems* have to be tackled by local actions that do not necessarily wait for the bureaucracies of institutions to gain inertia. Therefore, the question is: *how can we promote the integration between the local and the global scale?* To start with, education, information, and communication are needed in order to promote awareness, responsibility, and an ethical framework (Local Agenda 21) to address environmental issues at the local (community) level. Scientific activity including the providing of data, analysis, models, and interpretation needs to be more easily understood by scientists from other disciplines, and by the public at large. Further, rethinking and redesigning our *structure of life,* where people could be offered options for behavioral changes, will help sustain ecosystem functions juxtaposed against the current holding patterns of unsustainable behavior by humanity caught up in the urban rat race of a *global economy.*

Launching *global money* for *local projects* by setting up *environmental funds* could assist in beginning to finance the necessary process of change. For example, Gössling suggested an eco-tax for flying might be used to compensate the *want* of global mobility with the *need* to finance local wetland projects. National environmental funds in developing countries could be financed in part by governmental and non-governmental organizations in the developed world.

Advancements will come from cycles of *thinking* and *doing*. But how can the progress of *thinkers* and *doers* be more closely coupled? Doers and thinkers often have different value systems that may frustrate their ability to more closely interact. For example, for a *doer*, *thinkers* are too theoretical, not rooted in reality, and therefore incapable of translating the results of their research into practical directives. A *thinker* may feel that a *doer* lacks the depth of insight into the complex governing interactions that are necessary to fully appreciate the forces of nature at play. A thinker takes too long, while the doer may act without adequate knowledge of the problem. Mutual acknowledgement of differences (perhaps with some humor) and the potential for meaningful contributions can provide common ground for cooperation between thinkers and doers. Such cooperation will lead to synergistic assessments of ecosystem functions and more pragmatic design of management polices. Moderated meetings with professional facilitators could be organized for bringing *doers* and *thinkers* together. The role of social networks and local partnerships in both modeling ecosystems and designing management policies should be emphasized at these conferences.

Projects that provide the opportunities for interdisciplinary case studies, and pilot tests would be instrumental for training and communication links between sociologists, scientists and engineers, and thinkers and doers. In the Netherlands, a case study that integrated applied ecology and traditional civil engineering demonstrates how the design and construction of road infrastructure can be recommended to environmental planners as a strategy for network thinking and the development of Eco-Design-Tools (van Bohemen, 1998). A project in Latin America (Loiselle et al., 2000) examining the overall impacts of different development alternatives on resource quality and ecosystem integrity was presented to the working group as an example of the future of management tools for wetland resources. In this project, a multidisciplinary team of scientists from Europe, Argentina and Brazil will be monitoring activities with potential impact on a fragile wetland ecosystem. These investigations must include *thinkers* and *doers* in hydrology, meteorology, biology, and ecology together with public meetings with key provincial and local actors. Microeconomic models for understanding and analyzing *project-level* choices and impacts of rainwater retention strategies in Ottawa were highlighted (Trudeau). The ecosystem services working group also cited several other community projects demonstrating methods for sustainable solid waste, water and wastewater management in China, India, New Zealand, Sweden, and Switzerland (Yan, 1993; Heeb et al., 2000; Roy, 2000; Dakers, 2000).

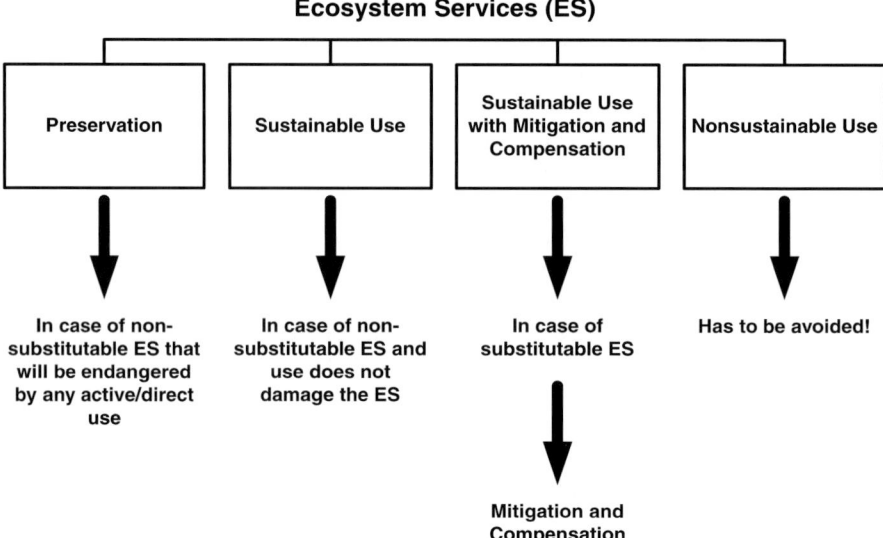

Fig. 1. A decision flow diagram for choosing the right approach for development plans that will impact on ecosystem services.

There would be global benefit of a compilation of such case studies that pool the learning from thinking *and* doing. For example, just imagine the resource base that could build if there existed a Journal of Ecosystem Services with an A series for *Doers* and a B series for *Thinkers*.

If we humans are a part of nature, then it is impossible for us to exist without influencing the surrounding nature (give and take). Assuming that we are gifted in the ability to predict potential outcomes for such giving and taking, what criteria do we follow to manage ecosystem functions servicing and emanating from our activities? Figure 1 illustrates the kind of decision tree that needs to be further developed *and* followed. It provides a framework for decisionmaking as follows:

(1) Identify the ecosystem services impacted by a proposed development plan.
(2) Identify the impacts of use and consequences of over-use.
(3) Link the impact study to an integrated assessment of the ecosystem services.
(4) Identify alternative development possibilities, which will have less impact.
(5) Identify mitigation or compensation strategies for the cases when no alternatives exist.
  • If impacts cannot be prevented, then they must be mitigated.
  • If effects exist, then compensation must take place to achieve no net loss.
(6) Build theoretical (integrated) models to assess strategies and outcomes.
(7) If the development plan is to be implemented, establish measurable indicators for impact and success or failure.

(8) Organize and implement an integrated monitoring plan.
(9) Establish an adaptive management strategy (*doers* to *thinkers* to *doers*) and agree *a priori* on actions for anticipated and unanticipated impacts that might be observed in the future.

This decision tree is a first iteration after mud wrestling, perspective sharing, and consensus building within a short, adaptive, self-organizing conference. More mud wrestling, and iterations for improvement need to be made. But while we *think*, we also need to *do*, and *do* locally. What are you *doing* today?

5. Conclusions

In conclusion, the key activities for *thinkers* in the area of ecosystem services are in the development of models to understand how ecosystems function and how they are affected by decision makers, engineers, economists, and others. At the same time, we need more pilot projects to link thinkers and doers, experts and stakeholders. Such models and the feedback (learning) from pilot *doing* must be channeled into education and information dissemination in order to create respect, recognition, and understanding of the value and the potential of ecosystem services (thinkers to doers to thinkers to doers). The evolution and development of international interdisciplinary institutions could at first serve to channel this information but must move to more progressively ensure that ecosystem services are managed for sustainable use (local to global to local). This assurance will come with the development and demonstration of guidelines and *best practices* for sustainable ecosystem management based on integrative decision trees that invoke stakeholder participation and embrace the concept, value, and role of ecosystem services to our very existence. There will be a continual evolution of best practices, meaning that adaptive management and the development of monitoring indicators must be well linked to policy and decisionmaking.

References

Daily, G.C. (Editor), 1997, Nature's Services – Societal Dependence on Natural Ecosystems (Island Press, Washington, DC) 392 pp.

Dakers, A.J., 2000, Ecological engineering: wastewater engineering, paper for Urban 2000 Conference, Duxton Hotel, Wellington, New Zealand, 15–16 June.

Daly, H.E. and Cobb, J.B., 1989, For the Common Good: Redirecting the Economy Toward Community, the Environment, and a Sustainable Future (Beacon Press, Boston, MA) 482 pp.

Heeb, J., Roux, M. and Dakers, A.J., 2000, Ecological engineering – three case studies. In: B.B. Jana, R.D. Banerjee, B. Guterstam and J. Heeb (Editors), Waste Recycling and Resource Management in the Developing World (University of Kalyani, India and International Ecological Engineering Society, Switzerland) pp. 15–25.

Loiselle, S., Rossi, C. and Gandini, M., 2000, The sustainable management of subtropical wetlands combining in situ monitoring and remote sensing technology, paper presented at EcoSummit 2000.

Roy, S., 2000, Ecological sustainability and metropolitan development – the Calcutta experience. In: B.B. Jana, R.D. Banerjee, B. Guterstam and J. Heeb (Editors), Waste Recycling and Resource

Management in the Developing World (University of Kalyani, India and International Ecological Engineering Society, Switzerland) pp. 293–302.

van Bohemen, H.D., 1998, Habitat fragmentation, infrastructure and ecological engineering. Ecol. Eng. 11:199–207. [Or, Ecological engineering and infrastructure; integration of ecological and civil engineering in the field of planning, design and construction of road infrastructure, paper presented at EcoSummit 2000.]

Wackernagel, M. and Rees, W.E., 1996, Our Ecological Foot-print: Reducing Human Impact on the Earth (New Society Publishers, Gabriola Island, B.C., Canada). German edition with updated data: 1997 (Birkhaüser, Basel).

Wulff, F. and Niemi, Å., 1992, Priorities for the restoration of the Baltic Sea – A scientific perspective. Ambio 21:193–195.

Yan, J., 1993, Advances of ecological engineering in China. Ecol. Eng. 2:193–215.

Background

Chapter 7

Science and Decisionmaking

V.H. Dale*

> Whoever, in the pursuit of science, seeks after immediate practical utility, may generally rest assured that he will seek in vain. All that science can achieve is a perfect knowledge and a perfect understanding of the action of natural and moral forces.
> *Hermann Ludwig Ferdinand von Helmholtz, Academic discourse, Heidelberg, 1862*

Abstract

Science is often viewed as an activity independent of society. There certainly is one pathway in science that is such a pure investigation of natural processes. Yet, in this age when the very sustainability of the Earth and its critical ecosystems are in question, key findings of environmental science must be communicated to and used by those who make decisions about the future of the Earth. The challenge is how scientists can effectively impart appropriate and useful information to decisionmakers.

Currently, many scientists are frustrated by the inefficiency of the existing communication processes and the unbalanced use of science in decisionmaking. Although science is an integral part of decisionmaking in some countries or for some issues, in general the connection is weak. Scientific results and model predictions are rarely expressed in terms of end points that have direct meaning or inherent value to decisionmakers. This leaves it up to the decisionmakers to extrapolate scientific results to the end points of true concern. Therefore, a number of questions arise as to how best to facilitate this interchange between science and decisionmaking. After a brief description of science and decisionmaking, this chapter presents three examples of how science has been used in making

---

* The submitted manuscript has been authored by a contractor of the US Government under contract No. DE-AC05-00OR22725. Accordingly, the US Government retains a nonexclusive, royalty-free license to publish or reproduce the published form of this contribution, or allow others to do so, for US Government purposes.

Understanding and Solving Environmental Problems in the 21st Century
Edited by R. Costanza and S.E. Jørgensen
Published by Elsevier Science Ltd.

decisions about land management and use. Based on the common elements of these examples, the chapter ends with a set of questions about appropriate ways to transmit science to decisionmakers.

1. Science and decisionmaking

Science covers the broad field of knowledge dealing with observed facts and the relationships among those facts. The word derives from the Latin word *scientia*, meaning knowledge. Systematic approaches are used to build the knowledge of science. Theory developed by one researcher is tested and often expanded by other researchers. As new understandings are attained, outdated knowledge is discarded. Thus, scientific knowledge is always expanding and being revised.

Scientific research uses a set procedure to develop this knowledge. The scientific method consists of observations, organization of data, development of a hypothesis, testing of the hypothesis via experiments, and expression of findings and new questions that may arise. Peer review and publication of findings is a key part of the process. The results of an investigation are shared with colleagues via scientific journals and presentations at meetings. Only rarely are findings reported in the popular press.

Sometimes science influences decisionmaking. In environmental areas, these influences have become more common since the first Earth Day, more than 30 years ago. That formal recognition of the importance of the Earth and its future led to the creation of government entities that strive to protect environmental resources (such as the Environmental Protection Agency in the USA). Today science often affects both environmental policy and day-to-day decisions. Yet, given the dire consequences of projected trends in sustainability, biodiversity, and the availability of ecosystem services, it is important that information from environmental science be included in the decisionmaking process.

2. Scientists' role in decisionmaking

Scientists have several roles in environmental decisionmaking. They communicate relevant information, work with interest groups to build consensus, maintain scientific dependence and integrity, and discover new options. Scientists communicate information via teaching, publications, the press, petitions, briefings, public meetings, expert testimony, user-friendly computer models, visualizations, and the Internet. Most often, this information is targeted to other scientists and may involve scientific jargon and technical terms. The format of delivery is specified in scientific communications to include an introduction, methods, results, discussion, and conclusions (although this organization is not the most effective way to communicate results to decisionmakers).

Scientists can build consensus by sharing information, teaching, and developing analyses that include all parties. They often participate in scientific workshops at which a consensus across a broad spectrum is developed (such as occurred at the EcoSummit that generated this book). Scientists are sometimes part of community gatherings, such as the Applegate Partnership in southwest Oregon and northern California, in which resource-management alternatives are explored and discussed among a broad group of stakeholders (www.mind.net/app/aphandou.htm). Scientists also build consensus by taking part in scientific advisory groups, such as the Committee of Scientists (1999), which provided suggestions to the US Secretary of Agriculture on how best to manage the US national forests and grasslands. Scientists need to better understand the complete spectrum of concerns that enter into the policy process. Science is often not the primary source of information used by decisionmakers to make decisions regarding environmental policy. However, by working with other stakeholders, scientists can often increase the use of science in decisionmaking.

Scientists maintain credibility by always questioning, discussing their ideas with others, and having their papers undergo peer review. The process of peer review is one of the mainstays of the scientific approach and bestows confidence on a result or mode of analysis. Scientists can enhance their credibility with decisionmakers by examining the relevancy of their work and using simple language to express their ideas (but these avenues are not always pursued).

Scientists can discover new options for environmental-change impacts by exploring future options (e.g., with computer simulation models), examining past situations, and determining causes of observed phenomena. The assumptions, constraints, and possibilities of the alternatives should be clearly set forth. These options have policy relevance to the extent that they are reasonable to implement and that they address concerns of the decisionmaker.

Although scientists can assist decisionmakers by providing accurate and relevant information, making good environmental decisions depends on dedicated government officials, an involved public, and concerned industry. The next section of this chapter illustrates some ways that scientists interact with decisionmakers in three example situations.

3. Three case studies

3.1. Mount St. Helens

Mount St. Helens is an active volcano in southeast Washington State. Although Crandell and Mullineaux (1978) warned of an impending eruption, no action was taken to protect the people and property in the area until after seismic activity in March 1980 gave the first signs of the eruption. Even with this warning, 57 people were killed, and 60 000 ha of land were devastated by the May 18, 1980 eruption.

This eruption was not a single event but created a new volcanic crater, a 15.5-km^2 pyroclastic flow, the largest debris avalanche in human history, massive mudflows, and a 550-km^2 area of trees that had been blown down bordered by 96 km^2 of scorched trees. It also gave rise to a living laboratory where scientists can study the process of succession.

Biologists realized that, in order to have a complete laboratory, a control site was needed as well as a diversity of disturbances. Therefore, they argued for the creation of a protected area that would include each type of disturbance as well as a control area that contained no volcanic impacts. Scientists became actively involved in the preservation efforts by lobbying for the creation of such a monument, writing letters to congressmen, endorsing petitions, and participating in the congressional hearings. When the Mount St. Helens National Monument was set aside in 1982, it not only protected the diversity of disturbance types and a control site but also established a scientific advisory board to review how science was conducted in the Monument.

However, the Monument had already been tainted by human influences. Immediately after the eruption, there was great concern about the potential for erosion in the areas downstream from the debris avalanche because the 1980 eruption had wiped out several bridges and destroyed homes. To address this justifiable concern, the Soil Conservation Service proposed aerial distribution of the seeds of several grasses and legumes in the hopes that the new plants would reduce the amount of erosion. The second International Congress of Systematic and Evolutionary Biologists was meeting in Vancouver, British Columbia during July 1980, and it adopted the resolution that "whereas the recent eruption of Mount St. Helens created unique habitats and a natural laboratory, resource availability limits plant establishment and growth, grasses inhibit recolonization and growth of trees; resolves the Congress opposes mass seeding of exotic species". This resolution was broadly distributed to members of the US Congress and administration. With this information in hand, the Soil Conservation Service restricted their seeding efforts so the entire devastation zone was not seeded. However, seeds were distributed over large portions of the Toutle River mud flow and subsequently moved into the lower debris avalanche within the monument.

Fortunately, of ninety-seven 250 m^2 plots that are distributed at 50-m intervals along two transects within the monument, only eleven were inundated by the exotic species (Dale, 1991). Unfortunately, these exotic species had no effect on the sheet erosion that was occurring on the debris avalanche but did affect the native ecosystem. Comparison of those plots with and without exotic seed showed that there was a reduction in the native plant diversity and in the survival of native coniferous trees, even though there was an increase in overall cover of the plots that contain the exotic species (Dale, 1991). It is most likely that the conifer trees died because there was an outbreak of the redback vole. Abundant seeds were produced by the introduced plants that established themselves on the mudflow and debris avalanche. During the summer and fall, these seeds fostered an explosion in the vole population (Franklin et al., 1995). But during the winter, snow covered the plants, and the voles had little to eat, so they ate the bark off the conifer trees, which were easy to locate under the snow.

In summary, the policy goals for the Mount St. Helens area were preservation of unique areas, reduction of erosion, and enhancement of the economic vitality of the region. Through two decades of research, scientists contributed to the resolution of these policy issues by communicating information, building consensus, maintaining credibility, and discovering options. Communication of information occurred via testimony, coverage by the press, and scientific papers. Consensus was created by scientists sharing their information and reviewing each other's papers. Also, the Forest Service and National Science Foundation supported field campaigns, called Mount St. Helens Pulses, at which a number of scientists not only visited their field sites and collected data but also met during the evenings to discuss their ideas and observations. Credibility was maintained because many publications have come out of the Mount St. Helens experience. In a review of papers on vegetation recovery following volcanoes, more than half of the papers have discussed the Mount St. Helens experience (Dale et al., 2002). Finally, the scientists have been effective in presenting options. For example, now the Natural Resources Conservation Service maintains native plant nurseries instead of only exotic-species nurseries so seeds of locally native plants are available for distribution when needed for recovery purposes.

3.2. Tennessee Cedar Barrens

The second example of how scientists can be involved in the decisionmaking process is the Oak Ridge Cedar Barrens in East Tennessee. Cedar barrens, which occur sporadically throughout the southeastern USA, are habitats of shallow soils underlain by limestone that are devoid of the native cedar trees (Baskin and Baskin, 1986). These Pleistocene-remnant habitats are noted for the rare plant species that they support, which are typical of Midwestern prairies, such as the tall larkspur and Earleaved false foxglove (DeSelm et al., 1969; DeSelm and Murdock, 1993). In January 1988, the citizens of Oak Ridge were surprised to read in their local paper that the city council had voted to sell a cedar barren that occurs adjacent to a school. Scientific information provides analysis that these communities are quite rare, and, in fact, only occur under unique edaphic conditions where the soils, geology, and land use are conducive to the habitat (Dale et al., 1998). However, the local decisionmakers were unaware of the unique ecological features of this habitat. Therefore, a group of concerned scientists organized a workshop at the school adjacent to the property in dispute. At the workshop, experts lectured on the scientific attributes of this habitat, and a field excursion took place. Many local citizens attended the workshop. A video was made of the workshop and mailed to the developers who were planning to build a mall at the site. On observing the video, the developers made the decision not to pursue construction on the entire site. The resolution was that 2.8 ha were set aside as a state conservation area. Although this may seem like a small area, it is large enough to protect the unique features of this edaphic condition (particularly since there are other cedar barrens in the local vicinity).

Thus, the policy issue for the Oak Ridge Cedar Barrens was development of this site and juxtaposition of inappropriate land uses (a mall and a school). The science issues were preservation of this rare habitat and the species it supported and maintenance of the network of cedar-barren habitats across the landscape. This was a situation in which the industry was concerned enough about the local response that they made the decision not to develop the land. Currently, the State of Tennessee is managing the site as a state natural area. Furthermore, the Oak Ridge Cedar Barrens serves as a field-study site for biology students from the adjacent school.

Scientists contributed to the resolution of this situation by communicating information about the rarity and value of the cedar barrens and by building consensus among the scientists and among the citizens. The scientists maintained their credibility throughout this process by referring to published papers on the value of these lands and the impacts of fragmentation. The scientists were also involved in discussing options. For example although these lands had been used as a trash dump, the scientists noted that cedar barrens thrive under occasional disturbances with fires helping to maintain them in the past and light dumping serving the purpose more recently.

3.3. The Brazilian Amazon

The last example of the relationship between science and decisionmaking is the Brazilian state of Rondônia in the southwestern Amazon. This area is undergoing a tremendous rate of deforestation that results in much of the land becoming agricultural tracts and then pastureland when the agriculture cannot be supported. This rapid change in land use is contributing to the global issue of climate change as well as to a local issue of soil degradation. The soil changes are quite detrimental to the people because the land becomes unable to support them. The problem has been so severe that between 1978 and 1988 there was an 18-fold increase in deforestation that correlates with the rate of road development (Dale et al., 1994a).

A model was developed that allowed the examination of the impacts of colonist families moving into the area (Southworth et al., 1991; Dale et al., 1993, 1994a). The model results indicate that the land is usually cleared within 18 years of the introduction of roads and about 65% of the carbon is released from the land and soil (Dale et al., 1994a). However, under a sustainable situation in which the farmers never burn the forest and use a diversity of the native crops, only about 40% of the land is cleared and about 30% of the carbon is released (Dale et al., 1994a). Most striking in the model results is that, under sustainable conditions, a high number of the immigrant families remain on the land even after 40 years. Currently, most of the immigrant families have moved on after 20 years (Dale et al., 1994a). Furthermore, with current practices, the model indicates that the land use and cover quickly becomes so fragmented that populations of many native animal species cannot be maintained. However, under sustainable conditions, a large number of animals can be supported (Dale et al., 1994b). In interviews with the local citizens, we found that 3 out of 90 farmers were practicing these sustainable practices (Dale and Pedlowski, 1992). As a result of these and other similar findings, the Rondônian Department of the Environment is now taking other farmers around to observe the farmers who were practicing sustainable agriculture and to learn about their techniques.

In conclusion, the decisions being made in the Amazon were how to encourage the farmers to make appropriate land-use decisions. The policy issues were reducing atmospheric CO_2 increase and subsequent climate change, reducing/minimizing soil degradation, and enhancing sustainable economic development. Scientists contributed to this resolution by communicating information about the situation. In fact, it was remote-sensing data first collected by scientists that alerted the world to this large impact that was occurring (Malingreau and Tucker, 1988). Scientists were also important in that they built a consensus that this rapid rate of deforestation was a problem and that a solution needed to be obtained. However, in all cases, scientists maintained their credibility by publishing their results in a series of articles (e.g., Brown et al., 1992; Fearnside and Ferreira, 1984; Hect, 1993; Moran, 1993; Skole and Tucker, 1993; Skole et al., 1997; Smith and Schultes, 1990). The scientists were also important in discovering options using model projections, as discussed previously.

3.4. Lessons learned

Together, these three examples illustrate how scientists can communicate relevant information, work with interest groups to build consensus, maintain scientific dependence and integrity, and discover new options. All four elements are necessary for the scientists to influence decisions. However, it is useful to determine who the decisionmakers and stakeholders are in each instance. For the Mount St. Helens situation, the decisionmakers are at the national level, whereas for the cedar barrens example industry leaders located in another state make the critical decision. For the Amazon example, there is a hierarchy of decisions makers. Local farmers make decisions about where to plant and land preparation procedures. The state department of the environment plans and carries forth outreach programs. Decisions about immigration programs for the Amazon are made at the national level, and international banks fund the development and paving of roads.

4. Characteristics of scientists and decisionmakers influence how they interact

These examples provide some insights into interactions between scientists and decisions makers. Scientists and decisionmakers come from two very different fields of endeavor, but their interests sometimes merge in the arena of environmental policy. However, scientists who wish to contribute to policy are a small subset of all scientists, although many scientific endeavors are motivated by a potential for application. But it needs to be recognized by scientists that the pipeline of information from science to decisionmaking is only useful when policy questions are addressed. Sometimes the decisionmakers are not even aware that science can pertain to a policy issue. In such a situation, regular and routine discussions between scientists and decisionmakers can enhance communications. Such discourse is enriched by mutual respect and the granting of authority. Communication between scientists and decisionmakers is a two-way street. Scientists need to listen to decisionmakers to understand what problems are in greatest need of additional research. This communication can be enhanced by a mutual understanding of the background of the two groups of people.

Like most professions, scientists have certain personality similarities. Application of the Myers–Briggs psychological types (Myers, 1987) groups most scientists into the "introverted, intuitive, thinking, judging" (INTJ) type (Tieger and Barron-Tieger, 1992). This grouping holds specifically for environmental scientists, for when scientists in the Environmental Sciences Division at Oak Ridge National Laboratory were tested about 90% fell into the INTJ category. Introverts are energized by spending time alone and think things through inside their heads. Intuitive people like new ideas and concepts for their own sake, build by inspiration and inference, and are future oriented. Thinkers tend to step back and apply impersonal analysis to problems. They value logic, justice, and fairness. Judgers tend to have a strong work ethic and work toward goals. INTJs are logical, critical, and ingenious people who are often perfectionists. They have a strong need for autonomy and personal competence and a firm faith in their own original ideas. They are natural brainstormers and work well with theoretical problems.

On examining the strengths of the INTJ personality type, it is clear why these people are good scientists. According to Tieger and Barron-Tieger (1992), the INTJ types
- Are visionary and excel at creating systems;
- Can understand complex and difficult subjects;
- Enjoy creative and intellectual challenges;
- Are good at theoretical and technical analysis and logical problem solving; and

- Work well alone and are determined even in the face of opposition.

In other words, these are people who can envision a solution to complex scientific questions; enjoy taking on challenges; excel at problem solving; are tenacious; and do not mind the many hours spent alone in the field, in the laboratory, or at the computer required to test a scientific hypothesis.

However, the INTJs also have a set of shortcomings associated with their personality type that, I believe, compromises their ability to relate to decisionmakers. According to Tieger and Barron-Tieger (1992), the weaknesses of the INTJ types are that they
- May be less interested in projects after creative problem solving is completed;
- Drive others as hard as they drive themselves;
- May be too independent to adapt to corporate culture;
- May have difficulty working with or for others they consider less competent; and
- Can be inflexible and single-minded about their ideas.

Thus, the typical scientist may have less interest in the application of science than in the discovery of it. Their hard drive may turn off the enthusiasm of others. Scientists' independence may jeopardize both their interest in and ability to transfer scientific information to decisionmakers. If they consider the decisionmaker as less competent than themselves, communication may be difficult. And their lack of flexibility may mean that, even when these deficiencies are pointed out, scientists will not change their ways.

Thus, when communicating with decisionmakers, scientists need to build upon their strengths and be wary of their weaknesses. Yet, scientists can use their observations, models, and understanding to anticipate trends and future needs, synthesize information, and create options for the application of science to decisions. Thus, scientists can use their keen ability to brainstorm to create options for applying science.

Scientists can avoid some possible pitfalls by attending to necessary and relevant facts (not just the novel ones), using tact and diplomacy to get others to buy into their ideas, avoiding appearing arrogant or condescending, and remaining flexible and open-minded (Tieger and Barron-Tieger, 1992). Thus, scientists need to consider the practical applications of their ideas and not just the unique implications. They should use persuasion rather than being unyielding and should consider the implications of their comments and perspectives. It is important to listen carefully to others' ideas and to consider incorporating other concepts into their own viewpoint.

Of course, some scientists are better at these communication skills than others and some have more interest in the application of scientific ideas. These communication abilities and interests do not necessarily correlate with being the best at developing new scientific ideas. In fact, the most common depiction of a scientist is as an introverted frazzled man who has little touch with day-to-day realities (i.e., the Einstein image). Yet, Einstein himself had a great influence on the application of science. His letter to President Roosevelt led to development of the atomic bomb, although subsequently Einstein argued for a ban on nuclear weapons.

Furthermore, introverts can become extroverts. One of my professors was a classic introvert. He had thick glasses, spoke in a meek, hesitant voice facing the blackboard, and was difficult to engage in conversation. But when I saw him fifteen years later, all of that had changed. Maybe it was the fact that after his divorce, he became an avid weight-lifter and transformed from a "90-pound weakling". But now he looks you straight in the eye and engages you with his strong voice and direct questions.

Such a dramatic physical transformation is not necessary for scientists to effectively transmit their science into applications. Instead, an attention to the needs of decisions makers and an interest in application is required. Each person has strengths and weaknesses that need to be recognized in their relations with others. The ability to communicate should be recognized as being just as important as science innovation. Unfortunately, this recognition does not always occur. For example, some of Carl Sagan's colleagues did not appreciate the knowledge and interest about astronomy that he imparted to the public.

Like scientists, decisionmakers have a general set of characteristics that influence how they relate to scientists and tend to use scientific information. For the purpose of illustrating this point, I use only the characteristics of politicians although decisionmakers also come from other fields of endeavor. According to the Myers–Briggs psychological types (Myers, 1987), politicians are most often extraverted, intuitive, feeling, and judging people (ENFJ) (Tieger and Barron-Tieger, 1992). These are people who tend to promote harmony and build cooperation, respect a variety of opinions, are decisive and organized, and are natural leaders. However, they also may have trouble dealing with conflict and tend to sweep problems under the rug, may not be attentive to factual accuracy, and may take criticism too personally. This set of weaknesses puts decisionmakers at odds with scientists who use analytic methods, such as statistics, to make decisions about conflicting data, interpret avoidance of a problem as cheating, and carefully adhere to facts. Thus, scientists may follow Lagrange's advice to "seek simplicity but distrust it", whereas decisionmakers may have no qualms about moving ahead with a simplistic solution to a concern.

This comparison of the characteristics of scientists and decisionmakers makes it clear that scientists and decisions makers are coming from quite different positions and world views. However, they can each use their strengths to enhance communication. The decisionmaker might focus on a harmonious solution to the issue, while the scientist might employ problem-solving skills to help resolve concerns.

5. Questions about the relationship between science and decisionmaking

There are clearly major differences between the processes of science and of decisionmaking. The differences deal with the value of science, the reward structure of science, and the role of uncertainty. Science only has value for decisionmaking when it has the potential to influence a decision. The quest for deeper theoretical "understanding" that routinely drives the scientific process is rarely of value to decisionmakers. Furthermore, the reward system of scientists does not recognize the value of contributing to the decisionmaking process. Schwartz (1999) points out that resource managers have been both blessed and cursed by the attention of biologists. Scientists are rewarded for novelty and thus often suggest and sometimes test new ideas for conservation. Managers are expected to employ the latest scientific developments, but application of untried ideas can result in failure, and often it is not clear which is the most appropriate technology. Therefore, Schwartz (1999) urges the testing of new ideas with empirical data before new methods are adopted. Finally, although public perception of scientific uncertainty can have a strong impact on the policy process, decisionmakers can typically accept a different level of uncertainty than can scientists.

This review leads to a number of questions about the relationship between science and decisionmaking. Some general questions are the following.

- Does "better" science translate to "better" decisionmaking?
- How much does science influence decisionmaking? How much should science influence decisionmaking? In some cases, decisionmakers do not want to know results that may conflict with their objectives and may therefore not ask the "right" or important questions. Is it the role of the scientists to present this information regardless? If so, how can this be done effectively?
- How can the linkages between science and environmental policy be facilitated?
- How much should decisionmakers' objectives and concerns drive the science (e.g., in terms of the temporal and spatial scale, resolution of studies, inputs and outputs, or science questions)?
- How can science appropriately influence decisionmaking? What determines what is appropriate? Similarly, how can decisionmaking appropriately influence science?
- Science sometimes contributes to specific decisions but more often contributes to general policies. What guidelines are useful in addressing specific decisions? (Here there is concern about one expert being played off against another in a regulatory hearing or similar circumstance.)
- How should scientists and decisionmakers be working together? What can enhance this interaction?
- How can scientists quantify the value of scientific information to decisionmakers? How can they use this information to make science more useful?

- Where do we draw the boundary between analysis performed by scientists and that performed by decisionmakers?

  In addition, there is a set of questions dealing with communications:
- How do decisionmakers receive scientific information (from staff, media, etc.)?
- What are the major barriers influencing the effective communication of science to decisionmakers? How do the attitudes and expectations of scientists affect their ability to communicate with decisionmakers?
- What aspects of the science should be communicated to decisionmakers (e.g., assumptions, uncertainties, range of applications)? How can scientists effectively convey scientific uncertainty to decisionmakers? An even more basic question is how can scientists convey the importance of scientific uncertainty to decisionmakers?
- How much should scientists be involved in communicating science to decisionmakers?
- Who are the decisionmakers to whom science should be communicated (e.g., administrative and agency decisionmakers at local to national levels, nongovernmental organizations (NGOs), the public)?
- Do these different decisionmakers want different kinds of science information or information communicated in different styles?
- The styles of doing science and decisionmaking are quite different. What changes need to be made in the style of communicating science so that the science is more readily received by the decisionmaker?
- What professional incentives/obstacles are there for scientists to improve communications with decisionmakers? How can these be improved?

Answering these questions requires a better understanding of the decisionmaking process and of the tools and instruments that scientists rely upon for analysis and communication. It calls for an explicit evaluation of what can make scientists become more effectively involved in the decisionmaking process. These are the topics of the following chapter that was developed at the EcoSummit by collaboration among 33 scientists representing 18 countries.

Acknowledgements

Ed Rykiel contributed helpful comments on the manuscript. I appreciate the editorial comments on the manuscript by Fred O'Hara and Linda O'Hara. The project was funded by contract from the Conservation Program of the Strategic Environmental Research and Development Program (SERDP) with Oak Ridge National Laboratory (ORNL). ORNL is managed by UT-Battelle, LLC, for the US Department of Energy under contract DE-AC05-00OR22725.

References

Baskin, J.M. and Baskin, C.C., 1986, Distribution and geographical/evolutionary relationships of cedar glade endemics in southeastern United States. Assoc. Southeast. Biol. Bull. 334:138–154.

Brown, I.F., Nepstad, D.C., de Pires, I., Luz, L.M. and Alechandre, A.S., 1992, Carbon storage and land-use in extractive reserves, Acre, Brazil. Environ. Conserv. 19(4):307–315.

Committee of Scientists, 1999, Sustaining the People's Lands: Recommendations for Stewardship of the National Forests and Grasslands into the Next Century (US Department of Agriculture, Washington, DC) 193 pp.

Crandell, D.R. and Mullineaux, D.R., 1978, Potential Hazards from Future Eruptions of Mount St. Helens Volcano, Washington, Bulletin 1383-c (US Geological Survey) 26 pp.

Dale, V.H., 1991, The debris avalanche at Mount St. Helens: vegetation establishment in the ten years since the eruption. Natl. Geogr. Res. Explor. 7(3):328–341.

Dale, V.H. and Pedlowski, M.A., 1992, Farming the forests. Forum Appl. Res. Public Policy 7:20–21.

Dale, V.H., O'Neill, R.V., Pedlowski, M.A. and Southworth, F., 1993, Causes and effects of land-use change in central Rondônia, Brazil. Photogramm. Eng. Remote Sens. 59:997–1005.

Dale, V.H., O'Neill, R.V., Southworth, F. and Pedlowski, M.A., 1994a, Modeling effects of land management in the Brazilian settlement of Rondônia. Conserv. Biol. 8:196–206.

Dale, V.H., Pearson, S.M., Offerman, H.L. and O'Neill, R.V., 1994b, Relating patterns of land-use change to faunal biodiversity in the Central Amazon. Conserv. Biol. 8:1027–1036.

Dale, V.H., King, A.W., Mann, L.K., Washington-Allen, R.A. and McCord, R.A., 1998, Assessing land-use impacts on natural resources. Environ. Manag. 22:203–211.

Dale, V.H., Delgado-Acevedo, J. and MacMahon, J., 2002, Effects of modern volcanic impacts on vegetation. In: J. Marti and G. Ernst (Editors), Volcanoes and Environment (Cambridge University Press, Cambridge).

DeSelm, H.R. and Murdock, N., 1993, Grass-dominated communities. In: W.H. Martin, S.G. Boyce and A.C. Ecternacht (Editors), Biodiversity of the Southeastern United States: Upland Terrestrial Communities (Wiley, New York) pp. 87–141.

DeSelm, H.R., Whitford, P.B. and Olson, J.S., 1969, The barrens of the Oak Ridge area, Tennessee. Am. Midl. Nat. 81:315–330.

Fearnside, P.M. and Ferreira, G.L., 1984, Roads in Rondônia: highway construction and the farce of unprotected reserves in Brazil's Amazonian forest. Environ. Conserv. 11:358–360.

Franklin, J.F., Frenzen, P.M. and Swanson, F.J., 1995, Re-creation of ecosystems at Mount St. Helens: contracts in artificial and natural approaches. In: J. Cairns (Editor), Rehabilitating Damaged Ecosystems, 2nd edition (Lewis Publishers, Boca Raton, FL) pp. 287–334.

Hect, S., 1993, The logic of livestock and deforestation in Amazonia. Bioscience 43:687–695.

Malingreau, J.P. and Tucker, C.J., 1988, Large-scale deforestation in the Southeastern Amazon Basin of Brazil. Ambio 17(1):49–55.

Moran, E.F., 1993, Deforestation and land use in the Brazilian Amazon. Hum. Ecol. 21(1):1–21.

Myers, I.B., 1987, Introduction to Type: A Description of the Theory and Application of the Myers–Briggs Type Indicator (Consulting Psychologists Press, Palo Alto, CA) 98 pp.

Schwartz, M.W., 1999, Choosing appropriate scale of reserves for conservation. Annu. Rev. Ecol. Syst. 30:83–108.

Skole, D.L. and Tucker, C.J., 1993, Tropical deforestation and habitat fragmentation in the Amazon: satellite data from 1978 to 1988. Science 260:1905–1910.

Skole, D.L., Chomentowski, W.H., Salas, W.A. and Nobre, A.D., 1997, Physical and human dimensions of deforestation in Amazonia. Bioscience 44(5):314–322.

Smith, N.J.H. and Schultes, R.E., 1990, Deforestation and shrinking crop gene-pools in Amazonia. Environ. Conserv. 17(3):227–234.

Southworth, F., Dale, V.H. and O'Neill, R.V., 1991, Contrasting patterns of land use in Rondônia, Brazil: simulating the effects on carbon release. Int. Soc. Sci. J. 130:681–698.

Tieger, P. and Barron-Tieger, B., 1992, Do What You Are: Discover the Perfect Career for You Through the Secrets of Personality Type (Little Brown and Company, Boston, MA) 330 pp.

Chapter 8

Science and Decisionmaking

E.J. Rykiel Jr., with J. Berkson, V.A. Brown, W. Krewitt, I. Peters, M. Schwartz, J. Shogren, D. Van der Molen, R. Blok, M. Borsuk, R. Bruins, K. Cover, V. Dale, J. Dew, C. Etnier, L. Fanning, F. Felix, M. Nordin Hasan, H. Hong, A.W. King, N. Krauchi, K. Lubinsky, J. Olson, J. Onigkeit, G. Patterson, K.S. Rajan, P. Reichert, K. Sharma, V. Smith, M. Sonnenschein, R. St-Louis, D. Stuart, R. Supalla and H. van Latesteijn

Abstract

The scientific community has an important role to play in public policymaking. However, several difficulties impede productive incorporation of science into decisionmaking. One difficulty is that most scientists lack essential knowledge and experience of decision and policymaking processes at all governmental levels, and therefore do not know how to participate effectively in these processes. If a scientist wants to be more effective, he or she has to take political discourse seriously. A second aspect is that a holistic scientific viewpoint is necessary to improve decisionmaking. Methods to understand behavior at the whole-system level – both natural and human dimensions – are required to replace the reductionistic approaches that attempt to solve problems piecemeal. The global extent of environmental problems is a clear signal that piecemeal problem solving does not result in effective decisionmaking. The challenge rests in developing a transparent process in which scientists and policymakers share information equally about scientific needs, political realities, and economic choices.

1. Introduction

1.1. Working definitions

For the purposes of this chapter, a *scientist* is defined as a person with expert knowledge of one or more sciences, who possesses an advanced degree in a scientific discipline, and who uses scientific methods for research. *Science* is the

Understanding and Solving Environmental Problems in the 21st Century
Edited by R. Costanza and S.E. Jørgensen
© 2002 Elsevier Science Ltd.   All rights reserved

particular way of knowing and reasoning that is based on scientific methods of systematic observation, experimentation, modeling, and testing. The definition of *decisionmaking* is the passing of judgment on an issue under consideration or the act of reaching a conclusion. A *decisionmaker* is defined as a person with the authority to make a judgement on an issue, especially the choice of one option among several, cause actions to be taken to implement the judgment, and promulgate policies that other persons are bound to follow.

Decisions come in all shapes and sizes from those we make in daily life to those that influence the course of history. Here the thrust is on environmental decisionmaking from saving the local wood lot from destruction to ratifying international treaties on issues of global concern. We acknowledge at the outset that many decisions do not have an ecological component. However, we point out that many decisions that do not appear to have an ecological impact when taken in isolation in fact have significant environmental effects when considered in their proper ecosystem context.

1.2. Multiple roles of science

Science is a part of society and not merely a neutral observer. Science and the political decisionmaking process interact in at least three major ways. First, scientific progress changes society as a whole and leads to new decisions; on the other hand, society influences directions of research in science. Neither can do without the other. Second, scientists have the duty to inform decisionmakers on current and potential problems when they become apparent, not only when politicians ask for or agree with the information. Third, scientists are essential contributors to political decision processes (fig. 1). In the ideal case, they try to contribute to both of the following two areas and make the procedures leading to their conclusions as understandable as possible.

(1) Scientists provide information on the current state of knowledge as objectively as possible by summarizing the current state of scientific understanding, by searching for alternative scenarios that try to address the gaps between the desired and the actual state, and by predicting consequences of different scenarios and providing uncertainty estimates for their predictions.

(2) Scientists inform decisionmakers on their interpretation and subjective valuation of the different scenarios.

If science is to fulfill its proper role in decisionmaking, scientists and decisionmakers must learn to communicate with each other. And each has to pay attention not only to what the other is saying, but also to the context from which each speaks. We must remember that communication is a multifaceted framework of interacting contexts constantly shifting in relative importance.

Ch. 8: Science and Decisionmaking

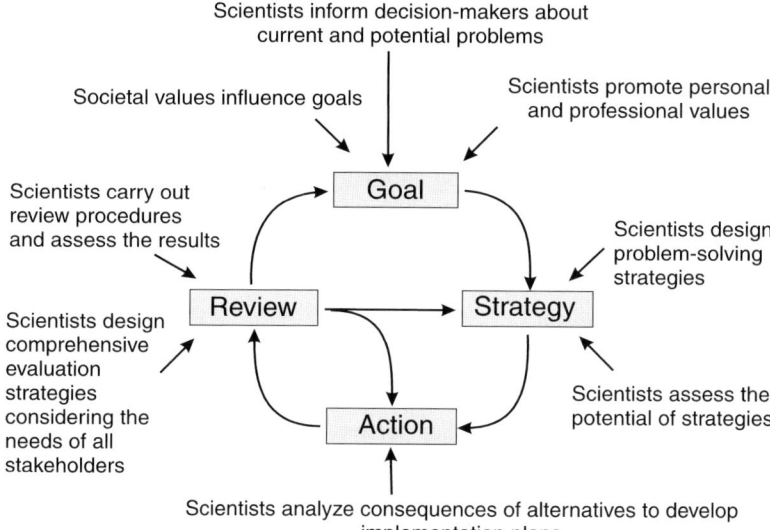

Fig. 1. Schematic diagram of decisionmaking with a special emphasis on the influence of scientists and science as a whole.

1.3. The role of scientists in controversial issues

Harf and Lombardi (2001) indicate that the term issue implies controversy, and list these four "dimensions of disagreement":
(1) whether a problem exists;
(2) what the characteristics of the problem are;
(3) what the preferred future alternatives or solutions are; and
(4) how preferred futures are to be obtained.

Scientists have a clear role to play in determining if a problem exists and defining its characteristics. Often scientists become aware of potential problems before other elements of society. However, we labor in relative obscurity, and, no matter how important, the knowledge we have is known only to a handful of colleagues. As demonstrated by the issue of global climate change, important problems are not addressed by decisionmakers until the media raise public consciousness and concern for the issue. Scientists are not educated to raise such concerns, or to facilitate public understanding of the issues, or even to participate in decisionmaking processes in which scientific information can have a significant impact.

Between the dimensions of identifying a problem and determining its characteristics is the no-man's-land that separates "objective" scientific information from a scientist's advocacy for a particular preferred future. The conflicts that may arise between scientists and other segments of society are in determining what the preferred futures may be, deciding which future to pursue, and determining how the chosen future is to be obtained. There is no surprise that virtually all scientists are advocates for using scientific information in all the dimensions of an issue where that information is relevant. Science can also assist by ruling out alternatives that are scientifically impossible. However, advocating a preferred future can result in the science and the scientist being summarily dismissed as "having a political agenda". Society perceives a strong tie between scientific credibility and neutrality in choosing among alternative futures.

1.4. Scientists and activism

A critical decision that each individual scientist must make is whether to become an advocate for a particular alternative. Science itself does not provide direction on whether to become an activist. Rather personal philosophy and value system are powerful determinants of the level of involvement a scientist chooses, as described below:
- **Never:** science and activism are incompatible, and to become an advocate devalues and may even corrupt the science.
- **Sometimes:** to be a scientist carries with it an obligation to advocate for science in general and for the essential nature of scientific contribution to ecological questions in particular.
- **Inevitably:** to be a scientist is to hold both personal and professional values about science which influence the selection of research topics, the choice of method and the application (and sometimes inevitably the interpretation) of results and these values should be clarified and stated.
- **Always:** to be a scientist and respect objectivity in science does not free the scientist from his citizenship obligations to relate the findings of science to their social context, and so to take a public stand on public issues involving science.

1.5. Education of scientists

Two glaring omissions in the education of scientists are apparent: (1) most scientists do not know how to communicate scientific knowledge to non-scientists, and (2) most scientists do not know how management and political decisions are made. Budding scientists are taught how to communicate with each other, but not how to communicate with non-scientists. Scientists who participate in government, especially at the federal level in the USA for example, generally express dismayed shock at how policy is formulated. An old saw says that the process of making sausage is sufficiently disgusting that it is better not to watch it being made. The making of policy bears a strong resemblance to sausage making. Perhaps it is time to educate scientists about how sausage and policies are made so they can make a more informed decision about their level of involvement.

3.1. Integrating science and economics for environmental policymaking in Europe

The European Commission's 5th Environmental Action Programme "Towards Sustainability" required the integration of the environmental dimension with other policy areas. One of the key elements of this Programme was to "get prices right" and to ensure that environmental externalities are accounted for in market mechanisms. To provide appropriate scientific and technical information, the European Commission together with the US Department of Energy launched in 1991 a joint research project to assess the environmental externalities of energy use. Energy engineers, natural scientists, health experts, ecologists, and of course economists joined the effort. The need for modeling the full impact pathway of pollutants from the power-plant stack through their interactions with the environment to a physical measure of impact, and where possible, a monetary valuation of resulting welfare losses forced the participating scientists to develop a common understanding and appropriate interfaces between the relevant scientific areas.

After the first phase of the project, an operational accounting framework for the assessment of external costs of energy technologies – named ExternE in Europe – was delivered (European Commission, 1995). At that time the US contribution was stopped; however, more than 50 teams from 15 countries in Europe participated in follow-up activities (European Commission, 1999). The ExternE label became a recognized brand; the scientific quality of the work was well accepted, and national and international organizations got used to referring to ExternE numbers as a standard source for external cost data.

This major research activity had a surprising twist, however. Striving to provide the most up-to-date science partly reduced public acceptance of the results because continuous integration of new findings could significantly change the results of analysis with the ExternE framework. At first, ExternE assessed local impacts only, which resulted in external cost estimates that were negligible compared to the private costs of electricity generation. The later consideration of effects from long-range trans-boundary pollution and of chronic mortality from long-term exposure to fine particles at low concentrations resulted in external cost estimates well above the private cost. The message from these new results to policymakers changed dramatically (European Commission, 1999) because the cost estimates went from negligible for local impacts to very high for long-range and long-term impacts.

ExternE failed to accomplish the original objective of quantifying the *total* environmental externalities resulting from a unit of electricity generation from different technologies. However, the process significantly improved understanding of environmental mechanisms across different scientific disciplines. Given the limitations of the approach, the ExternE accounting framework has nonetheless been used successfully in a large number of cost–benefit studies of environmental policy measures in Europe, including for example the economic evaluation of the European Commission's draft Incineration Directive, the Large Combustion Plant Directive, the EU strategy to combat acidification, and air quality standards under the EU Air Quality Directive (European Commission, 1997; Krewitt et al., 1999a,b; Olsthoorn et al., 1999).

3.2. Lake management and demand-driven research in the Netherlands

Even before 1250 AD, regional water boards were effective in water management in the Netherlands. The main tasks of these boards were to protect the land from seawater and allow surplus fresh water to drain towards the sea. Water quality has now become another task of regional water boards. In the case described here, a regional water board invited scientists and stakeholders to deliberate the future of a number of so-called 'border-lakes'.

These lakes cover an area of about 6000 ha, are shallow, narrow, and man-made by the creation of polders in the second half of the 19th century. For many years, the lakes were clear and covered by macrophytes. Near the end of the 1960s, lake quality deteriorated because of increased nutrient loading. For more than two decades, these lakes were characterized by an almost permanent bloom of cyanobacteria (predominantly *Planktothrix agardhii*), and Secchi disk transparency of two or three decimeters. About 1980, initiatives were undertaken to reduce the phosphorus loading to all lakes; later, additional measures were undertaken in some of the lakes, such as flushing and biomanipulation. These actions resulted in an improvement of the water quality, and since the 1990s, most of the lakes are clear for most of the year, reoccupied with macrophytes, and excellent habitat for a diversity of fish and bird species (Hosper, 1984; Jagtman et al., 1992; Meijer et al., 1994; Van der Molen and Boers, 1996; Van den Berg et al., 1998).

After experiencing the previous several decades of the deteriorated lake condition, regional water managers realized that precaution was required to prevent return of this undesired state. Scientists were invited to study the stability of the lakes' clear state and to forecast future developments that would affect the lake (Meijer et al., 1999). In addition, stakeholders were invited to discuss pros and cons of lake quality and to express their wishes. Several meetings were organized with both stakeholders and scientists to explore management options (Reeders and Helmerhorst, 1996). These meetings produced remarkable results: high investments will be undertaken to increase the efficiency of the phosphorus removal at a water treatment plant to counteract the anticipated population increase, and a relatively heavily polluted stream will be directed towards a less vulnerable lake.

This is an example of demand-driven research, which is conducted in response to a decisionmaker's request. This type of research is more likely to be influential in the decisionmaking process because it originates from the decisionmakers and stakeholders. However, the various motivations that lead to demand-driven research can weaken as well as strengthen the effectiveness of science in decisionmaking (Van der Molen, 1999).

4. Conclusions

The international nature of EcoSummit 2000 added a global perspective to the subject of science and decisionmaking, and demonstrated that the degree of influence that science exerts varies from country to country. European participants generally felt that science does influence decisions and policies in Europe, while North American and Asian participants felt that science is generally downplayed unless it serves a political agenda. As might be expected of a scientific meeting, the conclusion was that science does not yet have as much influence on decisionmaking as it should.

Following his year on the White House staff with the Council of Economic Advisors, Brennan (2001) listed 10 points that academics should know about how the US Federal Government "really works". Seven of these points are worth identifying here because they reflect the perceptions of national government operations raised independently in the Science and Decisionmaking Theme and the frustrations that scientists feel when they interact with the political system:

(1) The government is just large enough to be ineffective because most "time is spent in Big Meetings at which little happens".
(2) The quality of the analysis is inversely proportional to the importance of the issue.
(3) Presentations on important decisions should be simple; "If you can't write out an argument in 3 sentences or say it in 30 seconds, you may as well forget about it."
(4) When it comes to influencing policy, anecdotes trump data; "A good story usually beats a library of studies."
(5) With enough assumptions, any policy can be justified.
(6) Parochialism reigns supreme; however, "A clever analyst can design policies to benefit both a narrow interest group and the public at large."
(7) Action is unlikely until the newspapers and networks start to pay attention.

Brennan concludes, "Politics, not the merits of a position, often determines policy decisions, because politics is the only currency that everyone has in common."

Scientists generally have little or no education in politics and may therefore have difficulty relating to political arguments in the face of apparently compelling scientific information. Politicians can rely on minority and even extreme scientific positions if that suits their political agenda. A decisionmaker may understand the science quite well, yet for political reasons, not use the information.

Scientists are also human beings with values who may wish to express an opinion about what policies are best. Because of their unique status in society, scientists therefore have an obligation to distinguish clearly between factual information (what is) and their personal views of what decisions and policies should be made based on that information (what ought to be). In general, science students do not receive training in decision and policymaking and do not understand how decisions and policies are made.

Groves (1992), summarizing the early history of the conservation movement, noted that "... states will act to prevent environmental degradation only when their economic interests are shown to be directly affected." The lesson is that science is regarded as having value for decisionmaking only when it has an economic effect. While perception of risk to health has long been recognized as an important factor in decisionmaking, perception of risk to lifestyle has become a new third factor in environmental decisionmaking. In these days, when ecological scientists believe there is a greater awareness of the intrinsic value of functioning ecosystems, the approval of a voting public, who can translate their displeasure with environmental degradation into political and economic penalties, may be an increasingly important factor. The message for scientists who want to be involved in decisionmaking and particularly in high-level policy formulation is clear: make science relevant to economics, health, and quality of life concerns, and of those three, economics is primary.

The message from the Science and Decisionmaking Theme is that scientists have a critical role to play in society and need to be more involved in decisionmaking and policy formulation processes than ever before. But not every scientist needs to be deeply involved. Many scientists should be involved in decisions at the local level, and a few scientists so inclined should be involved at the state, national, and international levels. In addition, our students need to know more about how decisions are made and the ways in which they can participate as a scientist in the decisionmaking process. Finally, we must find the mechanisms to make all scientists more effective in asserting the proper role of science in society.

Acknowledgements

The ideas expressed in this chapter are the joint effort of the Science and Decisionmaking discussion group, which numbered approximately 33 individuals. It is understood that there may not be unanimous agreement with every point expressed in this chapter. The authors wish to express their appreciation to all who contributed one way or another and hope that all the participants will feel a sense of ownership of this chapter. In particular, we thank Rebekah Blok, Mark Borsuk, Randall Bruins, Kevin Cover, Virginia Dale, Jodi Dew, Carl Etnier, Lucia Fanning, Francesca Felix, Mohd. Nordin Hasan, Huasheng Hong, A.W. King, Norbert Krauchi, Henk van Latesteijn, Ken Lubinski, John Olson,

Janina Onigkeit, Gary Patterson, K.S. Rajan, Peter Reichert, Kamala Sharma, Val Smith, Michael Sonnenschein, Robert St-Louis, Deidre Stuart, and Ray Supalla.

References

Aikenhead, G.S., 1985, Collective decision-making in the social context of science. Sci. Educ. 69:453–475.
Bingle, W.H. and Gaskell, P.J., 1994, Scientific literacy for decision-making and the social construction of scientific knowledge. Sci. Educ. 78(2):185–201.
Brennan, T., 2001, An academics guide to the way Washington really works. Chron. High. Educ. (January 12), Section 2, B11.
Clark, T.W., 1997, Averting Extinction (Yale University Press, New Haven, CT) 270 pp.
DeBonis, J., 1995, Natural resource agencies: questioning the paradigm. In: R.L. Knight and S.F. Bates (Editors), A New Century for Natural Resources Management (Island Press, Washington, DC) pp. 159–170.
European Commission, 1995, Externalities of Fuel Cycles. European Commission, DG XII, Science, Research and Development, JOULE. ExternE – Externalities of Energy. Volume 2: Methodology (European Commission, EUR 16521).
European Commission, 1997, Economic Evaluation of the Draft Incineration Directive.
European Commission, 1999, Externalities of Fuel Cycles. European Commission, DG XII, Science, Research and Development, JOULE. ExternE – Externalities of Energy. Volume 7: Methodology, 1998 update (European Commission, EUR 19083).
Ford, E.D., 2000, Scientific Method for Ecological Research (Cambridge University Press, Cambridge) 564 pp.
Groves, R.H., 1992, Origins of Western environmentalism. Sci. Am. 7:42–47.
Harf, J.E. and Lombardi, M.O., 2001, Taking Sides: Clashing on Controversial Global Issues (McGraw-Hill/Dushkin, Guilford, CT) 380 pp.
Herreid, C.F., 1994, Case studies in science – a novel method of science education. J. Coll. Sci. Teach. 23:221–229.
Hosper, S.H., 1984, Restoration of Lake Veluwe, The Netherlands, by reduction of phosphorus loading and flushing. Water Sci. Technol. 17:757–768.
Hutchings, J.A., Walters, C. and Haedrich, R.L., 1997, Is scientific inquiry compatible with government information control? Can. J. Fish. Aquat. Sci. 54:1198–1210.
ICAFS, 1995, Why isn't science saving salmon? Fisheries 20(9):4, 48.
Jackson, J.A., 1994, The Red-Cockaded woodpecker recovery program: professional obstacles to cooperation. In: T.W. Clark, R.P. Reading and A.C. Clarke (Editors), Endangered Species Recovery: Finding the Lessons, Improving the Process (Island Press, Washington, DC) pp. 157–181.
Jagtman, E., Van der Molen, D.T. and Vermij, S., 1992, The influence of flushing on nutrient dynamics, composition and densities of algae and transparency in Veluwemeer, The Netherlands. Hydrobiol. 233:187–196.
Krewitt, W., Holland, M., Trukenmüller, A., Heck, T. and Friedrich, R., 1999a, Comparing costs and environmental benefits of strategies to combat acidification in Europe. Environ. Econ. Policy Stud. 2:249–266.
Krewitt, W., Heck, T., Trukenmüller, A. and Friedrich, R., 1999b, Environmental damage costs from fossil electricity generation in Germany and Europe. Energy Policy 27:173–183.
Lackey, R.T., 1997, Is ecological risk assessment useful for resolving complex ecological problems? In: D.J. Strouder, P.A. Bisson and R.J. Naiman (Editors), Pacific Salmon and Their Ecosystems (Chapman & Hall, New York) pp. 525–540.

Ludwig, D., Hilborn, R. and Walters, C., 1993, Uncertainty, resource exploitation, and conservation: lessons from history. Science 260:17–36.

Meffe, G.K., 1998, Editorial: conservation scientists and the policy process. Conserv. Biol. 12(4): 741–742.

Meijer, M.L., Van Nes, E.H., Lammens, E.H.R.R., Gulati, R.D., Grimm, M.P., Backx, J., Hollebeek, P., Blaauw, E.M. and Breukelaar, A.W., 1994, The consequences of a drastic fish stock reduction in the large and shallow Lake Wolderwijd, The Netherlands. Can we understand what happened? Hydrobiology 275/276:31–42.

Meijer, M.L., Portielje, R., Noordhuis, R., Joosse, W., Van den Berg, M.S., Ibelings, B., Lammens, E.H.R.R., Coops, H. and Van der Molen, D.T., 1999, Stabiliteit van de Veluwerandmeren, rapport 99.054 (Rijksinstituut voor Integraal Zoetwaterbeheer en Afvalwaterbehandeling, Lelystad, The Netherlands) 132 pp. In Dutch. ISBN 9036952832.

NCSE, 2001, Improving the Scientific Basis for Environmental Decisionmaking (National Council for Science and the Environment, Washington, DC) 32 pp.

Noss, R.F. and Cooperrider, A.Y., 1994a, Chapter 10: the task ahead. In: R.F. Noss and A.Y. Cooperrider (Editors), Saving Nature's Legacy: Protecting and Restoring Biodiversity (Island Press, Washington, DC) pp. 325–338.

Noss, R.F. and Cooperrider, A.Y. (Editors), 1994b, Saving Nature's Legacy: Protecting and Restoring Biodiversity (Island Press, Washington, DC) 416 pp.

NSF, 1996, Shaping the Future: New Expectations for Undergraduate Education in Science, Mathematics, Engineering, and Technology, NSF Publication No. 96-139 (National Science Foundation, Arlington, VA) 76 pp. Out of print; available on-line as ASCII or HTML file: http://www.nsf.gov/cgi-bin/getpub?nsf96139.

Olsthoorn, X., Amann, M., Bartonova, A., Clench-Aas, J., Cofala, J., Dorland, K., Guerreiro, C., Hendriksen, J.F., Jansen, H. and Larsen, S., 1999, Cost benefit analysis of European air quality targets for sulphur dioxide, nitrogen dioxide, fine and suspended particulate matter in cities. Environ. Resour. Econ. 14:333–351.

Reeders, H.H. and Helmerhorst, T.H., 1996, Op weg naar helderheid. Een heroriëntatie van BOVAR gericht op 2000, rapport 96.01 (Rijkswaterstaat Directie IJsselmeergebied, Lelystad, The Netherlands) 102 pp. In Dutch.

Shogren, J., 1998a, A political economy in an ecological web. Environ. Resour. Econ. 11(3–4): 557–570.

Shogren, J., 1998b, Do all the resource problems in the West begin in the East? J. Agric. Resour. Econ. 23(2):309–318.

Van den Berg, M.S., Coops, H., Meijer, M.L., Scheffer, M. and Simons, J., 1998, Clear water associated with a dense Chara vegetation in the shallow and turbid lake Veluwemeer. Ecol. Stud. 131:339–352.

Van der Molen, D.T., 1999, The role of eutrophication models in water management, Thesis (Agricultural University Wageningen). Report 99.020 (Rijksinstituut voor Integraal Zoetwaterbeheer en Afvalwaterbehandeling, Lelystad, The Netherlands) 167 pp.

Van der Molen, D.T. and Boers, P.C.M., 1996, Changes in phosphorus and nitrogen cycling following food web manipulations in a shallow Dutch lake. Freshw. Biol. 35:189–202.

Wondolleck, J.M., Yaffe, S.L. and Crowfoot, J.E., 1994, A conflict management perspective: applying the principles of alternative dispute resolution. In: T.W. Clark, R.P. Reading and A.C. Clarke (Editors), Endangered Species Recovery: Finding the Lessons, Improving the Process (Island Press, Washington, DC) pp. 305–326.

Yaffe, S.L., 1994, The Northern spotted owl: an indicator of the importance of sociopolitical context. In: T.W. Clark, R.P. Reading and A.C. Clarke (Editors), Endangered Species Recovery: Finding the Lessons, Improving the Process (Island Press, Washington, DC) pp. 4–71.

Background

Chapter 9

Ecosystem Health and Human Health

L. Vasseur, D.J. Rapport, J. Hounsell

Abstract

Ecosystems are complex and highly susceptible to human activities. The relationships between ecosystems and humans are very strong but unfortunately often misunderstood by decisionmakers and other parts of our societies. It is important, therefore, that we begin to examine the linkages between human activity and the health of our ecosystems, at both the local and global scales. The challenge is to ensure that all levels of society are informed and thus able to make more sustainable and healthy decisions. With increasing populations and impacts, not only the health of humans can be in jeopardy but also the health of the ecosystems that support us. In this chapter, we examine the connections between human health and ecosystem health and the reasons for ignoring these links. The arguments that we present target three important topics that reach all nations – climate change, biodiversity, and agrosystems and food production – as they are global in their impacts. From these examples, it is possible to extract the main challenges that will face humanity in order to improve the sustainability and health of all ecosystems.

1. Introduction

Over 35 years ago Rachel Carson's *Silent Spring* (Carson, 1962) brought critically needed attention to the problems of our environment. Since that time, political efforts such as the Brundtland Report, and protocols such as Agenda 21 and Kyoto, have attempted to improve the condition of our environment and the quality of human life. Communities around the globe are undertaking positive steps towards achieving sustainable development. However, even with these efforts, ecological crises are deepening, becoming more global in nature, and more intractable as human populations continue to grow. In a recent publication the United Nations Development Programme, the United Nations Environment Programme, the World Bank and the World Resources Institute (2000) agreed that the world's ecosystems are in dangerous decline because of increasing human demands and impacts. Declining ecosystem capacity translates into declining human health and well-being. The World Health Organization indicates that 25% of all preventable ill health in the world can be directly attributed to poor environmental quality (World Health Organization, 1997). Only effective, concerted, and co-ordinated human action can hope to avert or slow the downward spiral of irreversible damage to ecosystems, declining carrying capacity of the earth resources, and loss of options for supporting and improving human quality of life.

Although we have begun to recognise the importance of healthy ecosystems and the preservation of our resources for maintaining viable human futures, in general, we still have difficulties integrating these principles into our actions, legislation and policies. We are locked into a value system that places short-term gain over sustainability, and denies responsibility for future generations. It is important, therefore, that we begin to examine the linkages between human activity and the health of our ecosystems, at both local and global scales. The challenge is to ensure that all levels of society are informed and thus able to make more sustainable and healthy decisions. One method to strengthen political resolve is to more clearly articulate the relationship between human health and ecosystem health. Traditionally, these aspects are considered separate and independent, and while the public is highly concerned about human health, there is much less interest in the environment. There is a stronger emotional reaction towards the death of another human being (especially a close one) than towards the death/extinction of another species. Most cultures in today's society are not as emotionally, mentally, or physically linked to their environment as they are to other humans. However, it is becoming apparent that the boundary between health and the environment is no longer as concrete as once thought. Here we shall show that the two march hand-in-hand (albeit with time lags). The earth's ecosystems are now becoming overburdened and this poses serious threats to the future well-being of humankind (Brown et al., 1989; McMichael, 1993; Postel, 1994; Rapport et al. 1998a,b).

Can expressing the link between ecosystem health and human health instill the necessary urgency needed for immediate action? We recognise the risks for many human habits, such as overeating and smoking, yet we have more obese people now than ever and people still continue to smoke. Here, as with many ecosystem ills, the "system" fails only after years of neglect and abuse. The cumulative and long-term impacts on the system are not evident while the activity is occurring, only after subsequent health problems and often, death result.

2. Ecosystems, humans, and the concept of health

Prior to examining the links between human health and ecosystem health, the concepts of 'ecosystem' and 'health' should be defined.

2.1. Ecosystems

The entire planet, a landscape, or a lake can be considered as an ecosystem. An ecosystem is an astonishing assortment of species that interact with one another, as well as with the non-living environment surrounding them. Ecosystems are composed of essential elements for the survival of all organisms and their ability to function, and include all living components and their non-living physical and chemical environment (Draper, 1998). In all ecosystems, gradients of light, temperature, salinity, and nutrients exist where several communities of organisms will survive and interact. This is a complex matrix of reactions and interfaces between the biotic and abiotic components that allows organisms to survive, move, and interact. Since many of the elements composing organisms have global cycles, global ecosystems are always interconnected (Homer-Dixon, 1999). Although ecosystems have boundaries, they are somewhat arbitrary, as populations of organisms can vary in space and time. Furthermore, environmental factors (e.g., elevation, soil chemistry) may also vary continuously in time and space (Likens, 1992). Ecosystem boundaries are usually determined for the purpose of analysis rather than on the basis of some known functional discontinuity with an adjacent ecosystem.

An ecosystem provides a conceptual framework for the study of the interactions among species – including individuals, populations and communities – and their abiotic environments, and for the changes in these relationships over time. For many years ecosystems have been viewed as external systems where humans have little or no involvement. Humans are, however, only one of the many species on earth – albeit the most influential one, and exist as an integral part of many ecosystems (Draper, 1998). Only in more recent years, persuaded by overwhelming evidence, has this perspective begun to be commonly accepted.

In "natural" ecosystems (e.g., old-growth forests, oceans), humans are believed to have only a minimal or negligible influence. With very little direct human intervention, these systems are considered pristine or at least untouched. However, even in remote areas of the Arctic where there is little or no human presence, high concentrations of toxic chemicals have been found. Sadly enough, very few, if any, places on the earth today have not felt the impact of human activity. In "managed" ecosystems, humans play an integral role in the system, and are often predominant. Such ecosystems include urban areas, municipal parks, stream/river channels, etc. In all cases, humans have manipulated many of the components of these ecosystems in order to best utilise the resources. Because of this high degree of human influence, some of these managed ecosystems would now not be able to exist in their current state without the continued presence of humans. This is the case for several rivers (e.g., Rhine, Thames, Mississippi, etc.) that have been modified to a point that without human control, surrounding ecosystems would be at risk. They have lost a large part of their ecological integrity and most native species have been extirpated because of human modifications of these ecosystems.

2.2. Health

The concept of health has roots etched deeply in our lives and is concerned with survival and well-being. But it is also often defined negatively – for example in terms of the absence of disease (Haskell et al., 1992). Healthy humans are thus described as (1) free of diseases, and (2) capable of mentally and physically functioning/performing (Rapport et al., 1998b). Traditionally, the concept of health has been limited to the physical condition of the individual (i.e., humans, plants and animals). More recent applications of the concept however, lean towards more complex organisations, including the community and ultimately, the ecosystem level (Rapport et al., 2002). Ecosystem properties such as degradation under stress and distress syndromes (Rapport et al., 1995) suggest that health can also be a property at the whole-system level, and not merely used as a metaphor. Moreover, methods developed in risk and impact assessment as well as monitoring are applicable to the analysis of ecosystems under stress, and are therefore useful in assessing ecosystem health.

The concept of health is also shifting within traditional health (i.e., medical) professions. This is largely due to the recognition that ecological imbalances are increasingly the cause of many human illnesses. Traditional medicine is concerned with diagnosing the disease and curing it, while an ecosystem health approach emphasises linking the disease with its environment (human, ecological, social, etc.) to understand what caused it (Rapport et al., 2002).

The need to examine the meaning of ecosystem health arises from a growing concern that our ecosystem services are failing and the frustration to see the failure of current economic paradigms to protect the natural environment (Haskell et al., 1992). The idea of health of an ecosystem provides a link to sustainability. The main goal of ecosystem health is to develop policies that allow human cultures to thrive without changing the life support functions, diversity, resilience, productivity and organisation of ecosystems (Rapport et al., 1998a). Costanza et al. (1992) defined ecosystem health as follows: "An ecological system is healthy and free from 'distress syndrome' if it is stable and sustainable, that is, if it is active and maintains its organisation, and autonomy over time and is resilient to stress".

The concept of ecosystem health was foreshadowed in the writings of Aldo Leopold (1941), when he introduced the notion of "land health" (Callicott, 1992). Leopold recognised that not only individuals, but also whole ecosystems could show signs of disease or sickness as a result of stress from human activity. He also recognised that under such conditions, ecosystems would no longer be capable of supplying the ecosystem services that humans and many other

Table 1
Examples of ecosystem services for the benefits of all organisms (including humans)

Services	For all other species	For humans
Oxygen and air purification	×	×
Food	×	×
Fiber	×	×
Nutrient balance/cycling	×	×
Water purification	×	×
Climate control	×	×
Soil production	×	×
Religious expressions		×
Recreation		×
Psychological and spiritual development		×

species depend upon for their well-being. These services are numerous and their performance should also give us an idea of the state of the system (table 1). So, both ecosystem and human health can be based on two components: (1) absence of diseases (sickness) and (2) capacity to function optimally (also referred to as integrity). Humans label someone who is integrated into society as a person who is mentally and physically capable of functioning and providing services to society. Similarly an ecosystem functioning effectively and positively has a certain level of integrity. Thus it is the combination of attributes (i.e., resilience, productivity) that is critical, not one element alone. The journal *Ecosystem Health* incorporates these concepts of human health, by defining ecosystem health as: "a systematic approach to the preventive, diagnostic, and prognostic aspects of ecosystem management, and to the understanding of relationships between ecosystem health and human health. It seeks to understand and optimize the intrinsic capacity of an ecosystem for self-renewal while meeting reasonable human goals. It encompasses the role of societal values, attitudes and goals in shaping our conception of health at human and ecosystem scales."

Many ecosystems have already been transformed to a degraded state from which recovery is unlikely (Rapport and Whitford, 1999). It is often difficult to predict however, at what point an ecosystem looses it natural resilience. For example, we can examine the rapid decline of small-managed ecosystems, such as parks (e.g., Kejimkujik National Park and Point Pleasant Park in Nova Scotia, Canada; Vasseur, 2000a). With intensive management, parks seem relatively stable and capable of maintaining their functions for a long period of time. However, over years, changes occur rapidly and ecological integrity is lost. For example, young trees die, soil becomes compacted, ground vegetation disappears and tree

> **Box 1**
> Statistics on the state of our ecosystems and the environmental degradation due to human activities[a]
>
> - Half of the world's wetlands were lost last century.
> - Logging and conversion have shrunk the world's forests by as much as half.
> - Some 9 percent of the world's tree species are at risk of extinction; tropical deforestation may exceed 130 000 square kilometres per year.
> - Fishing fleets are 40 percent larger than the ocean can sustain.
> - Nearly 70 percent of the world's major marine fish stock are overfished or are being fished at the biological limit.
> - Soil degradation has affected two-thirds of the world's agricultural lands in the last 50 years.
> - Some 30 percent of the world's original forests have been converted to agriculture.
> - Since 1980, the global economy has tripled in size and population has grown by 30 percent to 6 billion people.
>
> [a] Source: United Nations Development Program et al. (2000).

regeneration drastically decreases. Like most managed ecosystems, in order to reduce the risk of liability, plants (mostly dead trees) are removed from the park because they can be dangerous to humans. With the increasing density of humans and demands on these resources, trampling and other types of damage occur. Compaction due to trampling leads to the removal of the ground vegetation and an increase in soil erosion. The removal of trees and other debris reduces the amount of organic matter and nutrients available through nutrient cycling. Since the roots of trees are also stressed, more trees die. The cycle accelerates and the ecosystem becomes less resistant and resilient. In fact, in many cases, restoration projects have to be initiated in the campground in order to improve the integrity of the ecosystem. A similar situation in which human intervention (trails, and restricted hunting of deer) have compromised the health of a park was recently reported for a small provincial park in Southwestern Ontario (Patel and Rapport, 2000). These examples are restricted to small-scale ecosystems, but evidence of large-scale ecosystem failing is abundant (box 1). Many are well known but little is done to resolve the problems. These cases add up at regional and even global scales (e.g., deforestation, desertification). The most difficult challenge is to admit the obvious: we are impacting on ecosystems. But how can we convince others? How can we show the links between ecosystem health and human health?

3. Ignoring the link

In most countries, policymakers and the general public have yet to be convinced of the link between ecosystem and human health. Although the present chapter examines examples and areas where the connections are clear, they are usually overlooked during the decisionmaking process. Reasons for such avoidance include (1) greed or enterprise and (2) ignorance or inattention. In the pre-industrial era the main goals of human society were subsistence activities. However, since industrialisation and improvements in living conditions and lifestyles, money and materialism have replaced subsistence as the key motivators of current modern society. With increasing populations, the demand for materials and supplies has risen to such a point that industries and corporations can now manipulate and control consumption patterns to their own means. Several industrialised countries encourage decisionmakers and politicians of developing countries to increase consumption, global trade, and production. The fundamental assumptions of our society however, are now being severely challenged, as witnessed by the riots that have accompanied the meetings of the World Trade Organization in Seattle (1999), the World Bank and International Monetary Fund in Budapest (2000), and the Free Trade Agreement of the Americas in Quebec City (2001).

In order to satisfy the increasing demands of our society, more natural resources are needed and without a change in our current attitudes the destruction of our ecosystems will continue unabated. This is particularly evident in developing countries, where populations and economy are both on the rise. Many of these countries see a Western lifestyle as the ultimate goal, a pressure that often leads to an overconsumption of resources – even though most of these countries do not have the technology to manage these consumption patterns. Economic growth, when unchecked, can accelerate the chain of impacts on the ecosystem leading to degradation and ultimately, declines in human health. The collapse of the fisheries in the North Atlantic represents a strong example where communities on the coast of both industrialised and developing countries have been greatly impacted by this resource loss (Kraft, 2001). While in some areas, such as in Nova Scotia (Canada), the loss of one resource has placed added demands on remaining resources: the forest industry has now replaced fisheries as the top resource industry (Tony Charles, presented at EcoSummit 2000). Figure 1 illustrates the links between economic activities and ecosystem health and human health in a resource based economy.

Given the gravity of the situation, one might conclude that lack of education, information, and communication are major barriers to corrective actions. Natural history, human health, and environmental science are often neglected in the high-school curricula in many of the high-income countries, and are practically non-existent in what little formal education is available to the children of developing nations. Similarly, environmental information is not always readily accessible in all countries of the world (Kraft, 2001). In general, even where such education is attempted, the material used for teaching is out of date and inadequate for better understanding of the interdependency of all ecosystem components (Sauvé, 1999). Although access to information seems to be possible in most democratic countries, the reality can be very different. In most cases, news media have tendencies to promote ideas that could lead to higher consumption. The public information system has become oversaturated with high sensation news. Green and environmental friendly media are still relatively rare and limited to a more

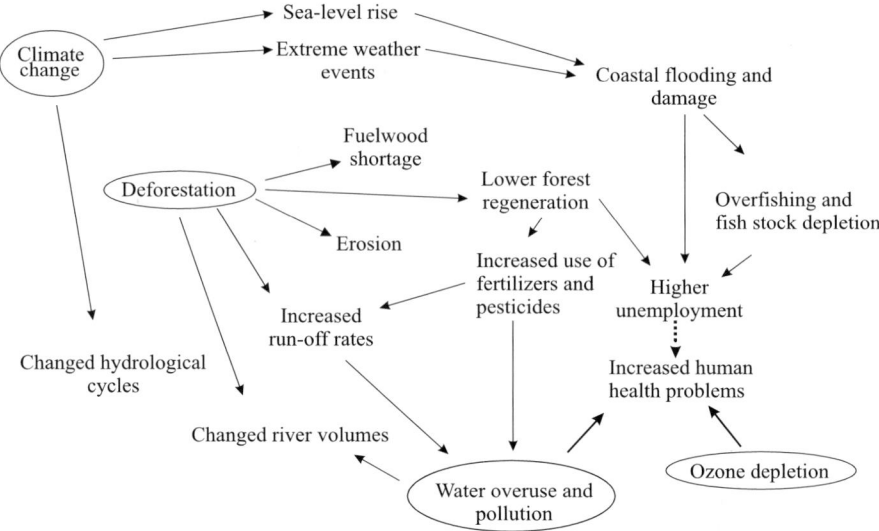

Fig. 1. Some possible links and effects of climate change and human overexploitation of natural resources on human health and ecosystem health in a Nova Scotia economic system of fisheries and forestry. Adapted from Homer-Dixon (1999).

educated and/or aware audience. With budget restrictions in most industrialised countries, questions related to human health care and employment are more on the agenda than the protection of species, and yet as we know, these aspects of life are all interrelated. In other countries where information is tightly controlled, the connections between ecosystem health and human health may be consciously neglected (comments of Dr. Le Dien Duc, one of the participants from a developing country at the EcoSummit 2000).

Inattention is also to blame for our lack of acknowledgement of the connections between ecosystem and human health. Concern over the environment is not new, nor is action to reduce the harmful effects of human activity on the environment (Mullin, 2000). In the 1990s several surveys were conducted in North America. In Canada, for example, 73% of respondents reported being very concerned about environmental issues (Krause, 1993), with more than half indicating that protecting the environment is more important than creating jobs. Most people are in fact very attentive to their environment. Yet, their behaviour suggests that the environment does not seem to be a priority. The main challenge is the establishment of a link between attitude and behaviour. While most people say that they are environmentally conscious, their attitude does not necessarily translate into behaviour. Behaviours however, are often dictated by status. For example, Sadalla and Krull (1995) showed that in a 'high status' society, conservation-related behaviours such as using a clothesline, taking the bus or recycling, are regarded as 'lower-class' activities and inappropriate.

Finally, religion, culture and other types of belief can also affect the perception and understanding of humans in relation to their environment. In the Judeo-Christian tradition, humans are assumed to have "dominion" over nature (Harper, 2001). This belief encourages people to consider humans as separate from the rest of nature, and as having dominance over nature. It is often this perception that has encouraged the degradation and demise of our life support systems. One may ask, however, why it is not recognised that this may also be the demise of our own species and culture. The answer might lie in the optimistic belief that new technologies will come to the rescue (as they have in the past). This undying faith in technology has very little evidence supporting it however, as often attempts to mitigate ecosystem degradation has simply resulted in transferring the problem elsewhere (e.g., applying fertilizers to compensate for declines in soil fertility can often lead to contaminated ground water, rivers, lakes, etc.) (Rapport et al., 2001).

It is now essential to describe some examples showing how human activities affect ecosystems, which in return influence human health. In the next three sections, global problems related to human and ecosystem health and the links between humans and ecosystems are explored. The topics discussed here are climate change, agrosystems and food production, and biodiversity and declining productive capacity. Although there are many more issues related to ecosystem health and human health, we have chosen to consider these as they have significant global implications. They pose challenges however, not only at the global policy level, but as well, at the regional and local scales.

4. Climate change

Human dependency on the burning of fossil fuel has increased in the past century. With greater population size and increasing economic activities, emissions into the atmosphere have reached levels never recorded before in human history. For example, atmospheric CO_2 concentration is now 29% above its pre-industrial level, a concentration higher than at any time in the last 160 000 years (United Nations Development Program et al., 2000). One thing is also known; industrial countries are responsible for 76% of the world's cumulative carbon emissions since 1950. The increased concentration of these "greenhouse gases" is predicted to have a significant impact on the earth's climatic conditions (Serreze et al., 2000). Although there is much debate on what changes will occur when and at what rate, it is undeniable that the earth's global temperature has been steadily rising since the early 1900s and will continue to do so.

To better predict the potential impacts of climate change, computer models have been developed. The most elaborate of these models is the General Circulation Models (CGMs) (Government of Canada, 1996; Hengeveld, 2000). These models integrate physical parameters that govern the global ecosystem and simulate the interactions of the sun, atmosphere, oceans, vegetation and other lands, under various scenarios. Although these models cannot precisely predict the potential impacts of climate change at the regional scale, they do agree that several major global changes will occur with the doubling of CO_2 in the atmosphere (Government of Canada, 1996; Draper, 1998). For example, the GCMs predict an increase of global temperature by 1–3.5°C. This warming would not be distributed equally around the globe however, leading to more unpredictable climatic conditions at regional and time scales (Sachs et al., 2001).

Several ecosystem components will be affected by climatic change. While temperature is commonly discussed in the news, precipitation patterns are also predicted to change, leading to desertification in some areas and more rainfall in others. With an increase in average global temperature, sea level could rise at a rate of 1 to 10 cm per decade over the next century as polar ice continues to melt in the warmer climate (Draper, 1998; United Nations Development Program et al., 2000). Are these levels rising as fast as predicted? We do not really know. But, since a large number of people live along the world's coastlines, the number of environmental refugees could reach millions of people per year over the next century (United Nations Development Program et al., 2000). Where will these people go? In the case of Trinidad-Tobago, concerns are expressed repeatedly in international meetings (Lester Forde, participant to the Hemispheric Trade and Sustainability symposium organized prior to FTAA, April 2001, Quebec City). Impacts on health could be more severe than predicted with a larger number of deaths, starvation, and the occurrence of infectious diseases. Displaced refugees will have difficulties finding new arable lands and conflicts for these lands are therefore highly probable. Under degraded and poorer conditions, cases of infectious diseases (such as cholera, typhoid, etc.) are expected to explode due to lack of hygiene and poor living conditions.

Forests may benefit in the short term by this increase in CO_2 availability. Warmer temperatures may extend the optimal growing conditions of North-American forests further north, however this warming may also push drier, inhospitable southern climates north as well. If tree species cannot migrate northward at the same rate as the optimal growing conditions, certain species distribution may be significantly reduced. Similarly in agriculture, where photosynthetic rate can increase, unpredictable climatic conditions can reduce crop production in important agricultural regions. Other impacts of climate change include warmer weather conditions in the Arctic with a shorter ice period and decay of permafrost (Government of Canada, 1996; Hengeveld, 2000). A slight change in ice break will have drastic impacts on migrating animals and nomadic communities trying to reach safe lands in the spring.

With these ecosystem changes, human activities will undoubtedly be affected. Stresses from lower food production, less forest cover, and unpredictable weather could lead to increasing human health issues. Extreme heat conditions can become more severe, such as the 1995 heatwave in Chicago where more than 700 excess deaths occurred in the metropolitan area (Patz, 2000). In cities around the world, with increased temperatures, heat strokes are more frequently reported and heat "emergencies" are being declared (e.g., Toronto, Canada, August 2001). Ground-level ozone smog can also intensify with increased temperatures, and lead to more cases of asthma and other respiratory problems. Warmer temperatures can mean fewer cold-related deaths; however, in the USA twice as many people die from the heat as from the cold (EPA, 2001).

The increase of global temperatures may also increase the northward spread of many infectious diseases currently limited to subtropic and tropic regions, particularly those spread by mosquitoes and other insects. Diseases such as malaria, and dengue fever, may become more prevalent in northern areas. In fact, malaria has already reappeared north and south of the tropics (Epstein, 2000).

Extreme weather conditions can also threaten water quantity, and increase the incidence of water-borne diseases such as cholera (Epstein, 2000). In industrialized countries where excellent health-care systems are present and resources for effective prevention measures are available, these types of diseases can be mitigated. In countries where such resources are not readily available, however, climate change can bring havoc to the health of human populations. Due to lack of sanitation and degraded environmental conditions, infectious diseases are usually one of the main concerns that medical practitioners have to deal with in developing countries. Under these conditions, lack of hygiene, food, potable water, and shelter, people tend to be more susceptible to diseases. Their vulnerability to climate change is therefore greater than in industrialised countries (Colwell et al., 1998; Lindgren, 1998; Woodward et al., 1998).

Unpredictable climatic events (e.g., hurricanes, droughts, etc.) associated with climate change can also impact human health directly and indirectly. With hurricanes, the number of deaths can increase as a result of ecosystem degradation. For example, when Hurricane Mitch ravaged Central America in October 1998, over 10 000 people were killed. Many of these deaths were the result of mudslides, which were triggered by the torrential rainfalls impacting on the weak slopes of deforested hillsides. While heavy rainfall can have similar impacts in a healthy ecosystem, in these densely populated, marginal, and overexploited lands, impacts are often amplified. Under such conditions, shortages of food and drinking water are another factor leading to starvation, malnutrition, diseases, and deaths.

Other examples include changes in precipitation and water shortages. Already in some areas of the world, water shortages from changing climatic conditions have occurred over the past decades. This can lead to violence and political fighting between regions or countries. For example, Israel has limited Arab access to groundwater, which has fuelled tensions. Syria and Iraq are in a similar situation with Turkey because of hydropower developments on the Tigris and Euphrates Rivers, leading to political conflicts (Postel, 1996).

The issue of climate change has been in the news for the last 20 years. However, despite scientific predictions, international protocols (e.g., Kyoto) and the desires of some policymakers, little progress has been made. The main reason for such inertia rests on the lack of understanding of the problem and the consequences in the short and long term. Governments have argued that the human cost of climate change is very difficult to estimate due to uncertainties in projection models and a poor understanding of the human response process (Hengeveld, 2000). These uncertainties, influenced by the idea that humans are external to the environment, have led some economists and others to believe that solutions to climate change will be found and impacts on human health will be minimal (Harper, 2001). Although this may be possible for some industrialised countries (for the short term), developing countries could be faced with devastating impacts. The precautionary principle is rarely applied even though the risk to the global ecosystem is potentially extremely high. Climate change will impact the global community indiscriminately, and only co-ordinated human efforts across all borders can hope to mitigate or possibly avert the damage.

5. Agrosystems and food production

Agrosystems are managed ecosystems built for the purposes of producing food and other agricultural products (Draper, 1998). These ecosystems are complex and encompass the land and the species grown on it, along with the farmer's management techniques and technology. These systems are controlled by natural functions, government policies, and the global economy. The level of sustainability of these ecosystems is directly related to food production, as well as to the long-term maintenance of other ecosystem functions such as nutrient cycling and biodiversity. With an increasing human population, the need for agrosystems is increasing to a point that without new technologies, most of the earth's surface would need to be cultivated in order to feed the world. Food production economic value in 1997 reached US$1.3 trillion. Although these numbers seem positive, the reserves of grains are low and declining (Brown, 1998).

Agrosystems cover more than 25% of the global land area but 75% of this land is of poor quality (United Nations Development Program et al., 2000). The main challenge is how to increase food production without further degrading these existing ecosystems. Most of the agrosystems that are located on poor lands are in developing countries where climatic conditions are more conducive to unpredictable events such as drought or flooding. Under these conditions, agricultural development should follow precautionary principles. Unfortunately, with large populations to feed and global economic incentives, marginal lands are often overexploited. Although these initiatives should encourage the development of more agrosystems, crop production in Africa and Central America for example, has declined. There are several reasons for this, the main one being the decline in the health of these ecosystems. In developed countries, the amount of land used for agriculture is also decreasing, and can be linked to changes in social, economic, and technological conditions. In all cases, agrosystems are either lost or converted for other usage (e.g., suburban areas), or lost through desertification and degradation.

The use of unsustainable practices such as a heavy reliance on fossil fuels for production and the application of chemicals, fertilisers, and pesticides, have led these ecosystems to be highly managed but fragile to any environmental change. While they affect sustainability of surrounding ecosystems by impacting soil, water, biodiversity, and even the air through increases in greenhouse gases, their own survival is limited. Current practices, such as monoculture, intensive row cropping, rapid fallowing, and up-and-down slope cropping, have led to rapid loss of the physical matrix. Water runoff and soil erosion due to soil compaction and intense agricultural practices are common problems in all agrosystems. With the overuse of fertilisers, soil nutrient imbalance is becoming another major limiting factor for crop production. Overuse of chemical pesticides has also reduced soil diversity. Microorganisms and invertebrates usually present in soil are instrumental for the reconstruction of the physical matrix and the cycling of nutrients in soil. With the complete removal of organic matter and addition of chemical complexes in the soil, these organisms have declined to a point that nutrient cycling is no longer possible in most intensively farmed lands (Homer-Dixon, 1999).

Agrosystems are complex and interrelated to other ecosystems. Thus water erosion, irrigation, and heavy pollution from the overuse of fertilisers and pesticides impact on them as well. Studies show that heavy pollution from chemical use on agricultural lands has been reported in surrounding waterbodies and forests (Freedman, 1998). The runoff of sediments and fertilisers in waterbodies has led to anthropogenic eutrophication and a reduction in water quality. Since more modern crops are susceptible to droughts, irrigation has become an important farm task.

In areas where water is already limited, shortages are becoming more frequent, resulting in more violent conflicts. The region of Gaza offers a very strong example of the impacts of increasing water demands and the overuse of fertilisers and pesticides, not only on agrosystems, but also on adjacent ecosystems and human health, overall leading to conflicts and violence between communities or nations (Homer-Dixon, 1999). To meet the demands of an increasing population, agricultural practices in Gaza have dramatically increased over the years, maintained by an increasing use of chemicals, fertilisers, and pesticides. As conditions in this area are sometimes extreme, water is also an important resource for the maintenance of these agrosystems. The population of Gaza however, depends almost exclusively on water drawn from an aquifer underneath the territory for their irrigation and drinking water uses. Over the years, increasing pressures on the aquifer have created new environmental problems for the Gaza region. The use of water now exceeds the recharge rate of the aquifer by about 60 million cubic meters a year, and salt intrusion and chemical contamination also threaten the aquifer. The pressure from agricultural practices and an increasing population has become a major problem in Gaza, as the aquifer is quickly becoming unsuitable for irrigation and not fit for human consumption. A report by the World Bank suggests that Gaza's water contributes to the high frequency of gastrointestinal and parasitic infections in the region (Homer-Dixon, 1999).

Food production in aquatic systems can also lead to increasing ecosystem and human health issues. In many countries, overfishing has led to a major decline in the biodiversity of aquatic systems, and in some cases to the destruction of complete ecosystems (Kraft, 2001). In developing countries, agrosystems and food production still represent the main subsistence activities. However, with rapid population growth and increasing demands on food production, these activities have begun to greatly contribute to the decline of ecosystem health and human health in these countries. Cambodia offers a typical example of such problems. Fish size and diversity have decreased in most areas of the world due to overfishing. In Cambodia, the reduction of the mesh size of the fishing nets currently helps sustain the fisheries commercially. However, this short-term view has important implications on the survival of villagers relying on this resource for subsistence (Vasseur, 2000b). In addition, along the shore of the Mekong River, rural inhabitants have continued the practices of slash-and-burn agriculture. With the increase in population and land limitation, farmers have been forced to use fertilisers and pesticides with a fallow rotation non-existing. Overexploitation and degradation of the land leads to soil and water erosion. At the same time, fisheries have intensified since the 1980s and medium and large-scale operations are more common. In most cases, fishers and their families will live on floating houses. Fish are either captured or farmed underneath the floating houses in cages (Ahmed et al., 1998). Due to the lack of sanitation, both on shore and on boats, all wastewater is directly dumped in the river. With the increasing use of detergent and other chemical agents, wastewater is highly contaminated. With high densities of fish in cages and high level of pollution, the health of the aquatic ecosystem has declined rapidly. For example, algal diversity has drastically changed and is mainly composed of blue-green algae. Under environmental stress, fish mortality has increased. Fish diversity and individual size have also been altered. Under such stressed conditions, fish also become more susceptible to parasites and other diseases. These changes in the Mekong River ecosystem have drastic impacts on human populations, mainly their health. Levels of infectious diseases remain high despite humanitarian help. While water is unfit for human consumption, due to the high level of poverty, people still drink directly from the river. There is a chronic occurrence of gastrointestinal problems and parasites in most of these populations. The current lack of data (due to lack of research activities in the country) does not allow for the completion of this picture. But it is hypothesised that due to the overuse and unregulated use of chemicals and water erosion, fish might be highly contaminated. Since fish represents more than 75% of protein uptake in rural communities, it is expected that human health could gradually be affected by this increase of chemicals in the food chain (Vasseur, 2000b).

6. Biodiversity and declining productive capacity

In the Convention on Biological Diversity, biodiversity is defined as "variability among living organisms from all sources including *inter alia,* terrestrial, marine or other aquatic ecosystems and the ecological complexes of which they are part" (Convention on Biological Diversity, 1994, p. 4). This definition is relatively vague and driven more by political agenda than scientific thinking. In a more scientific context, biodiversity is defined as "the variety of life and its processes and encompasses genetic, species, assemblage, ecosystem, and landscape levels of biological organisation and their structural, compositional and functional components" (Primack, 1993). This means that, if all components of an ecosystem are protected, the ecosystem as a whole is functioning well or it is healthy. The word biodiversity encompasses the various aspects of ecosystem services and structure.

Biodiversity consists of five integral components: genetic diversity, species diversity, ecosystem diversity, ecological functions, and the physical matrix. Because most species have several populations that are at least in part isolated, individual populations are locally adapted and thus may carry differences in genetic diversity. When one talks of conserving biodiversity, it is well to bear in mind that whole organisms, not the genes within them, are the functional entities. This is because only organisms (not genes) can live in real ecosystems. Genetic information within individuals is like a great library of books that was created over a long time. It represents potential opportunities for the evolution of new lineage in the future, and potential for better crops, domesticated animals, cures for diseases, and biotechnological production of useful substances. It also offers the potential for adaptation to long-term natural and anthropogenic changes in the environment. Loss of genetic variability reduces capacity to adapt to changes. Unfortunately under the current forestry and agricultural practices, genetic diversity is reduced due to strong selection for more productive races or varieties. While more than 80 000 plant species are edible, humans have in the past used only 7000 of them and currently rely heavily on only 20 species such as maize, rice, and wheat (Wilson, 1990). Plant breeders have exhausted their traditional techniques to improve crops. With biotechnology and gene transfer, it is expected that some options are still available but only up to a certain limit, which is close to the physiological limit of the plant (Brown, 1998).

Close to one million species have been described from the terrestrial environment and estimates of 5 to 50 million species on earth are often offered (Wilson, 1988; World Resources Institute, 1986; Freedman, 1998). The vast majority of terrestrial species are arthropods, and most of them are believed to live in tropical forests (Raven and Johnson, 1992). However, due to human activity, global species diversity is now decreasing at its fastest rate in 65 million years (Hayes, 2000), and may represent the sixth great extinction of earth's history. There are three major causes for the loss of biodiversity: (1) habitat destruction, (2) invasion of exotic species, and (3) overutilization pressures from global trade and population growth. Economic development has brought the most important threat to diversity with the destruction of habitats. Many of the richest areas for endemic plants and animals are located in prime locations for urban or agricultural development, or forest harvesting. Most biodiversity hotspots (Myers et al., 2000) across the world are part of these systems targeted for exploitation. However, there are major consequences in destroying these habitats. Reduced biodiversity leads to reduced productivity and stability of the natural system. Plants and animals containing potential new medicines may be lost forever (Chivian, 2001). However, our utilitarian view of the world's resources pushes developers and policymakers to adopt strategies that can be detrimental to the survival of other species on earth. The amount of tropical forests lost every year is so great that it has an impact on the global circulation system and influences the long-term rates of precipitation and cycling of CO_2. Without vegetation (and tropical forests have a major role to play in this), the level of CO_2 in the atmosphere could be increasing more dramatically than expected (Freedman, 1998). Human activities have altered the atmosphere and ozone layer and thus affected the capacity of the other species to function or even survive.

When species abundance changes, the effect of its contribution to the overall community also shifts, affecting other species and the whole ecosystem as a result. For example, the carnivorous sea otters along Canada's West Coast were overexploited almost to extinction by hunting. Species of herbivorous sea urchins, unchecked by otter carnivory, multiplied and reduced shoreline kelp forests, causing drastic changes to species dependent upon kelp. The return of sea otters on the West Coast is now bringing forth a resurgence of the many ecological functions supported by the great kelp beds (Draper, 1998).

The loss of species diversity can have major impacts on human health. The ecology of the Argentinean Hemorrhagic Fever (AHF) provides a good illustration of the links between biodiversity, ecosystem health, and human health (Morse, 1993). During the First World War, a decrease in grain import from Europe prompted Argentina to become self-sufficient in its grain production. A program to convert the pampas grasslands of Argentina to grain production was initiated. In this effort, the pampas grasslands, a biodiverse and established natural ecosystem, were reduced to maize-only production. The canopy of the maize reduced the balance of solanophilic and shade-loving grasses to one of shade-loving grasses only. The field mice species that grazed on shade-loving grasses expanded rapidly and essentially became the dominant species (previously the solanophilic grass feeding species were dominant). However, these field mice carried a virus that was capable of causing a highly virulent disease in humans, and the population explosion exposed field workers to the disease and outbreaks began to occur in fieldworkers and nearby villages. It took approximately 7 years to analyse the situation and discover these linkages. In this case, because of a reduction in biodiversity and complexity of the ecosystem, an imbalance between species led to an outbreak of a highly infectious virus in humans. This example dramatically shows the level of interdependency between ecosystem health and human health.

From the discovery of new medicines, food, or fibres to the conservation of primitive or wild genes of economically important species, humans and ecosystems will require biodiversity to maintain the long-term health of human populations. In areas where environmental conditions are degraded or naturally extreme, modern agricultural systems can prove inadequate for the survival of human populations. In India, for example, traditional tribes are returning to their primitive varieties of seeds, to reduce starvation and death from hunger (Norberg-Hodge, 1999). In other countries, without a certain level of productivity from the natural ecosystem, populations would not be able to survive. These ecosystems produce their food and traditional medicines. But mainly due to current trends towards global trade and market pressures, policymakers have tendencies to over-exploit their natural resources and ecosystems. The approval by the Brazilian government to cut half of its remaining tropical rainforests shows how politics of special interest and greed as well as a lack of support for biodiversity as a function of human health can lead to dramatic ecosystem degradation. While this decision will bring economic benefits in the short term, it will lead to the destruction of ecosystems and higher levels of poverty for the local inhabitants in the longer term. Under those conditions, it is expected that infectious diseases and starvation will become severe problems for these local people.

7. Discussion

We are all aware of examples of how humans are degrading their natural environment. In recent years, this concern has grown and spread throughout much of the world (Mullin, 2000; Harper, 2001). While Rachel Carson (1962) raised our consciousness about the devastating impacts of pesticides, the position of most governments regarding the environment does not seem to have changed. The rapid acceleration of technological growth associated with industrial and post-industrial society poses a risk to the health of our ecological systems. These small but incremental negative impacts result from our use of the environment strictly for our needs with no regard for the sustainability of the global ecosystem. There are three main reasons why humans should value the environment: instrumental, aesthetic, and ethical. Until people realise that overuse of ecosystems leads to their degradation and unhealthy conditions, and appreciate ecosystems as having an ethical and aesthetic "raison d'être", ecosystem health and sustainable development will be unachievable goals.

Maintaining healthy ecosystems will require responsive policy frameworks and infrastructures conducive to corrective actions and environmentally friendly management choices. A comprehensive assessment of ecosystem health requires simultaneous consideration of at least six major elements of the regional ecosystem: biophysical integrity, human health, economic factors, innovative technologies, social acceptability, public awareness, and public policy and ethics. All these aspects must be viewed not as competing interests but as integral components of a single complex system characterized by continual novelty, contradiction, surprise, and complex dynamics (Funtonowicz and Ravetz, 1994).

Although the general public and policymakers are not always looking at their decisions and actions as affecting the health of ecosystems, it is obvious that the links are there. In fact, since humans are so intimately linked to the environment, any type of action that humans take will invariably have an impact on the environment and influence the health of the system (Figure 1, above).

In order to maintain a healthy ecosystem, development and management activities should be capable of maintaining all the properties previously mentioned. This seems relatively easy and attainable. However, too little is done and measures to maintain ecosystem health remain a global priority. The debate concerns the entire planet as a whole especially for the billions of less fortunate humans whose children face a possible future of dying or staying alive by destroying the last resources. An often-quoted statistic is that some 40000 people die of starvation every day. However, because these deaths are almost exclusively in developing countries, little public attention is brought to this matter, except in extreme cases where famine grips a large area in a short period of time and millions teeter on the verge of starvation (as has happened recently in the Sudan, and in Ethiopia). The problem of starvation and lack of space in several (mostly developing) countries will not disappear over night. While we can claim that maintenance of healthy ecosystems is essential, we miss the point if we forget the current issue of population growth. Can we really have effective policies for ecosystem health without curbing population growth? Solutions (such as redistribution of populations, sterilisation campaigns, or one child per family policy) presented or enforced by scientists, decisionmakers, and others are not popular and remain arguable by various groups. But, this will still come back to the agendas of most countries.

The challenge that decisionmakers have to face and that literally thousands of reports and books have pointed out is the need for greater sharing between the rich and the poor to help maintain a healthy ecosystem. The sharing should go beyond money to include scientific knowledge, technology, management skills, education, etc. An informed, environmentally committed public can provide the incentives policymakers and industry need to move towards healthier lifestyles and ecosystems (Shabecoff, 2000). Pressures coming from the public, as shown in the past few years at world meetings can push world leaders to change the global agenda. However, there are other opportunities to improve world environmental decisions, environmental literacy and local initiatives being good examples. One of the problems is that the wealthier countries are expected to donate and the poorer countries dislike being told how to manage the money that they receive from the wealthy countries. The Group of 8 (G8) however needs the resources coming from the Group of 77 (in fact there are 130 poor countries in the world). How can these challenges be resolved? Commissions after commissions and the Rio de Janeiro (Brazil) Summit were different attempts to resolve the global debate. The Stockholm Conference in 1972, and later work such as the Brundtland Commission (in 1983; The World Commission on Environment and Development, 1987) produced the first drafts of what sustainable development should be and how the problem could be solved. The Rio Summit was the largest world conference in human history and from this conference, Agenda 21, the blueprint for a sustainable society on Earth, was produced. Are we meeting the expectations? The second chapter of this theme examines the various gaps towards a better connection between ecosystem health and human health, the source of solutions, and the priority actions that should be undertaken for better living on a healthy planet.

Acknowledgements

The authors would like to thank the members of the Working Group Organizing Committee, for their contribution and commitment to the development of this theme: Mohi Munawar, Diane Malley, Andrew Hamilton, Ken Minns, Sharon Lawrence and David Rapport. As well, invited contributions from Robert McMurtry, John Howard, Robert Lannigan, Dieter Riedel, William Fyfe, John Cairns, Jr., and Thomas Edsall were central to the development of the theme discussions.

References

Ahmed, M., Navy, H., Vuthy, L. and Tiongco, M., 1998, Socioeconomic Assessment of Freshwater Capture Fisheries of Cambodia – Report on a Household Survey (Mekong River Commission, Phnom Penh, Cambodia) 186 pp.

Brown, L.R., 1998, Struggling to raise cropland productivity. In: L.R. Brown, C. Flavin, H.F. French, J. Abramovitz, C. Bright, S. Dunn, G. Gardner, A. McGinn, J. Mitchell, M. Renner,

D.M. Roodman, J. Tuxill, and L. Starke (Editors), State of the World 1998 (W.W. Norton and Company, New York, for Worldwatch Institute) pp. 79–95.

Brown, L.R., Flavin, C. and Postel, S., 1989, Foreword. In: L.R. Brown, A. Durning, C. Flavin, L. Heise, J. Jacobson, S. Postel, M. Renner, C. Pollock Shea and L. Starke (Editors), State of the World 1989 (Earthscan Publication Ltd, New York) pp. xv–xviii.

Callicott, J.B., 1992, Aldo Leopold's metaphor. In: R. Costanza, B.G. Norton and B.D. Haskell (Editors), Ecosystem Health: New Goals for Environmental Management (Island Press, Washington, DC) 269 pp.

Carson, R., 1962, Silent Spring (Houghton Mifflin, Boston) 368 pp.

Chivian, E., 2001, Environment and health: 7. Species loss and ecosystem disruption – the implications for human health. CMAJ 164:66–69.

Colwell, R.R., Epstein, P.R. and Gubler, D., 1998, Climate change and human health. Science 279:968–969.

Convention on Biological Diversity, 1994, Convention on Biological Diversity. Text and annexes (Interim Secretariat of the Convention on Biological Diversity, Geneva) 34 pp.

Costanza, R., Norton, B.G. and Haskell, B.D. (Editors), 1992, Ecosystem Health: New Goals for Environmental Management (Island Press, Washington, DC) 269 pp.

Draper, D., 1998, Our Environment: A Canadian Perspective (ITP Nelson, Scarborough) 499 pp.

EPA, 2001, Global warming – health impacts, www.epa.gov/globalwarming/impacts/health/index.html.

Epstein, P.R., 2000, Is global warming harmful to health? Sci. Am. 283 (August 2000): 50–57 (available at http://www.sciam.com/2000/0800issue/0800epstein.html).

Freedman, B., 1998, Environmental Science: A Canadian Perspective (Prentice-Hall Canada, Scarborough) 568 pp.

Funtonowicz, S. and Ravetz, J.R., 1994, Emergent complex systems. Futures 26:568–576.

Government of Canada, 1996, The State of Canada's Environment – 1996 (Supply and Services Canada, Ottawa) 825 pp.

Harper, C.L., 2001, Environment and Society. Human Perspectives on Environmental Issues (Prentice Hall, Upper Saddle River) 467 pp.

Haskell, B.D., Norton, B.G. and Costanza, R., 1992, What is ecosystem health and why should we worry about it? In: R. Costanza, B.G. Norton and B.D. Haskell (Editors), Ecosystem Health: New Goals for Environmental Management (Island Press, Washington, DC) 269 pp.

Hayes, D., 2000, Mobilizing to combat global warming. World Watch 13:6–7.

Hengeveld, H.G., 2000, Projections for Canada's Climate Future. (Meteorological Service of Canada, Environment Canada, Downsview, Ont.) 54 pp.

Homer-Dixon, T.F., 1999, Environment, Scarcity, and Violence (Princeton University Press, Princeton, NJ) 280 pp.

Kraft, M.E., 2001, Environmental Policy and Politics (Addison-Wesley Longman, New York) 286 pp.

Krause, D., 1993, Environmental consciousness: an empirical study. Environ. Behaviour 25:126–142.

Leopold, A., 1941, Wilderness as a land laboratory. Living Wilderness 6:3.

Likens, G.E., 1992, The Ecosystem Approach: Its Use and Abuse (Ecology Institute, Germany) 259 pp.

Lindgren, E., 1998, Climate change, tick-borne encephalitis and vaccination needs in Sweden – a prediction model. Ecol. Model. 110:55–63.

McMichael, A.J., 1993, Global environmental change and human population health: a conceptual and scientific challenge for epidemiology. Int. J. Epidemiol. 22:1–8.

Morse, S.S. (Editor), 1993, Emerging Viruses (Oxford University Press, Oxford) 317 pp.

Mullin, D., 2000, Environmental Psychology: Human Behaviour and the Natural Environment (NSEIA, Halifax, NS) 147 pp.

Myers, N., Mittermeier, R.A., Mittermeier, C.G., da Fonseca, G.A.B. and Kent, J., 2000, Biodiversity hotspots for conservation priorities. Nature 403:853–858.

Norberg-Hodge, H., 1999, Reclaiming our food: reclaiming our future. Ecologist 29:209–214.

Patel, A. and Rapport, D.J., 2000, Assessing the impacts of deer browsing, prescribed burns, visitor use and trails on an oak-pine forest: Pinery Provincial Park, Ontario, Canada. Nat. Areas J. 20:250–260.

Patz, J.A., 2000, Climate change and health: new research challenges. Ecosyst. Health 6:52–58.

Postel, S., 1994, Carrying capacity: earth's bottom line. In: L. Brown, C. Flavin, S. Postel and L. Starke (Editors), State of the World 1994 (Earthscan Publication Ltd, New York) pp. 3–21.

Postel, S., 1996, Forging a sustainable water strategy. In: L.R. Brown, J. Abramovitz, C. Bright, C. Flavin, G. Gardner, H. Kane, A. Platt, D.M. Roodman, A. Sachs and L. Starke, State of the World 1996 (Worldwatch Institute) pp. 40–59.

Primack, R.B., 1993, Essentials of Conservation Biology (Sinauer Associates, Sunderland) 569 pp.

Rapport, D.J. and Whitford, W.G., 1999, How ecosystems respond to stress: common responses of aquatic and arid systems. BioScience 49:193–202.

Rapport, D.J., Gaudet, C.L. and Calow, P., 1995, Evaluating and Monitoring the Health of Large-Scale Ecosystems (Springer, Berlin).

Rapport, D.J., Costanza, R. and McMichael, A.J., 1998a, Assessing ecosystem health: challenges at the interface of social, natural, and health sciences. Trends Ecol. Evol. 13:397–402.

Rapport, D.J., Christensen, N., Karr, J.R. and Patil, G.P., 1998b, The centrality of ecosystem health in achieving sustainability in the 21st century: concepts and new approaches to environmental management. In: D.M. Haynes (Editor), Human Survivability in the 21st Century, Trans. R. Soc. Canada Ser. VI, IX:3–40.

Rapport, D.J., Fyfe, W.S., Costanza, R., Spiegel, J., Yassi, A., Bohm, G.M., Patil, G.P., Lannigan, R., Anjema, C.M., Whitford, W.G. and Horwitz, P., 2001, Ecosystem health: definitions, assessment and case studies. In: M.K. Tolba (Editor), Our Fragile World: Challenges and Opportunities for Sustainable Development (Eolss Publishers, Oxford) 2300 pp.

Rapport, D.J., Howard, J., Lannigan, R., McMurtry, R., Jones, D.L., Anjema, C.M. and Bend, J.R., 2002, Thinking outside the box: introducing ecosystem health into undergraduate medical education. In: A.A. Aguirre, R.S. Ostfeld, C.A. House, G.M. Tabor and M.C. Pearl (Editors), Conservation Medicine: Ecological Health in Practice (Oxford University Press, New York) in press.

Raven, P.H. and Johnson, G.B., 1992, Biology (Mosby Year Books, St. Louis, MO) 1192 pp.

Sachs, J.P., Anderson, R.F. and Lehman, S.J., 2001, Glacial surface temperatures of the Southeast Atlantic Ocean. Science 293:2077–2079.

Sadalla, E.K. and Krull, J.L., 1995, Self-presentational barriers to resource conservation. Environ. Behaviour 27:328–353.

Sauvé, L., 1999, Environmental education between modernity and postmodernity: searching for an integrating educational framework. Can. J. Environ. Educ. 4:9–35.

Serreze, M.C., Walsh, J.E., Chapin III, F.S., Osterkamp, T., Dyurgerov, M., Romanovsky, V., Oechel, W.C., Morison, J., Zhang, T. and Barry, R.G., 2000, Observational evidence of recent change in the northern high-latitude environment. Clim. Change 46:159–207.

Shabecoff, P., 2000, Earth Rising: American Environmentalism in the 21st Century (Island Press, Washington, DC) 224 pp.

The World Commission on Environment and Development, 1987, Our Common Future, Bruntland, G.H. (Chair) (Oxford University Press, Oxford) 383 pp.

United Nations Development Program, United Nations Environment Program, World Bank and

World Resources Institute, 2000, World Resources 2000–2001. People and Ecosystems: the Fraying Web of Life (Elsevier, New York) 400 pp.

Vasseur, L., 2000a, Impact of recreational activities in the Kejimkujik campground. Recommendations for future management and conservation, Research report for Kejimkujik National Park (Parks Canada, NS).

Vasseur, L., 2000b, Sustainability of the freshwater aquatic ecosystem in Cambodia: should we change the conditions? Presented at EcoSummit 2000, Halifax, June 18–22.

Wilson, E.O. (Editor), 1988, Biodiversity (National Academy Press, Washington, DC) 521 pp.

Wilson, E.O., 1990, Threats to biodiversity. In: Managing Planet Earth: Readings from Scientific American (W.H. Freeman, New York) pp. 49–59.

Woodward, A., Hales, S. and Weinstein, P., 1998, Climate change and human health in the Asia pacific region: who will be most vulnerable? Clim. Res. 11:31–38.

World Health Organization, 1997, Health and Environment in Sustainable Development: Five years after the Earth Summit: Executive Summary. Available at http://www.who.int/environmental_information/Information_resources/htmdocs/execsum.htm.

World Resources Institute, 1986, World Resources 1986: An Assessment of the Resources Base that Supports the Global Economy (Basic Books, New York) 582 pp.

Consensus

Chapter 10

Ecosystem Health and Human Health: Healthy Planet, Healthy Living[*]

L. Vasseur, P.G. Schaberg, J. Hounsell, with P.O. Ang Jr., D. Cote, L.D. Duc,
J.S. Ebenezer, D. Fairbanks, B. Ford, W. Fyfe, R. Gordon, Y. Guang, J. Guernsey,
A. Hadi Harman Shaa, A. Hamilton, W. Hart, H. Hong, J. Howard, B. Huang,
Y. Huang, D. Karnawati, R. Lannigan, S. Lawrence, Z. Li, Y. Liu, D. Malley,
L. McLean, R. McMurtry, V. Mercier, N. Mori, M. Munawar, M.A. Naragdao,
K. Okamoto, D. Rainham, D. Riedel, E. Rodriguez, M. Saraf, H. Savard, N. Scott,
A. Singleton, R. Smith, H. Taylor, N.T. Hoang Lien, S. Xing and H. Xuan Co

Abstract

The links between human health and ecosystem health are clear for many people but inaction to bring a balance between the two is still omnipresent among decisionmakers and certain parts of our societies. There is a need for concerted efforts to first educate and inform all people in the world about these links and the fragility of the ecosystems in which we live. While some ecosystems might be able to restore their health without human interventions, it is clear that others may need our help. There are several potential solutions, the challenge being to engage the world in the implementation of these actions. This chapter explores some of these solutions and potential actions. These priority actions were in many cases proposed by the discussion group at the EcoSummit. There was a strong recognition of these priorities. The discussion group passed a resolution on conserving, protecting, and enhancing ecosystem health and human health. The main goal of our society should be towards a healthy planet and healthy living.

---

[*] We would like to dedicate this chapter to Andrew Hamilton, our working group rapporteur, who has worked on linking human health and the environment for many years. Thank you for your eloquent summary of our discussions, found in the resolution at the end of this chapter. It brought all our thoughts together in a common vision.

Understanding and Solving Environmental Problems in the 21st Century
Edited by R. Costanza and S.E. Jørgensen
© 2002 Elsevier Science Ltd. All rights reserved

1. Introduction

Humans are altering the earth's environment in dramatic and increasingly pervasive ways. Through a variety of enterprises (e.g., industry, agriculture, recreation, and international commerce), humans are transforming fundamental natural processes such as climate, biogeochemical cycling, and even the biological diversity upon which evolutionary changes depend (Vitousek et al., 1997). Accompanying mounting evidence of human domination of the earth's ecosystems is growing recognition that ecosystem health is a critical prerequisite of human health and well-being (Cortese, 1993). Clearly, without key ecosystem services such as providing clean air, clean water, food, and a hospitable climate, life (human and other) cannot exist. In addition, the role that other ecological functions (e.g., disturbance regulation, biological control, and organic matter decomposition) play in supporting long-term biological health and stability has gained broader recognition (Costanza et al., 1997). The better understanding of human linkages to natural and altered ecosystems has been fostered, in part, through a growing number of research and case studies from around the world. Collectively, efforts like these have helped to define the risks posed by a myriad of long-recognized (e.g., asbestos, heavy metal, organic solvent exposures, etc.) and emerging (e.g., endocrine disruption, antibiotic resistance, interactive toxicity, etc.) ecosystem and human health issues.

Scientific evidence has helped illustrate some of the human and ecological costs of environmental degradation. However, it is unlikely that contemporary awareness fully incorporates all future threats posed to ecosystem and human health. In fact, current environmental issues may well be just a prelude of the consequences yet to come. The ever-increasing human population, along with an explosive increase in consumption, exacerbate and enlarge the footprint pressing on aquatic and terrestrial ecosystems as a result of land use from agriculture and human settlement, natural-resource use, activities such as transportation and recreation, and wastes from domestic, municipal, and industrial development. Even disregarding the possibility that interactions or synergies among the drivers and effects of anthropogenic ecosystem disruption could hasten system degradation, it seems clear that further erosion of ecosystem and human health is likely barring meaningful human response. Less certain, however, is how possible concurrent reductions in biological stress response systems might alter current trends. For example, evidence is accumulating that the widespread dispersal of synthetic chemicals may be altering animal hormone and immune systems that are needed for normal growth, development and disease resistance (Soto et al., 1996; Nilsson, 2000). Other work suggests that anthropogenic disturbance of ecosystem nutrient relations may diminish the capacity of plants to sense, respond to, and/or survive an array of environmental stresses (DeHayes et al., 1999). In addition, reductions in biodiversity could eliminate keystone species,

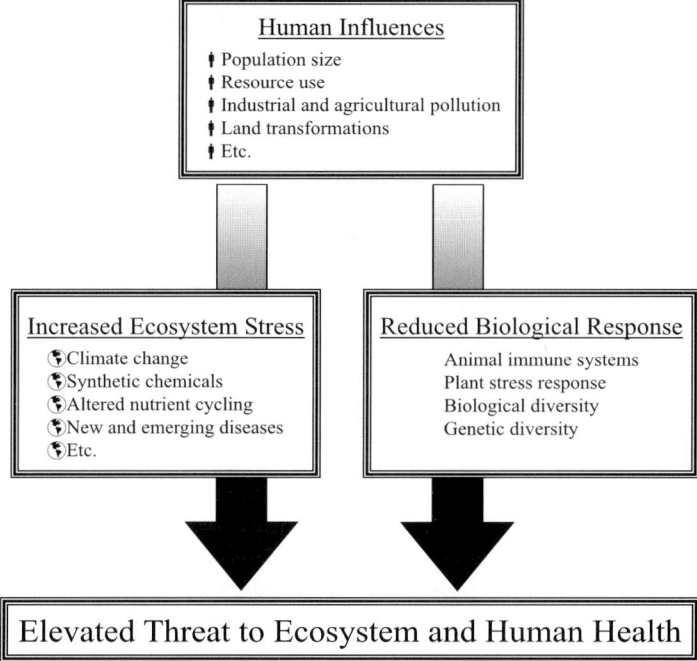

Fig. 1. Conceptual model of the dual threat (increased stress and reduced response/resistance that human population growth and activity pose to ecosystem and human health.

create discontinuities in ecosystem nutrient and energy pathways, and enhance vulnerabilities to significant losses of ecosystem services following natural or man-made disturbance (Tilman, 2000). Losses of genetic diversity within species would also reduce the capacity of populations to successfully adapt to environmental change (DeHayes et al., 2000). Especially if environmental perturbations are pronounced (e.g., as predicted under some climate-change scenarios), disruptions in the ability of biological systems to accommodate change might have dramatic ecological, health, economic, and social impacts.

In essence, human activity may be subjecting individuals and ecosystems to unprecedented levels of environmental stress by diminishing biological mechanisms to successfully respond to them (fig. 1). In addition, social conflicts (e.g., growing inequities in income distribution and consumerism) can unfold to exacerbate detrimental impacts to vulnerable populations. These combined factors are likely to threaten the function, health, and sustainability of ecosystems worldwide with a speed and intensity never before realized. Our knowledge of the causes and consequences of unprecedented human intervention in the natural world is growing. Yet, much may remain unknown until the current "experiment" with global health progresses further.

The threat to ecosystem health and human health posed by anthropogenic activities is real. Contemporary examples of damage are well documented, and current predictions of future impacts are supported by sound science. However, predictions of escalating harm are likely valid only in the face of inadequate human response. But what are the *appropriate* responses to the complex environmental problems we face?

This chapter summarizes the discussion and analysis of contemporary ecosystem health problems that occurred during the EcoSummit 2000 in the Working Group on Ecosystem Health and Human Health. The working group consisted of a heterogeneous amalgamation of participants from over 12 countries and a wide range of disciplines (including scientists from social, economic, biological, engineering, environmental, and medical specialties, as well as clergy, policymakers, and students representing a wide range of disciplines), who collectively brought a diverse mix of perspectives to the discussions. The aim of this EcoSummit was to encourage integration of both the natural and social sciences with the policy- and decisionmaking community for the purpose of developing a deeper understanding of complex environmental problems. A foundational premise was that this enhanced understanding is a necessary prerequisite for the debate and action needed to build a sustainable future that protects ecosystem and human health.

This chapter's structure mirrors the progression of discussion for the working group. To initiate the cycle of discussion, and help focus interactions, five broad questions were sequentially examined. These questions were:

(1) What are the linkages between human health/disease and ecosystem health?
(2) What are technological, social, political, and economic sources of solutions to the problems?
(3) What are the priority actions that should be taken to protect, preserve, or restore the health of ecosystems and growing human populations?
(4) What are the barriers to effective action?
(5) What are useful measures, indicators, or metrics of progress?

While the first question overlaps the general overview presented in the previous chapter on ecosystem and human health, several arguments are included here to reinforce the conceptual and functional linkages between ecosystem and human health. Because the questions discussed involved material of a highly intricate nature and vast scope, additional focus was provided by primarily examining these questions within the context of three global issues that all participants could relate to: climate change, agrosystems and food production, and biodiversity. As a capstone to discussions, a statement was written to summarize the positions and priorities that emerged, and to provide a unified call to improve communication and actions that foster ecosystem health and human health. This statement, written in the form of a resolution, closes the chapter[1].

2. Linkages between ecosystem health and human health

Underpinning group discussion was the shared view that human health and ecosystem health are inexorably linked and interdependent. "The atmosphere, fertile soils, freshwater resources, the oceans, and the ecosystems they support, play a key role in providing humans with shelter, food, safe water, and the capacity to recycle most wastes" (World Health Organization, 1997). This section underscores some of the linkages between key natural resources (i.e., air, water, food, soil, and biological diversity) and ecosystem/human health targeted during group discussions. Although examples of the health costs of ecosystem degradation are many and can be found throughout the world, this section highlights examples from Canada (the host country for EcoSummit 2000) and other regions most familiar to discussion participants. In addition, to adequately address the complexity and diversity of the environmental problems the world faces, the group proposed that new paradigms for understanding the linkages between ecosystem and human health were needed. Accordingly, this section on linkages closes with a call from the group for an expanded perspective when considering models for integrating environmental and biological health.

2.1. Air quality

Air quality can be examined at the local or global level, as airborne pollutants can travel long distances linking ecosystems from various regions of the world. Unfortunately, there are ecosystems that are more vulnerable to air pollutants than others because (due the vagaries of air currents and/or topography) they act as sinks for the residues of atmospheric contamination. As a result, these ecosystems can suffer elevated health burdens. For example, the Arctic ecosystem is highly vulnerable to long-range transport pollutants, which have contaminated all nodes of the food web. First Nations and Inuit peoples are two groups whose homelands have been disproportionately affected by this ecosystem contamination (Commoner et al., 2000). Generally in Canada, however, air quality has been improving. And, although average annual levels of ground-based ozone increased 29% between 1979 and 1993 (Health Canada, 1997), the number of days for which ground-level ozone posed a high health risk generally declined in Canada over the last 25 years (although there is a large fluctuation from year to year due to

---

[1] A note to the reader: although citations occasionally appear in this chapter, they are included simply as a supplement to the review of group discussion presented herein.

weather-related and other sources of variation). Despite this improvement, high ozone levels still occur frequently in certain parts of Canada when episodes of hot, stagnant weather coincide with periods when local and transported emissions of precursor contaminants are high (Environment Canada, 1999).

Air pollution poses an overwhelming global threat to ecosystem and human health. Worldwide, an estimated 3 million deaths are attributed to air pollution (World Health Organization, 1997). In Canada, the number is approximated at 5000 annually (Environment Canada, 2001). Through a myriad of ways (including reduced lung capacity, eye, nose, and throat irritation, aggravation of lung and heart disease, etc.) most Canadians can expect to be personally impacted by poor air quality at some time in their life (Health Canada, 1997).

Although poorly quantified, air pollution has also altered the health of sensitive forest ecosystems (e.g., see Mickler et al., 2000), perhaps forewarning a broader disruption in ecosystem structure and function as the native resiliency and buffering capacity of systems are depleted. The impacts of air pollution can be direct (e.g., ozone injury to leaf physiology and tree health), or can involve chemical transformations in the environment that can lead to broader ecosystem disruption. For example, pollutants such as sulfur dioxides and nitrogen oxides can react with atmospheric moisture to create sulfuric and nitric acids, which precipitate to the earth as acid deposition. These acidic inputs can lower the pH levels of lakes and other surface waters, increase the solubility of metals such as aluminum and mercury that usually have low biochemical availability, reduce the number of plant and animal species that aqueous ecosystems can support, and can thereby disrupt the energy and nutrient relations of surrounding environments. Similarly, pollutant-induced depletion of the protective ozone layer is now allowing more UV radiation to reach the earth's surface. Because DNA absorption of UV increases the rate of genetic mutation, ozone depletion increases the risks of skin cancer among humans and may be one factor raising mutation rates in vulnerable wildlife species such as frogs (Wardle et al., 1997). In these and many other ways, human disruption of one resource can "spill over" and influence other critical life-support systems.

2.2. Water resources

Canada contains 9% of the Earth's fresh water supply (Environment Canada, 2000). Although most Canadians have access to one of the safest drinking water supplies in the world, water reserves here are not uniformly free of contamination. For example, as noted for air quality, First Nations and Inuit communities have a disproportionate exposure to contaminated water supplies. In fact, the incidence of waterborne diseases is several times higher for Aboriginal communities than for the broader Canadian public (Federal, Provincial and Territorial Advisory Committee on Population Health, 1999). However, the greater tendency for rural native peoples to experience tainted water supplies does not indicate that water systems in the rest of Canada are free from problems. For example, an *E. coli* outbreak in the water supply of Walkerton, Ontario, Canada – a small rural town – left 7 dead and over 2000 ill. This tragedy prompted an increased analysis of ground water throughout Canada, and showed that, for a variety of reasons, microbial and chemical ground water contamination in the entire country was far worse than previously assumed. On a global scale, it is now predicted that limitations of fresh potable water will be a focal point of human health concerns and a significant contributor to regional conflicts as disputes over water resources escalate (Postel, 1999). According to the WHO, "over 1000 million people do not have access to an adequate supply of safe water for household consumption" (World Health Organization, 1997). Although already extremely high, the number of people living with inadequate clean water supplies will likely rise dramatically in the near future as human population growth and increased water use outstrip supply capacities. Water deficits are predicted to be particularly large in certain locations including India, China, parts of the USA, and Africa (Postel, 1999). Water shortages increase the diversion of dwindling natural reserves to supply human demands, with urban/manufacturing applications typically favored over agricultural uses. Whatever the use, a greater human sequestration of water results in a reduction of supplies left to support ecosystem function and health.

Dams, levees, and other forms of hydraulic infrastructure exist because they serve real human needs (e.g., increased water supplies for drinking, irrigation and industry, flood control, and hydroelectric power). However, these same waterways have historically served important ecological functions such as the buffering of floodwaters, assisting with nutrient cycling and dispersion while maintaining salt and sediment balances, protecting wetlands and their ability to absorb pollutants, and providing critical habitat for a diverse array of aquatic species (Postel, 1999). Unfortunately, as currently designed and operated, engineered water control systems almost exclusively fulfill human needs and rarely accommodate other ecosystem services. This apparent schism between meeting both human and ecosystem needs has increased relevance in light of the substantial and growing control of humans over freshwater resources (Postel, 1999). For example, it has been estimated that about 77% of the river systems in the USA, Canada, Europe, and the former Soviet Union are moderately to strongly altered by dams, reservoirs, diversions, and irrigation (Dynesius and Nilsson, 1994). In fact, human diversion and exploitation of water resources is so great that little or no fresh water reaches the sea during parts of the year for numerous river systems worldwide including the Yellow (China), the Colorado (North America), the Ganges and Indus (South Asia), the Amu Darya and Syr Darya (Central Asia), and the Nile (northeast Africa) rivers (Postel, 1999). Among its many impacts, overexploitation of freshwater resources has severely threatened dependent plant and animal species, and helps to account for the elevated risk of extinction for fish species worldwide relative to other life forms (Postel, 1999).

2.3. Food resources

Worldwide, food is one of the major routes of exposure for many pathogens and toxic chemicals (World Health Organization, 1997). The increased interdependency and complexity of the world's food supplies has helped to strain production and distribution systems, and contributed to recent increases in foodborne disease (World Health Organization, 1997). In Canada, the food supply is generally safe. However, as a result of food-borne bacterial contamination, an estimated 10 000 cases of food-related illness are reported every year (Health Canada, 1997). Other forms of contamination, although currently rare, are also of increasing concern. Food also accounts for 80–95% of our total daily intake of persistent organic pollutants, including PCBs, dioxins, furans, and PAHs (Health Canada, 1997). As with air and water resources, First Nations and Inuit people suffer disproportionately from food-borne pollutant exposures. In particular, their traditional diet of fish and marine mammals places them more at risk of exposure to environmental contaminants such as PCBs and mercury (Health Canada, 1997). For instance, PCB concentrations in breast milk of Inuit women in Northeastern Quebec are five-fold higher than levels of PCBs in breast milk of women in Southern Quebec (Health Canada, 1997; Commoner et al., 2000). The most probable source of airborne contaminants such as PCBs and dioxins is long-range atmospheric transport, with Arctic ecosystems being major sinks where pollutants bioaccumulate in food chains affecting entire ecosystems from algae to humans (Commoner et al., 2000). However, fish and game contamination is not only of concern to people living in polar ecosystems, but is relevant in many regions where diets include game species (e.g., Langlois et al., 1995; Tsiji et al., 1999), and has been documented for some commercial foods as well. For example, the US Food and Drug Administration recently advised that shark, swordfish, king mackerel, and tilefish (all long-lived, large fish that feed on smaller fish) can accumulate high levels of methyl mercury. Consequently they recommended that vulnerable segments of the population (i.e., pregnant women, women of childbearing age, nursing mothers and young children) not eat these commercially available fish species (US Food and Drug Administration, 2001).

Identified contamination results in consumer advisories that hopefully limit direct health risks to human consumers. But what about contamination that is not officially documented? Are humans cautioned against these possible risks? And what risks do heavy-metal and synthetic-chemical accumulations pose to ecosystem health and function? The potential ecological damage of chemical bioaccumulation in aquatic food chains is well documented, and research is now assessing the risk to terrestrial systems (e.g., Lasorsa and Allen-Gil, 1995). Still, information on the impacts of certain forms of contamination on ecosystem function and health has long been reported. Indeed, Rachael Carson's groundbreaking book *Silent Spring* helped instigate the modern environmental movement by exposing the untold threat that pesticides like DDT pose to ecosystem food chains and the health of associated wildlife populations. More recent works have expanded the scope of concern about widespread pesticide contamination to include a number of emerging issues including the possible destruction of beneficial natural predators and parasites, and the ecological consequences of pesticide resistance in pests (Pimentel et al., 1992). Even with recent gains in understanding, the long-term, cumulative, interactive, and potentially synergistic impacts of pesticides and other forms of chemical/heavy metal contamination on ecosystem health remain far from resolved.

2.4. Soils

Soil plays a critical role in the Earth's life support system. Canada has approximately 5% of its total land area that is suitable for agricultural purposes, and only one-half of this is considered prime agricultural land (Acton and Gregorich, 1995). There are many factors reducing the quality of the soil in Canada, including erosion, loss of organic matter, compaction, as well as urban sprawl, the increased use of agricultural chemicals, and unsound waste management practices (Acton and Gregorich, 1995). An extreme example of land degradation is desertification. While droughts and fires can be considered natural causes for desertification, increasing human density, livestock ranching, fuelwood harvests, and deforestation are more important drivers of desertification than was previously recognized (Barrow, 1991). Grassland ecosystems in all regions of the world are subject to overexploitation and unfavorable weather patterns, which can also promote desertification. Desertification induces an obvious and dramatic alteration to ecosystem structure, function, and health, which translates into dire problems (e.g., possible malnutrition, starvation, refugee and migration issues, associated increases in communicable disease, and death) for dependent human societies.

2.5. Biodiversity

Human activity is driving the extinction of plant, animal, and microbial life at a rate thousands of times above estimated natural levels (Chivian, 1993). Indeed, human alteration of the global environment has likely triggered the sixth major extinction event in the history of life on earth (Chapin et al., 2000). Currently, in the USA, the major threats to biodiversity are thought to be habitat destruction and degradation, alien species, pollution, over-exploitation, and disease (in descending order of importance) (Wilcove et al., 1998). Locally and even regionally, the relative influence of these drivers of species loss often varies. In Canada, for instance, the impact of alien species may be somewhat elevated: it is estimated that about 25% of plant life in Canada is now of non-native origin (Vitousek et al., 1997). Examples such as the proliferation of zebra mussels and lamprey eels in the Great Lakes also indicate that this threat is not limited to the plant kingdom. Furthermore, as humans continue to alter the planet, additional factors could become important drivers of biodiversity loss. By the year 2100, climate change, nitrogen deposition, and increased atmospheric carbon-dioxide levels are predicted to become major influences on the viability of species worldwide (Sala et al., 2000). To complicate matters more, human populations are often concentrated in regions with unusually high levels of native biodiversity, putting particular strain on these biological reserves. In fact, it is estimated that nearly 20% of the world's population now lives within biodiversity hotspots (Cincotta et al., 2000).

The consequences of this enormous loss of global biodiversity to ecosystem and human health are likely to be staggering. Because biological activity integrated among a vast array of life forms helps mediate energy and material fluxes through ecosystems and even influences important abiotic conditions (like resource limitations, disturbance, and climate), reduced biodiversity could measurably alter fundamental ecosystem services that sustain life (Chapin et al., 2000). It could endanger the continued productivity of natural systems (including ones that humans rely on for food and fibre generation) by reducing their resilience to disturbance and increasing overall susceptibility to disease outbreaks (United Nations Development Program et al., 2000). In addition to its disruptive effects on ecosystem equilibrium and productivity, dramatic losses of biodiversity would undoubtedly diminish human health by hindering potential medical gains. Through the loss of important medical models (e.g., species with unusual physiological characteristics such as elevated cancer resistance which are studied to better understand human physiology and disease), and the elimination of plant, animal, and microbial species that could be sources of important new medicines, the impact of depleted biodiversity on medical science could be tremendous (Chivian, 1993).

The consequences of depleted biodiversity and associated impacts on ecosystem function are increasingly evident worldwide. One vivid example described during group discussions involved the effects of mangrove destruction in tropical regions. Here, anthropogenic forces (e.g., urban development, intensive shrimp culture, charcoal/firewood harvesting, high sedimentation rates from coastal development, and pollution) have resulted in the rapid disappearance of mangrove trees, with more than 50% of original mangrove areas destroyed in many countries (United Nations Development Program et al., 2000). The depletion of these keystone plants has reverberated through mangrove ecosystems, endangered remaining native species through the disruption of energy and nutrient cycles, and impacted human settlements as the protective buffer provided by mangrove stands has diminished, increasing damage to human settlements from storms.

2.6. Other models

Although much progress has been made toward a better understanding of ecosystem and human health, it is certain that much more has yet to be learned. To foster this learning process, new paradigms may be needed to expand traditional views and understandings of the linkages between all systems that contribute to biological function and health. As pointed out during group discussions, one source for new models of understanding may be the human body itself. While ecologists/environmentalists view biodiversity at a large scale, medical pathologists look at it at the human scale. For example, the human gut includes a rich "ecosystem" of bacteria and other biotic life forms. This diversity is a necessary part of the function and health of the digestive tract and is essential to nutritional well-being. However, this diversity can be disrupted by the use of antibiotics. Although a linchpin of modern medical therapy, many antibiotics (especially broad-spectrum agents) not only reduce populations of target organisms (pathogenic bacteria), but can also deplete populations of beneficial organisms in the gut (Rabsch et al., 2000). As a result, one problem (a pathogenic infection) can be substituted by another (digestive distress and potential nutritional imbalance). As such, this example highlights the perspective of the human body as an internal ecosystem, dependent on balance and integrity within itself, but integrally tied to external factors that can disrupt this balance and alter function and health.

No matter what models of interaction are used, only when a broader understanding of the basic interrelationships of all life forms and environments become widespread, will humans devise and support management options that truly restore and sustain ecosystem and human health. Whatever specific strategies are developed, the working group emphasized that the goal should be the attainment of health as defined by the World Health Organization (1948): the state of complete physical, mental, and social well-being, not merely the absence of disease.

3. Sources of solutions

Through the course of discussion, the group outlined several fundamental human and ecosystem conditions that it felt were prerequisites for the attainment of this more comprehensive definition of human health. These prerequisites included peace, shelter, education, clean water and air, nutritious food, adequate income, ecosystem stability, sustainable resources, social justice, and equity. True ecosystem health would include complete maintenance of all ecological functions and services, many of which directly support human sustenance needs such as food production, climate control, etc. Because human activities threaten ecosystem and human health on an unprecedented scale, it was argued that decisive action is needed to safeguard the existence and well-being of all biota. There is much at stake, but there is also a mind-boggling array of specific environmental issues to address. Clearly, priorities need to be established to focus the human and financial resources needed to advance ecosystem and human health initiatives. Although the establishment of such priorities is an important societal responsibility, the

Box 1
Sources of solutions

It was the intent of organizers that the discussions and subsequent actions of participants in the working group on Ecosystem Health and Human Health would help contribute sources of solutions to ecosystem and human health problems. By encouraging the integration of the natural, social, and health sciences with policy- and decisionmaking communities, it was hoped that a deeper understanding of complex environmental problems could be developed. Furthermore, there was an expectation that the debate and discussion conducted could stimulate participants into taking meaningful action following the conference. There was also some promise that the professional contacts made at the EcoSummit would lead to productive collaborations that could help address ecosystem health issues in new and innovative ways.

Indeed, although the discussions of the working group were centered on the five questions that are the focus of this chapter, there was a constant undercurrent of desire that the group's efforts lead to something more constructive than just more talk. Clearly, discussions reinforced participant beliefs that ecosystem and human health are co-dependent and jointly in peril. However, this belief grew so strong that members also felt compelled to act. The publication of this chapter was one action intended to help build a better understanding of, and catalyze solutions for ecosystem health problems. In addition, members developed a list of broad goals and specific objectives to guide their own efforts to improve ecosystem and human health following the conference. These goals focused on

(1) improving the knowledge about the linkages of human and ecosystem health,
(2) supporting the dissemination of this knowledge through education and better communication, and
(3) using this knowledge to direct and promote meaningful change.

To be most effective, it was proposed that these activities be applied across a spectrum of scales, ranging from personal, to groups of targeted individuals, to society at large. The specific objectives outlined explicit actions (e.g., the development of local conferences or courses on ecosystem health, trying to convince one opinion leader of the importance of ecosystem health to human health, making personal changes in lifestyle to reduce consumption of non-renewable resources and energy, etc.) that participants could engage in to further prioritized goals. Some of these actions are now being realized. Several participants from the Faculty of Medicine & Dentistry at The University of Western Ontario (London, Canada) were so compelled by the discussions at the EcoSummit, that they organized a Primer Course in Ecosystem Health that was hosted by their University in June 2001 (see www.med.uwo.ca/ecosystemhealth). This course provided a strong background in the fundamentals of ecosystem health to a range of professionals and students who might not otherwise connect the need for ecosystem integrity with the sustenance of human health.

Only time will tell if participants will actualize the goals and fulfill specific objectives outlined during EcoSummit 2000. However, if the enthusiasm generated by group discussions is any indication, it is reasonable to expect that members will work to instigate meaningful personal and societal change and improve the health of world ecosystems and human populations.

working group proposed a series of considerations that members felt should be at the center of the continuing debate. Participants concluded that ongoing efforts to solve ecological/human health problems should focus on,
(1) Maximizing *global* human well-being;
(2) Ensuring *long-term* ecological sustainability/integrity;
(3) Preserving *all* aspects of biodiversity; and
(4) Creating the necessary *linkages/connections* for sustainable development.

Due to the geographic diversity represented by group members, issues of place and culture were always at the forefront of discussions especially when discussing solutions to environmental problems. Members recognized that solutions should be tailored to the communities in which problems occur. Throughout discussions, whenever solutions to environmental problems were addressed, a primary consideration raised was whether or not they were appropriate to local biological and social systems. One after another, group members recounted examples of well-intended "solutions" to environmental problems that failed because they did not accommodate local needs. As a result, a consistent theme emerged: flexibility in approach was often a cornerstone to successful environmental problem solving. Whether defining the appropriate level of technology needed, pertinent social constraints or enhancement opportunities, or even deciding the appropriate scale needed (e.g., neighborhood? district? region?) to capture "local" variation, it was agreed that planning should be flexible and solutions adapted to the situation at hand. It was emphasized that, although stringent goals concerning environmental quality are needed to prevent the continued degradation of ecosystem services, these goals might be achieved more rapidly, efficiently, and effectively if flexible approaches to implementation are employed.

Another offshoot of the considerable diversity embodied within the working group was that issues of equity were often raised. Admittedly, the human enterprises that help drive environmental degradation often impart some benefit (often economic) to certain groups or individuals. Clearly, neither the benefits of healthy ecosystems nor the harmful externalities of environmental damage impact all people or nations equally. The distribution of environmental "goods" and "bads" is unequal, and this inequity is itself a powerful driver perpetuating environmental harm and ecosystem decline. In effect, if individuals, businesses, or governments *perceive* that the net impact (goods – bads = net) of their actions is favorable to them, the incentive to change is often limited. However, assessments of net impact are traditionally incomplete and often undercount the full costs of resulting ecosystem dysfunction. Considering this, it was suggested that one promising source of solutions to environmental problems could be an effort to educate individuals about their connection to associated degradations of ecosystem and human health. Enhanced awareness of personal impacts may prompt individuals to adjust their "internal accounting systems" (and the perceptions/actions they support) to better reflect the true costs of existence. But what kind of educational experience has enough personal relevance to actually alter ingrained perceptions of how the world works? For some people, the source of this inner growth is "experiential leaning" (hands-on learning that results from personal experience).

Several members of the working group had considerable experience with programs that personally involved individuals with the day-to-day battle to improve local environments. A good example of this was the urban farming project established by Reverend Joseph Ebenezer in Chicago, Illinois, USA. Reverend Ebenezer described to the group his ambitious program that used abandoned lots, backyards, and rooftops as locations for community agriculture and aquaculture projects. These projects reap multiple benefits to the communities they serve (mostly poor or marginal communities): they provide nutritious food, they invigorate a sense of self-reliance and empowerment, and they help build a positive community identity. Importantly, the organic gardening techniques employed also teach valuable lessons in biology and ecology that transform the fundamental perceptions of food and health for these urban participants who otherwise might have little connection to the land.

Another prominent example of experiential learning discussed was the trend for colleges to require students to perform community service as a requirement for graduation (e.g. Canada, the USA and South Africa). The goal of these programs goes beyond a desire for students to contribute to community welfare. Implicit as well is the realization that service to others can be a life (and perception) altering experience. What better way is there to learn about new and sometimes very different realities than to live them? Indeed, perhaps if more of us experienced the net environmental impacts that others are forced to live with, our perceptions and associated actions would evolve to reduce pervasive environmental destruction.

4. Priority actions

It was emphasized repeatedly throughout group discussions that the seeds of change originate at the individual level. Indeed, group dialog focused on the fundamental importance of personal change as a prerequisite for broader societal reform. At some point, however, individual beliefs and action must translate into accomplishments at the institutional and societal levels in order to be fully effective. Governmental and corporate policies and management can have a pronounced and widespread influence on ecosystem health and human health. Clearly they cannot be ignored. In fact, the road to a healthier future will likely be built on an assortment of very specific changes in public policy and management. In addition, once an array of management options has been identified, a global consensus will be needed. Past agreements like those made in Stockholm, Rio, and Kyoto have proved to be insufficient. Restoring ecosystem and human health will require that thorough measures be prescribed and enacted globally. Although at one time many considered zero population growth to be "the answer" to ecosystem health problems, it is now broadly recognized that comprehensive reforms including reductions in resource consumption (particularly in developed nations that disproportionately deplete world reserves) must accompany population control.

As outlined earlier, the detrimental environmental and health consequences of growing populations and escalating human activity are many. Even fundamental

Box 2
Prioritizing issues: group analysis versus systematic review

As one means of focusing group discussion and guiding subsequent individual or collective action, participants compiled a list of serious threats to ecosystem and human health. Over the course of the conference, participants individually prioritized these issues from most to least "important", and toward the end of deliberations a synthesis of individual rankings was produced to create a composite tally of the relative importance of identified issues. The resulting composite listed 17 issues that ranged from #1 – climate variability and associated natural disasters, to #17 – the potential ecological threat of genetically modified organisms. The group recognized that there were many imperfections in this somewhat hastily prepared collective prioritization. Clearly it was redundant. For example, 6 of the 17 issues had interconnected ties to climate change (e.g., climate variability, rising atmospheric CO_2, rising global mean temperatures, etc.). In addition, in retrospect there were many important existing or emerging threats to ecosystem and human health (e.g., heavy-metal pollution, the growing resistance of microorganisms to antibiotics, potential declines in human immune and ecosystem stress response systems, etc.) that were missing from this list. It also seemed that there might have been some confusion on what criteria to use for judging importance. Was importance based on the current threat to ecosystem and human health, or was it based on the potential threat at some time in the future? Was the importance level based on the scope of the problem (e.g., worldwide vs. local or regional), or perhaps it was influenced by the sensitivity of the issue to timely remediation (e.g., because a policy decision now could make a real difference in alleviating or preventing a problem later)?

Despite the missteps inherent in this informal attempt to rank problems, the group openly recognized that some prioritization of ecosystem health issues is needed to guide research directions and policy actions. There are limited funds available for ecosystem/human health issues, and policymakers justifiably want advice on how to get the most "bang for the buck" toward resolving those problems that are most threatening. Comprehensive cost–benefit analyses should be conducted that account for all the costs (including alterations in health, finances, societal, community, and personal impacts, etc.) and benefits (including broad definitions of ecosystem and human health) of existing actions and proposed changes. Such analyses would likely integrate a range of pertinent criteria (e.g., considerations of problem scope, timely remediation, etc.) that would influence assessments of relative risk. A recent book may provide an example of the kind of integrated analysis that needs to be done. For their book *The Consumer's Guide to Effective Environmental Choices*, Michael Brower and Warren Leon (1999) from the Union of Concerned Scientists synthesized a wide array of scientific information on the environmental impacts of various consumer practices to produce a virtual "top 10 list" of activities that most harm the environment. Importantly, they outline specific ways for consumers to lessen their contributions to these pressing threats against ecosystem integrity and human health. The resulting analysis and advice establish a documented and well-reasoned set of priorities that serve as a practical guide for consumers that could also be used by policymakers to help evaluate policy alternatives. Of course, even this relatively comprehensive assessment has many significant limitations to its use and relevance. For example, because it specifically focuses on the effects of consumer actions, it avoids a vast area of pertinent influence on ecosystem health: the impact of the government sector. In addition, like any report, it is an informational "time capsule", in this case reflecting a rational cost–benefit analysis at the time of publication (1999), but with an unknown duration of validity. Despite limitations like these, careful prioritizations of the relative risks of the numerous threats to ecosystem and human health need to be undertaken. Although opinion surveys like the one conducted within the working group can help to highlight perceptions of risk, more stringent analyses should regularly be undertaken to more objectively rank threats to health, track the progress of existing corrective actions, and better define emerging problems.

earth processes like climate regulation are at risk. Considering the nature and scope of the threat to all life, it is likely that a broad range of actions will ultimately be needed to reduce anthropogenically-driven environmental destruction, restore earth ecosystems, and ensure human survival as one species amongst others on earth. Still, it was the consensus of the group that some specific policy priorities seem evident. For example, to cope with continuing global environmental changes, industrialized countries such as Canada, the USA, and those in the European Union will have to contribute their share to international greenhouse-gas emission reductions, and make other meaningful investments to monitor and maintain ecosystem and human health. A large part of the necessary investment in mitigative and adaptive measures may justifiably come from economic sectors most directly associated with the causes or consequences of environmental disruption. Of course, individuals will have a pivotal role in instigating and supporting policy changes by adopting a new personal ethic oriented less towards consumption and more towards conservation and healthy lifestyles. In addition, the health sector (including local and national health agencies) will need to provide leadership by enhancing public awareness of health issues related to environmental degradation, by helping to identify public health priorities, and by helping to shape appropriate prevention and response policies. Importantly, actions should be implemented in collaboration with neighboring nations because threats to ecosystem and human health do not respect geopolitical boarders.

A strong-held belief among many participants was that the equitable development and consumption of resources within and among regions and countries should be a policy priority. For example, Apartheid in South Africa included among its many discriminative practices a dramatic inequality of resource distribution. Although for many years the World Bank measured progress by assessing earned annual income, they now evaluate the distribution of key resources, such as the number of homes and villages that are supplied with water and sanitation. In South Africa, the hope is to provide "the basics" and then let the community co-evolve with the environment. The South African constitution proclaims: "Some water, for all, forever." This simple statement captures a basic premise that surfaced time and again in our group discussions: that issues of equity and the environment are integrally related. With this in mind, our group concluded that a better collaboration between rich and poor nations is badly needed. Although this has been repeatedly requested in the past, it seems ever more evident that long-term improvements in the economic and environmental health of developing countries will not occur without a halt to the overexploitation of world resources by industrialized nations.

Even concerning issues of international equity, the group agreed that small-scale (individual and community) actions figure prominently as instigators of change. In many cases, the statement: "Think globally, act locally" is still vitally relevant. While most environmental impacts can have global implications, many preventative and corrective actions have their roots at the local level. This is where individuals and communities can have the most control. This is where each person has a real chance to make a difference.

5. Barriers to effective action

Even though there is a growing awareness that human activities pose a mounting threat to natural and human populations, it seems unlikely that the changes needed to safeguard ecosystem and human health are imminent. Numerous attempts to reshape individual lifestyles and public policies (ranging from Earth Day to the Rio Summit) have occurred with increased frequency as worldwide awareness of environmental problems has grown. Still, tangible evidence of meaningful action to halt human-induced environmental harm and rebuild/nourish ecosystem and human health is scarce. Unfortunately, there are many reasons for inaction. Some were reviewed in the previous chapter. However, through group discussions it was suggested that overcoming certain crucial barriers might have enhanced importance because they disproportionately deter action and impede change. These five barriers are described below.

5.1. Sustenance needs

In many developing communities, the primary and immediate goal is short-term survival. Understandably, if fundamental sustenance needs for food, water, and shelter are not met, long-term concerns about education and environmental quality take on a low priority. For example, in Vietnam, deforestation is fast occurring. As explained by one of our Vietnamese participants, "There is not enough water for rice production or drinking: there is a clear relationship between all these activities. You can't ask a hungry person to save something for tomorrow. People have to eat every day ... What is a solution here?" Indeed, this is one of the reasons why equity issues figure so prominently in the debate concerning the environment and health worldwide. Social inequities that deprive people of the basic necessities of life often force them to establish in marginal, fragile, and less productive ecosystems where local patterns of resource use and exploitation (e.g., rapid rotations of slash and burn agriculture) can be harmful to both ecosystem and human health in the long term. Perspective is everything. When one is well fed and has the other basic necessities of life fulfilled, it can be all too easy to condemn unsustainable resource use. But what would *you* do if faced with uncertain prospects for short-term survival? Clearly, a major obstacle hindering long-term planning and management in many regions is the persistent inability of impoverished, often rural peoples to meet basic sustenance needs. In these regions, cycles of poverty and human deprivation must be broken if long-term ecosystem health is to be improved.

5.2. Little connection to the land

Another obstacle to change discussed by the group was a perceptual one that may have its roots in modern western culture. It was proposed that, as life in industrialized lands has become more urbanized and professions more specialized, people have "forgotten" that they are reliant upon and a part of nature. They have lost their connection to nature and the natural cycles under which their ancestors evolved. Freed from the perception that they are "of the earth", many contemporary cultures have undertaken a range of activities that are, at times, destructive to it. Changing perceptions of agriculture and food may exemplify this. The group observed that people in urban centers typically seem oblivious to the "details" (e.g., ecological, environmental, health, and other ties) of food production. Yet, they also often demand that food prices stay low. Here, food acquisition and consumption is habitually a hurried "chore", accomplished on the run amidst a busy schedule of "more important" commitments. The role of food in the health and sustenance of individuals, families and communities has faded. And with this withered social value, food has experienced a diminished monetary value as well. Consumers seem unwilling to pay much for food, and depressed commodity prices make it difficult for farmers to make a living. Subsidies are then needed to support farming, but this causes farmers to have less and less influence on the market and the type of agriculture they practice. Indeed, the incentive of farm management becomes economic survival, not the production of nutritious food or the stewardship of the land.

5.3. Resistance to change

Inertia and an unwillingness to change are additional barriers to resolving persistent environmental problems. Despite a growing number of vivid examples of ecosystem decline and the associated health impacts, attitudes and actions often remain entrenched in destructive patterns until critical biological thresholds are surpassed. Examples of the cycle of resource exploitation, depletion, ecosystem decline, site abandonment, and renewed exploitation elsewhere abound, and many were recounted by members of the group. For example, in the mid-1940s the anchovy fishery off the California coast collapsed. Although it is now recognized that overfishing was only partially responsible for this decline, this episode served as one of many wake-up calls to the limits of seemingly boundless resources. In Nova Scotia, such a decline in fish populations pushed fishing communities to move towards another ecosystem and resource – forests for timber production – that resulted in an inland migration, which escalated the clearing of interior woodlands. The current decline of commercial fisheries worldwide indicates that such depletions can occur on a global scale. Even following repeated examples of overexploitation, irresponsible patterns of resource development have been slow to change. Perhaps as a holdover of the "frontier mentality" humans have clung

Ch. 10: *Ecosystem Health and Human Health: Healthy Planet, Healthy Living* 207

to the perception that there would always be some new resource or reserve to tap next. Indeed, the struggle with the notion of limits seems stronger now than ever as population growth and explosive consumption drive an ever-growing demand for resource development. Certainly it is hard to accept restraints on resource extraction, especially when additional stocks are sometimes readily evident (as in wilderness reserves, etc.). However, it was clear to the group that greater restraint is essential if the structure and function of ecosystems are to be safeguarded and human health preserved.

5.4. Ignorance

Another prominent reason that individuals and groups seemingly ignore the integral connections between ecosystem and human health is that they simply are not aware of them. A lack of environmental awareness and a low functional knowledge of earth and human ecology are important barriers to productive action. Although it is often contended that "ignorance is bliss", a lack of understanding about ecosystem stability and human health might better be portrayed as an example of "what you don't know could kill you". Obviously, broad-based educational outreach is needed to inform all citizens of the basic co-reliance between the environment and biological health. This understanding could be the basis for the informed public debate needed to guide policy changes. In addition, however, there are innumerable aspects of environmental health that still remain a mystery. An expanded research effort is also sorely needed to provide a firm scientific basis for the critical decisions ahead.

5.5. Low critical mass

New ideas require some level of general acceptance before they gain the sufficient "critical mass" to demonstrably influence public opinion and policy. In fact, some contend that it is not just the level of acceptance that is important, but also *who* embraces an idea that matters. For example, in any mixed group of people, not everyone in that group will have the same reaction to a new concept, nor will everyone equally embrace and propagate new ideas. In group discussions it was suggested that approximately 10–15% of the people in any group tend to be innovators (people who are more likely to accept and spread new ideas); about 35% are inclined to be early changers (who tend to wait only a little while before embracing new ideas), around 35% tend to be later changers, and roughly 15% are apt to be dinosaurs (who never change). If this contention is correct, then when a new concept (like the importance of ecosystem integrity to human health) is first presented to the public, it might be best to target a small but highly influential group (i.e., the innovators) to most effectively disseminate the idea and develop the critical mass needed to influence societal change. This strategic targeting could be valuable at any scale of endeavor (from local to international). Indeed, several participants recounted examples of how focused outreach and education had helped to accelerate positive change within their communities.

6. Measures, indicators, or metrics of progress

Presently, there is a veritable industry blossoming around the development of indicators of ecosystem health. Indeed, a variety of indicators are now being evaluated to assess how well they depict the health of managed ecosystems (e.g., Ecosystem Health Indicators – Environment Canada, www.ec.gc.ca / cehi/en/indic_e.htm; Pilot Analysis of Global Ecosystems (PAGE) – United Nations Development Program et al., 2000). The hope is that, if the right indicators of environmental quality are developed, then adaptive management could be used to avoid harm (when the indicators showed it was immanent) and accommodate resource use and development (provided indicators showed that environmental quality was unimpaired).

In principle, the group agreed that a meaningful indicator of ecosystem health should be representative, sensitive to change, relevant to public policy, easily understood, and scientifically credible/transparent. Data used to generate the indicator should also be measurable and available at appropriate scales. Ideally, the indicator would also be associated with larger benchmarks to allow for global, long-term monitoring possibilities. Furthermore, participants felt that meaningful data on the status of ecosystems was vitally needed for effective risk assessment and to monitor alterations in environmental quality associated with changing policies and practices. While exploring this idea, the group initially generated a long list of measures/indicators of environmental quality. This list included many standard environmental indicators already measured or proposed for use worldwide. Although many of these indicators are clearly useful for evaluating specific aspects of ecosystem health, they might not have intuitive meaning to policymakers or the general public. The group concluded that what was needed was a simple indicator of ecosystem health that had ready relevance to human health; an indicator that provided a steady gauge of whether environmental quality was improving or declining, something that indicated whether human health was more or less at risk as time and conditions changed.

Of course, the relevant challenge is not to define indicators, but to relate them to ecosystem and human health. A pertinent indicator would be a useful tool to convince policymakers and the general public of the necessity to reduce pressures that threaten ecosystem and human health. However, because it seemed highly unlikely that any one environmental parameter would provide a truly comprehensive view of ecosystem integrity, the group concluded that any

Box 3
Achieving ecosystem health

Achieving and sustaining ecosystem health will undeniably require that human populations change the fundamental perceptions, attitudes, and actions that propel ecosystem degradation. Although many people express a general concern about the health of ecosystems, doing something about it is an entirely different matter. Still, some communities *are* making the effort to re-evaluate ingrained attitudes and behaviors and pioneer productive responses to environmental problems. Sharing examples of such "success stories" helps to highlight the idea that, despite the pervasive threat humans pose to ecosystem health, positive community action can make a meaningful difference.

One such success story can be found in the host province of the EcoSummit 2000: Nova Scotia, Canada. A small province on the East Coast of Canada, Nova Scotia has traditionally been a resource-based economy, more concerned with using their environment than with conserving it. However, in 1996, in accordance with established national priorities the province set a goal to reduce the amount of waste it generated by 50% before the year 2000. This decision, prompted in part by a growing shortage of landfill sites, required a concerted effort throughout the society to change waste generation and disposal patterns. Through a comprehensive strategy that included recycling, composting, reduction, and re-use, Nova Scotia surpassed the national average (22%) and nearly reached its ambitious goal of a 50% waste reduction. The province has even begun to view their wastes as a "resource", as the business of waste reduction (including waste recycling and composting facilities) has generated over 3000 jobs in the province.

Highlights of the Nova Scotia system include:
- A deposit/refund system for beverage containers. Nova Scotia now returns over 160 million containers annually, with a return rate on beverages sold of over 80%.
- 100% of Nova Scotia residents have access to curbside recycling.
- 100% of businesses in Nova Scotia also have access to recycling programs.
- Compostable organic materials were banned from landfills in 1998. 72% of residents now have access to curbside organic collection, and 100% of all municipalities promote backyard composting for leaf and yard wastes.
- 900 000 tires are re-used or recycled annually through a Used Tire Management Program conducted through a partnership with approximately 900 registered tire retailers.

(Source: Nova Scotia Department of the Environment, 2001)

Although waste reduction is only one step towards a healthier ecosystem, the province of Nova Scotia has shown that communities *can* be successful in making changes. Importantly as well, communities that have realized the benefits of change first-hand may also seek out and enact other projects that comprehensively promote overall ecosystem health. By gradually building on past successes while experimenting with new options, locations like Nova Scotia could help lead the way towards broad societal reform that generates a healthy planet through healthy living.

meaningful indicator would likely need to integrate some defendable but short list of environmental quality parameters. But what sort of measurements would make the most sense to include? The scientific debate on this decision is currently in full swing. The devil is always in the details. Still, whatever components are chosen, the group repeatedly emphasized that resulting indicators need to be focused, strategic, and attractive enough to policymakers to assure that they are incorporated into the decisionmaking process.

Even if a scientifically justified index of ecosystem health were devised, its use and interpretation would initially be fraught with problems. Innumerable questions of use and interpretation would need to be answered. Among them basic questions such as, what values were "normal" and what magnitude of change was indicative of a significant alteration of environmental quality? Most likely, the index would need to be calibrated for a variety of sites and circumstances. Questions of index interpretation across sites, temporal and spatial scales are among the many issues that would need to be addressed prior to use.

Despite the group's acknowledgement of the many technical problems facing the development and deployment of a useful index of ecosystem integrity and health, the promise of such a tool was too great to ignore. Therefore, further discussions focused on two approaches for indicator development: an ecological index and a bio-social integrated index. The ecological index would use environmental data to evaluate ecosystem status, especially as it related to qualities important to human health (e.g., air and water quality). The bio-social index would follow the model of commonly used indices like the Gross Domestic Product (GDP), but would be supplemented to include measures of human health and welfare. This latter approach might be more attractive to policymakers who already have a ready familiarity with indices like the GDP. The Quality of Life working group at EcoSummit 2000 discussed such a system in considerably more detail (see chapters 11 and 12).

Considering the high priority that group members placed on individual and community action, it was also suggested that some measurements of environmental progress be developed that were pertinent at the community level. For example, the number of individuals or communities involved could be an appropriate "performance indicator" of progress (at least regarding informational and educational elements). A community-based ecological indicator could also be used to more directly rate a community's impact on local and global ecosystem health. For instance, a community's "ecological footprint" could be measured and compared to other communities, national averages, or evaluated over time to track progress and highlight areas of needed improvement. As stated previously, once people or communities are better educated and aware of local environmental problems (and the global impacts of local actions), they often are more willing to promote change and work toward solutions. Perhaps as a result of this, community-based management projects appear to be gaining popularity throughout many countries. Experience has shown that environmental actions at the community level using volunteers are often more sustainable and effective than those made through political decisions alone. Since the motivation of volunteers is frequently high, participation can be maintained for long periods of time. Especially if volunteers sense that their efforts help improve environmental conditions within their own community, they are more likely to continue participation and even enroll others to help. With this in mind, the group concluded that community involvement is likely critical to the sustained progress of local ecosystem health programs. The involvement of volunteers and community-generated input can improve the transparency of the decisionmaking process, enhance the sense of community empowerment, and increase levels of trust for resulting decisions.

A key intent behind the development of ecosystem health indicators is that they be used to help monitor, promote, and improve ecosystem and human health. But who would use an indicator of ecosystem health? One obvious answer is that data on environmental quality would be extremely useful to policymakers, especially if presented as a short list of key indicators pertinent to the specific interest/jurisdiction of evaluating officials (e.g., CO_2-related data for climate-change policymakers). However, there is also a need for more comprehensive indicators that combine both biological and social data, and provide a more integrated picture of ecosystem health and societal wellbeing. For example, a bio-social index could be a useful educational tool that could help educate the public about ecosystem and human health issues, and might even find acceptance by some policymakers as a means of tracking "the big picture" (e.g., are current policies improving overall health or diminishing it?).

How could these data and their interpretation be best communicated? While scientists are traditionally involved in data management, analysis, and interpretation, discussion participants proposed that subject-specific experts are often not the best people to communicate data to the general public and policymakers. Clearly more effective avenues of communication with the public are needed. One option discussed was the possibility that doctors could act as a conduit of information on the linkages of ecosystem and human health. Doctors have established access to community members and are respected leaders regarding issues of health. In many ways, doctors have been the primary interface between science professionals and the broader community for generations. It could be argued that the interdependence of ecosystem and human health issues would automatically require greater doctor–patient communication concerning this topic. However, more immediate development of this dialogue could serve an important preventative role as an educated and activated public would help promote personal lifestyles, and corporate and governmental policies that foster improved ecosystem and human health. Undoubtedly, physicians and other health-care professionals are underutilized links between environmental scientists and the general public. However, it is uncertain how pressures to keep health-care costs low (often at the expense of doctor-to-patient time) could impact their availability as conveyors of ecosystem and human health information.

It was also suggested that, in at least some countries (e.g., South Africa), religious leaders could become effective messengers of the important connections between ecosystem and human health. Religious leaders (ministers, shamans, etc.) are often among the most trusted and influential members of communities, and therefore frequently impart an enormous influence on public opinion. By enlisting the help of these leaders, governments and others could tap into an established communication network and more effectively spread environmental health information (including indicators of ecosystem health) to local communities. Group members cautioned, however, that religious leaders are not uniformly viewed as unbiased protectors of the public good. Especially when aligned with repressive governments or movements, religious leaders might not retain public confidence. Open affiliation with religious leaders under these circumstances could actually hinder the promotion of ecosystem health programs.

Educators represent another likely user group for ecosystem health data. Indeed, through the course of discussion it became evident that the linkage between scientists and primary and secondary school educators has already been successfully developed in some locations. One participant from Vietnam reported that scientists in his homeland have long recognized that the school children of today are the citizens and decisionmakers of tomorrow. As a result, scientists there initiated an active program of involvement in community-based environmental education. Similarly in Canada, several programs have targeted school children and citizens of all ages for education efforts (e.g. Adopt-A-Stream; Naturewatch, a community-based monitoring program launched in 2002 by the Ecological Monitoring and Assessment Network of Environment Canada and the Canadian Nature Federation, which includes programs such as Plantwatch, Frogwatch and Icewatch). The involvement of school children within the USA in acid rain monitoring networks provides another noteworthy example. In recognition of the successes of past but thematically more limited efforts, it was suggested that a more comprehensive program be developed to collect broad-based ecosystem health data using area schools. Local students and teachers would collect data that could be used to generate ecosystem health indicators that over time could provide community, regional, national, and world leaders a detailed gauge of ecosystem health trends. Although an ambitious undertaking, the development of such a program could be fostered through concerted grass-roots efforts instigated by science professionals and educators. For example, in industrialized countries, many of us serve on school boards where we can influence science curricula and the hiring of teachers who would utilize such a program.

Finally, another group that should have an interest in environmental health data is the media. Their professional roles (to inform the public of issues that influence their lives) make them obvious target groups. Furthermore, journalists often hold the interest of policymakers, likely facilitating communication to this important group. However, it was cautioned that journalists are not always "scientifically literate" and that scientists are frequently not "media smart". The nuances and limitations of scientific information are often difficult to relate, especially in instances where "details" seem to obscure or confuse a "story". Still, it is evident that clear communication and cooperation among a range of professions and users will be needed to fully utilize the potential value that indicators of ecosystem health might someday provide.

7. Conclusions

Through the course of discussion, the working group covered a broad spectrum of topics within the overarching subject of ecosystem health and human health. The five questions the group sequentially addressed, provided a needed structure, and helped guide joint progress. However, even with this guide and overlying structure, discussions were far ranging. In particular, the considerable diversity of participants greatly enriched the breadth and scope of debate.

Despite the broad theme, open debate, and the rich diversity of participants, the fundamental level of consensus among participants was astounding. Anchoring this consensus was the knowledge that ecosystem and human health are intricately interdependent and that human activity is increasingly threatening both. Indeed, participants spent a considerable amount of time detailing the mounting evidence of anthropogenically induced ecosystem dysfunction and the associated impacts on human health. However, as the guiding questions for debate and this resulting chapter indicate, an even greater proportion of our attention was occupied with considerations of appropriate response. The group was uniformly convinced that unbridled human activity was threatening the intricate web of biological and ecological processes that support all life. Participants seemed equally certain that this threat was so grave and immediate that broad-based and timely action was not only warranted, but imperative.

Many participants commented that "we don't manage the ecosystem, we can only manage ourselves." Because local and even global environmental issues ultimately result from the additive effects of our individual but cumulative decisions and actions, solutions to these problems should address individual beliefs and behaviors as well as cultural norms and public policies. Although the group shared this unified perspective of the serious threat posed to ecosystem and human health, members were also struck by the realization that many other inhabitants of the earth did not share this perspective. Participants openly wondered: could this be the root of the environmental problems we face?

As detailed earlier, the barriers to change are many and also span a range of scales. However, time and time again, it was the group's conclusion that education and communication were the primary tools needed to overcome these barriers and foster meaningful personal and societal change. Because the goal of education and communication efforts is ambitious (i.e., to expand perceptions of the interdependency of ecosystem and human health), innovative methods will need to be deployed because the goal is not just to communicate facts, but rather to help individuals reevaluate ingrained attitudes and behaviors.

Many shifts in attitude are needed. However, some fundamental starting points seemed evident. For example, Public NIMBY (Not In My Back Yard) perceptions prevalent in many industrialized countries will have to change in order to reduce human impacts on the environment. In its place, a new philosophy, SIMBY (Start In My Back Yard) (Vasseur, 1997), could be implemented to highlight the role of the individual in fostering meaningful change. Of course, this change has to start small but eventually permeate all levels of society. Still, there are already many successful pilot programs that have helped people become active stewards of their ecosystems, health, and future. As recounted throughout group discussions, numerous examples (e.g., prisoners working in community restoration projects, homeless people becoming responsible for container gardens, and development of new green industries) have shown that it is possible to change public attitudes and lifestyles while maintaining economic prosperity in a healthy ecosystem. But more needs to be done to broadly foster change and affect all communities at the global ecosystem level.

The working group was not alone in its concern for ecosystem health and human health. Prior to the EcoSummit 2000, students from several countries gathered under the same theme to discuss their views of the future for ecosystem health and human health. As prospective leaders and stakeholders, they recommended numerous changes, which although more specific than our group's findings, were nonetheless quite similar:
- Ethics and compassion to replace greed;
- Quality versus quantity of education in environment;
- Environmental policy should be an umbrella for trade policy;
- Environmental accountability is needed; and
- With respect to economic measures, GPI (Genuine Progress Index) or the United Nation's Human Development Index (HDI) should replace GDP because the latter does not address distributive justice.

As the students concluded: "Concerns about ecosystem health and human health should not only target policymakers. While the current generation makes the mess, the future generation already knows that they will have to deal with the clean up" (Singleton, 2000). These last statements have reinforced the ideas that the statement from this theme should be linked to possible immediate actions.

A growing number of people from around the world now realize that human population expansion and resource consumption increasingly threatens ecosystem and human health. The dangers are real and the stakes remarkably high. Action is clearly needed. But what to do, and where to start?

As leaders from a range of professions across the globe, participants felt responsible to help initiate the public debate and personal growth needed to guide humanity through the quagmire of environmental problems that we have all helped to create. To summarize the group's findings and to help instigate substantive change, members offer the following resolution, prepared by group Rapporteur, Andrew Hamilton.

Healthy planet, healthy living

A RESOLUTION ON CONSERVING, PROTECTING, AND ENHANCING ECOSYSTEM HEALTH AND HUMAN HEALTH

RECOGNIZING that our human health and well-being are utterly dependent on the health and life support system of the earth;

RECOGNIZING that the earth is a dynamic and ever-changing living system made up of interconnected components of air, water, land and living organisms, including humans;

CONCERNED with the status and trends in the life sustaining capacities of the earth, including local, regional, and global air pollution, climate change and climate variability, deteriorating quality and diminished supplies of fresh water, reduced soil productivity, declines in wild stocks of fish, and the rapid loss of biological diversity at the genetic, species, and ecosystem levels;

NOTING that many of these undesired changes in the earth's life-sustaining capacities have directly and indirectly caused serious acute and chronic health effects in humans, especially in those most vulnerable, and most exposed that in many instances are unborn and young children, the poor and undernourished, and the elderly;

ACKNOWLEDGING that many of these undesirable changes in the earth's life sustaining capacities are directly and indirectly attributable to the demands and stresses placed on them by humans and human activities;

TROUBLED by the human tendency to take human and ecosystem health for granted, to ignore clear warning signals and to delay taking early and timely action to prevent undesired stresses, outputs and outcomes until major disasters have taken place;

CONCERNED, similarly, that our individual and collective actions to reduce, alleviate or eliminate undesired stresses, outputs and outcomes with respect to human and ecosystem health are often directed almost entirely at reducing symptoms rather than addressing the root causes;

CONVINCED that effective and lasting measures to restore, maintain and enhance the life sustaining capacities of the earth will not become a reality without parallel measures to address the inequitable distribution of power, influence and resources within and between countries and regions of the world;

REALIZING that decisions and actions need to be appropriate to the scope, scale, and time frame of the issue being addressed and that they will inevitably be influenced by our ability to anticipate and prevent or alleviate undesired stresses, outputs, and outcomes as well as on our ability to respond and adapt to both predicted and unanticipated situations and events;

APPRECIATING the need and opportunity to build on continuing local, national, and regional efforts to reduce our individual and collective stresses and unsustainable demands on the earth's life-support systems;

OPTIMISTIC that individuals, societies, and institutions are capable of making more informed, responsible, and enlightened decisions to reduce our individual and collective stresses and unsustainable demands on the earth's life support systems;

We THEREFORE have agreed to:

ENDEAVOUR to work in a network of scientists, elected officials, government officials, managers, and concerned citizens, including indigenous people, and as members of organizations, communities, and the global society, to understand and communicate our concerns and understandings to one another and to others;

CHALLENGE politicians, economists, sociologists, ecologists, members of the media, and others to comprehend and communicate those aspects of the global, regional, and local free market and trading systems that are undermining the earth's life-support system;

FOSTER and ENCOURAGE the development and use of measures of progress, which, unlike the Gross Domestic Product (GDP), would also include indicators of ecosystem health and human health and well-being;

PROMOTE the development and use of environmentally appropriate policies, practices, technologies, and human behavior so as to better conserve and protect the life support system of the earth;

PROVIDE, through the furnishing of information and expertise and by other means as appropriate, opportunities to help empower individuals, as well as local and environmental groups, to influence or make decisions to protect and enhance ecosystem health and human health; and

ENCOURAGE local, national, and international opinion setters, leaders, and funding bodies to recognize and work individually and together to address, with a sense of urgency and priority, the erosion of the earth's life-support system.

Andrew Hamilton, Rapporteur

Working Group on Ecosystem Health and Human Health
EcoSummit 2000
Halifax, Nova Scotia
22 June 2000

Acknowledgements

This chapter is the result of a collaborative effort from many individuals over the course of discussions both before and during the EcoSummit 2000. The authors would like to thank the members of the Working Group Organizing Committee, specifically Andrew Hamilton, Jennifer Hounsell, Sharon Lawrence, Diane Malley, Ken Minns, Mohi Munawar, David Rapport and Liette Vasseur for their contribution and commitment to the development of this theme. As well, earlier contributions from John Cairns Jr., Thomas Edsall, William Fyfe, John Howard, Robert Lannigan, Robert McMurtry and Dieter Riedel were central to the development of the theme discussions. A special thank you to Liette Vasseur, working group chair, Diane Malley, working group facilitator, and Andrew Hamilton, working group rapporteur.

References

Acton, D.F. and Gregorich, L.J., 1995, The Health of Our Soils: Toward Sustainable Agriculture in Canada, Publication 1906/E (Centre for Land and Biological Resources Research, Research Branch, Agriculture and Agri-Food Canada). Available on-line: http://sis.agr.gc.ca/cansis/publications/health/.

Barrow, C.J., 1991, Land Degradation (Cambridge University Press, Cambridge) 295 pp.

Brower, M. and Leon, W., 1999, The Consumer's Guide to Effective Environmental Choices: Practical Advice from the Union of Concerned Scientists (Three Rivers Press, New York) 292 pp.

Chapin III, F.S., Zavaleta, E.S., Eviner, V.T., Naylor, R.L., Vitousek, P.M., Reynolds, H.L., Hooper, D.U., Lavorel, S., Sala, O.E., Hobbie, S.E., Mack, M.C. and Díaz, S., 2000, Consequences of changing biodiversity. Nature 405:234–242.

Chivian, E., 1993, Species extinction and biodiversity loss: the implications for human health. In: E. Chivian, M. McCally, H. Howard and A. Haines (Editors), Critical Condition: Human Health and the Environment (MIT Press, Cambridge, MA) pp. 193–224.

Cincotta, R.P., Wisnewski, J. and Engelman, R., 2000, Human population in biodiversity hotspots. Nature 404:990–992.

Commoner, B., Woods Barlett, P., Eisl, H. and Couchot, K., 2000, Long range air transport of dioxin from North American sources to ecologically vulnerable receptors in Nunavut, Arctic Canada, Research report (North American Commission for Environmental Cooperation, Montreal). Available on-line: http://www.cec.org/files/PDF/POLLUTANTS/dioxrep/_EN.pdf.

Cortese, A.D., 1993, Introduction: human health, risk, and the environment. In: E. Chivian, M. McCally, H. Howard and A. Haines (Editors), Critical Condition: Human Health and the Environment (MIT Press, Cambridge, MA) pp. 1–11.

Costanza, R., d'Arge, R., de Groot, R., Farber, S., Grasso, M., Hannon, B.M., Limburg, K., Naeem, S., Paruelo, J., O'Neill, R.V., Raskin, R.G., Sutton, P. and van den Belt, M.J., 1997, The value of the world's ecosystem services and natural capital. Nature 387:253–260. See http://www.floriplants.com/news/article.htm.

DeHayes, D.H., Schaberg, P.G., Hawley, G.J. and Strimbeck, G.R., 1999, Acid rain impacts calcium nutrition and forest health. BioScience 49:789–800.

DeHayes, D.H., Jacobson, G.L., Schaberg, P.G., Bongarten, B., Iverson, L. and Dieffenbacher-Krall, A.C., 2000, Forest responses to changing climate: lessons from the past and uncertainty for the future. In: R.A. Mickler, R.A. Birdsey and J. Hom (Editors), Responses of Northern U.S. Forests to Environmental Change (Springer, New York) pp. 495–540.

Dynesius, M. and Nilsson, R., 1994, Fragmentation and flow regulation of river systems in the northern third of the world. Science 266:753–762.

Environment Canada, 1999, Urban Air Quality, SOE Bulletin No. 99-1. Available on-line: http://www.ec.gc.ca/ind/English/Urb_Air/Bulletin/uaind1_e.cfm.

Environment Canada, 2000, Environmental Priority – Clean Water. http://www.ec.gc.ca/envpriorities/cleanwater_e.htm.

Environment Canada, 2001, Clean Air webpage. http://www.ec.gc.ca/air/introduction_e.cfm.

Federal, Provincial and Territorial Advisory Committee on Population Health, 1999, Toward a Healthy Future: Second Report on the Health of Canadians, Prepared by the Federal, Provincial and Territorial Advisory Committee on Population Health for the meeting of Ministers of Health, Charlottetown, PEI, September 1999 (Ministry of Public Works and Government Services Canada, Ottawa) Available on-line: http://www.hc-sc.gc.ca/hppb/phdd/report/toward/eng/index.html.

Health Canada, 1997, Health and Environment: Partners for Life (Ministry of Public Works and Government Services, Ottawa, Canada) 208 pp.

Langlois, C., Langis, R. and Pérusse, M., 1995, Mercury contamination in Northern Québec environments and wildlife. Water Air Soil Pollution 80:1021–1024.

Lasorsa, B. and Allen-Gil, S., 1995, The methylmercury to total mercury ratio in selected marine, freshwater, and terrestrial organisms. Water Air Soil Pollution 80:905–913.

Mickler, R.A., Birdsey, R.A. and Hom, J. (Editors), 2000, Responses of Northern U.S. Forests to Environmental Change (Springer, New York) 578 pp.

Nilsson, R., 2000, Endocrine modulators in the food chain and environment. Toxicol. Pathol. 28:420–431.

Nova Scotia Department of the Environment, 2001, Nova Scotia: Too Good to Waste, Status Report 2001 of Solid Waste-Resource Management in Nova Scotia. Available on-line: http://www.gov.ns.ca/envi/wasteman.

Pimentel, D., Acquay, H., Biltonen, M., Rice, P., Silva, M., Nelson, J., Lipner, V., Giordano, S., Horowitz, A. and D'Amore, M., 1992, Environmental and economic costs of pesticide use. BioScience 42:750–760.

Postel, S., 1999, Pillar of Sand (Norton & Company, New York) 313 pp.

Rabsch, W., Hargis, B.M., Tsolis, R.M., Kingsley, R.A., Hinz, K.-H., Tschape, H. and Baumler, A.J., 2000, Competitive exclusion of *Salmonella* Enteritidis by *Salmonella* Gallinarum in poultry. Emerg. Infect. Dis. 6(5):443–448. Available on-line: http://www.cdc.gov/ncidod/EID/vol6no5/rabsch.htm.

Sala, O.E., Chapin III, F.S., Armesto, J.J., Berlow, E., Bloomfield, J., Dirzo, R., Huber-Sanwald, E., Huenneke, L.F., Jackson, R.B., Kinzig, A., Leemans, R., Lodge, D.M., Mooney, H.A., Oesterheld, M., Poff, N.L., Sykes, M.T., Walker, B.H., Walker, M., Wall, D.H., Walker, M. and Wall, D.H., 2000, Global biodiversity scenarios for the year 2100. Science 287:1770–1774.

Singleton, A., 2000, Report from the Student Forum *Ecosystem Health and Our Future*, presented to the working group on Ecosystem Health and Human Health at the EcoSummit 2000, Halifax, Nova Scotia, Canada, June 21, 2000.

Soto, A.M., Sonnenschein, C. and Colborn, T.E., 1996, Endocrine disruption and reproductive effects in wildlife and humans. Comm. Toxicol. 5:315–506.

Tilman, D.G., 2000, Causes, consequences and ethics of biodiversity. Nature 405:208–211.

Tsiji, L.J.S., Nieboer, E., Karagatzides, J.D., Hanning, R.M. and Katapatuk, B., 1999, Lead shot contamination in edible portions of game birds and its dietary implications. Ecosyst. Health 5:183–192.

United Nations Development Program, United Nations Environment Program, World Bank and

World Resources Institute, 2000, World Resources 2000–2001. People and Ecosystems: The Fraying Web of Life (Elsevier, Amsterdam) 400 pp.

US Food and Drug Administration, 2001, FDA Announces Advisory on Methyl Mercury in Fish, FDA Talk Paper T01-04, January 12, 2001. http://vm.cfsan.fda.gov/~lrd/tphgfish.html.

Vasseur, L., 1997, Environmental Science. Lecture notes ENV 300.1 (Saint Mary's University, Halifax, Nova Scotia, Canada).

Vitousek, P.M., Mooney, H.A., Lubchenco, J. and Melillo, J.M., 1997, Human domination of the earth's ecosystems. Science 277:494–499.

Wardle, D.I., Kerr, J.B., McElroy, C.T. and Francis, D.R., 1997, Ozone Science: A Canadian Perspective on the Changing Ozone Layer (Environment Canada, Ottawa).

Wilcove, D.S., Rothstein, D., Dubow, J., Phillips, A. and Losos, E., 1998, Quantifying the threat to imperiled species in the United States. BioScience 48:607–615.

World Health Organization, 1948, Preamble to the Constitution of the World Health Organization as adopted by the International Health Conference, New York, 19–22 June, 1946; signed on 22 July 1946 by the representatives of 61 States (Official Records of the World Health Organization, no. 2, p. 100) and entered into force on 7 April 1948, http://www.who.int/m/topicgroups/who_organization/en/index.html.

World Health Organization, 1997, Health and Environment in Sustainable Development: Five years after the Earth Summit: Executive Summary. Available on-line: http://www.who.int/environmental_information/Information_resources/htmdocs/execsum.htm.

Background

Chapter 11

Quality of Life and the Distribution of Wealth and Resources

R. Costanza, J. Farley and P. Templet

Abstract

Enhancing and sustaining the quality of human life is a primary goal of environmental, economic, and social policy. But how do we define and measure quality of life (QOL)? How is QOL distributed among people in the current generation and among the current and future generations? How do we model the dependence of QOL on the full range of environmental, economic, and social variables? Answering these questions is fundamental to understanding and solving environmental problems in the 21st century.

1. How is Quality of Life (QOL) defined?

If we are to assess the impact of distribution of wealth and resources on QOL, we must have some clear idea of what QOL actually is. Is it synonymous with satisfaction? With happiness? With human well-being? With consumption? A quick perusal of the literature shows that QOL is a topic of research in a broad range of disciplines. In fields as disparate as advertising, economics, engineering, industry, medicine, politics, psychology, and sociology, improving QOL is often claimed as a primary goal. However, real paradoxes in interpretations of QOL exist. For example, a substantial motivation behind the environmental movement is to improve human QOL, and the same motivation can be argued for the industries (e.g., logging, mining, auto) that are so often its foes. Farquhar (1995, cited in Haas, 1999) claims that the term may be one of the most multidisciplinary in common use, yet even within a discipline there seems to be little consensus regarding its actual definition. In fact, a common criticism against the phrase 'Quality of Life' is that "the concept lacks specificity; it has as many meanings as life has aspects" (Schuessler and Fisher, 1985).

It seems that improving human QOL should be the dominant policy objective of any government (Schuessler and Fisher, 1985), yet over the past 50 years

Understanding and Solving Environmental Problems in the 21st Century
Edited by R. Costanza and S.E. Jørgensen
© 2002 Elsevier Science Ltd. All rights reserved

this priority has been given to increasing the production of goods and services for consumption (Ekins and Max-Neef, 1992). It would seem to follow that policymakers implicitly assume that consumption is a suitable proxy for QOL. How far is this from the truth in modern consumerist society? Consumerism has been defined as a cultural orientation that holds that "the possession and use of an increasing number and variety of goods and services is the principal cultural aspiration and the surest perceived route to personal happiness, social status and national success" (Ekins, 1991). In 1990, nearly three-quarters of entering college students in the USA believed that being 'very well off financially' was 'essential', presumably for their QOL (Durning, 1992). Bloom et al. (2000) assert that "few statements in the development literature command as much universal assent as the claim that higher incomes lead to higher human development", where 'human development' implies QOL. Some believe that our economy depends on consumption to the extent that we must *make* it the vehicle by which we improve our QOL, if it is not already. In the words of retailing analyst Victor Lebow, "our enormously productive economy ... demands that we make consumption our way of life, that we seek our spiritual satisfaction, our ego satisfaction, in consumption ... We need things consumed, burned up, worn out, replaced and discarded at an ever increasing rate" (quoted in Durning, 1992, p. 22). However, the term QOL first came into common usage in the 1960s to address the issue of increasing crime and violence in the midst of growing material prosperity (Haas, 1999), explicitly distinguishing QOL from consumption. The American Heritage Dictionary defines QOL as "the degree of emotional, intellectual, or cultural satisfaction in a person's everyday life *as distinct from the degree of material comfort*" (The American Heritage Dictionary of the English Language, 1992; emphasis added). In addition, while people may believe that greater consumption would increase their QOL, psychology studies find little correlation between consumption and happiness (Durning, 1992).

The definition of QOL has evolved through time since the phrase first became widely used in the early 1960s. Early researchers often sought objective definitions (e.g., Mishan, 1967). However, empirical studies generally find poor correlation between objectively measured and subjectively assessed QOL. Hence, since the late 1970s, there has been a growing consensus that QOL is not an objective condition at all, but is rather a subjective one, concerned with people's own estimations of their individual welfare. Evidence suggests that individuals subjectively interpret their own QOL relative to an ideal standard or to a reference group (Haas, 1999). Thus, communities with low standards of living are sometimes found to rate their QOL as the same or even better than communities with higher standards of living, presumably because they aspire to less or compare themselves to others of similarly modest circumstances (Schuessler and Fisher, 1985). Encompassing these various considerations, Haas (1999) defines QOL as "a multidimensional evaluation of an individual's current life circumstances in the context of the culture in which they live and the values they hold. QOL is primarily a subjective sense of well-being encompassing physical, psychological, social, and spiritual dimensions. In some circumstances, objective indicators may supplement or, in the case of individuals unable to subjectively perceive, serve as a proxy assessment of QOL".

There are four noteworthy elements in Haas's definition of QOL. First, while much of the literature emphasizes the subjective nature of QOL, this definition allows for objective indicators. Two goals of this and the following chapter will be to suggest policies for improving QOL, and to suggest objective indicators for determining the success of such policies. While QOL may be primarily subjective, it is easier to advance and assess the success of policies that have measurable objective goals rather than subjective ones. Second, the emphasis on the subjective nature of QOL opens the door to policies designed to influence people's perceptions of their own QOL. Third, in this definition, the physical dimension of QOL (i.e., wealth and resources) is only one element of many and is the only one that has physical limits. Since the concern of this chapter is the relationship among wealth, resources, distribution, and QOL, these latter two elements of Haas's definition open up the possibility that we can distribute wealth and resources more equitably without compromising the QOL of those who currently possess the lion's share. The fourth element of interest in Haas's definition is that QOL is determined in the context of culture and values. Many economists argue that preferences are fixed and given. The economist's goal is simply to determine how those preferences can most efficiently be satisfied, and the policymakers' goal is to create the conditions to facilitate this. However, since cultures and values can change, it follows that the specific determinants of QOL can as well.

If the second and last elements of Haas's definition are correct, they suggest that a society could increase the QOL of its citizens by purposefully changing their preferences. The idea of purposefully changing people's preferences may seem patronizing and against the liberal view that it is the inalienable right of the individual to have sovereign preferences. However, the reality is that one person's actions do have impacts on others' well-being, and the advertising industry is actively devoted to changing our preferences every day. If we are concerned with the QOL of the entire world and of future generations, then it seems reasonable to argue that we are justified in changing preferences in such a way that maintaining or enhancing the QOL of one country or generation does not compromise that of others. Since wealth and resources are the only components of QOL that can be physically depleted, they are the only components whose excessive consumption (and dissipation into waste) can threaten the QOL of others.

The laws of thermodynamics ensure that the ultimate source of wealth and resources, and the ultimate recipient of the waste products from their use, is our environment (fig. 1). Therefore, we must closely examine the relationship

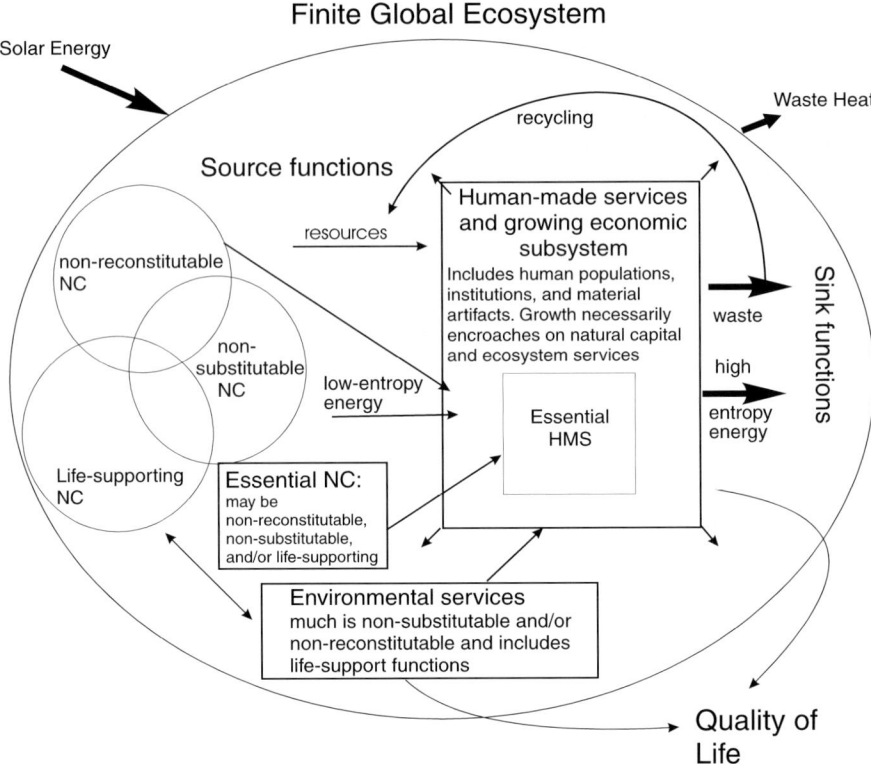

Fig. 1. Conceptual model of the dual threat (increased stress and reduced response/resistance that human population growth and activity pose to ecosystem and human health. Adapted from Collados and Duane (1999).

between QOL and the natural environment. Collados and Duane (1999) provide an appropriate framework. The natural environment (whose resource stocks and the services they generate will hereafter be referred to as natural capital) generates numerous environmental goods and services that enhance QOL in three ways. First, they provide the materials used by the human economy to produce all human-made products. Second, they directly provide humans with benefits of a type that cannot be imported from elsewhere. Third, they are essential for the reproduction of additional environmental goods and services. Of the human-made products produced from natural capital, some are essential for human life (though 'essentiality' may be culture-specific) while all others are non-essential. Natural capital can be divided into four classifications according to its ability to produce environmental services. First, natural capital required to make essential human capital is itself essential. Second, natural capital required for the reproduction of itself is life supporting. Third, natural capital for which no human-made substitutes exist is non-substitutable, and fourth, that which cannot be regenerated once it is destroyed is non-reconstitutable. Specific stocks of natural capital may exhibit none, any, or all of these properties. Clearly then, the relationship between QOL and the natural environment is critical.

In summary, QOL is a complex, multi-dimensional concept that may be largely subjective but whose enhancement is probably facilitated by certain objective factors. Policy goals for improving or maintaining QOL can seek to create the objective conditions associated with a superior QOL, or attempt to change people's subjective assessment of conditions in a way that improves their QOL. As the only element of QOL that can be physically depleted, the use of wealth and resources by some countries or generations can affect the QOL of others. Further, the depletion of wealth and resources can threaten natural life-support functions without which human life itself is threatened. The appropriate distribution of wealth and resources is therefore a critical element in any effort to sustain and improve the QOL we now enjoy. If we are to pursue policies towards this end, however, we must first be able to measure the outcomes of these policies, a topic to which we now turn.

2. How has Quality of Life been measured?

2.1. Economic income, economic welfare, and human welfare

If improving QOL is indeed the goal of social policies and programs, it follows that appropriate national aggregate accounting systems should attempt to measure the extent to which policies actually improve QOL, and this is arguably a fair statement of what they are intended to do. Although QOL is largely a subjective assessment, in practice, it must be measured with objective proxies. The number of existing and proposed aggregate accounting systems reflects considerable disagreement over what are the most suitable proxies, and the requirements for different systems differ depending on what proxies are used. Such proxies include (1) the level and pattern of economic activity, (2) sustainable economic income – the amount that can be consumed without depleting capital stocks (Hicks, 1946), (3) economic welfare – the net economic component of total welfare (Daly and Cobb, 1989), and (4) human welfare – the degree to which human needs are fulfilled (Max-Neef, 1992). This range of proxies is arrayed in fig. 2 and table 1.

2.2. Level and pattern of economic activity: gross national product

The simplest objective for an aggregate accounting system is to develop an indicator of the production of goods and services in the economy for comparisons either across space or time. In order to avoid double counting, one can focus only on "final" goods and services (i.e., those which attain their final point of use during the accounting period, and are not intermediate in the sense of being destined for

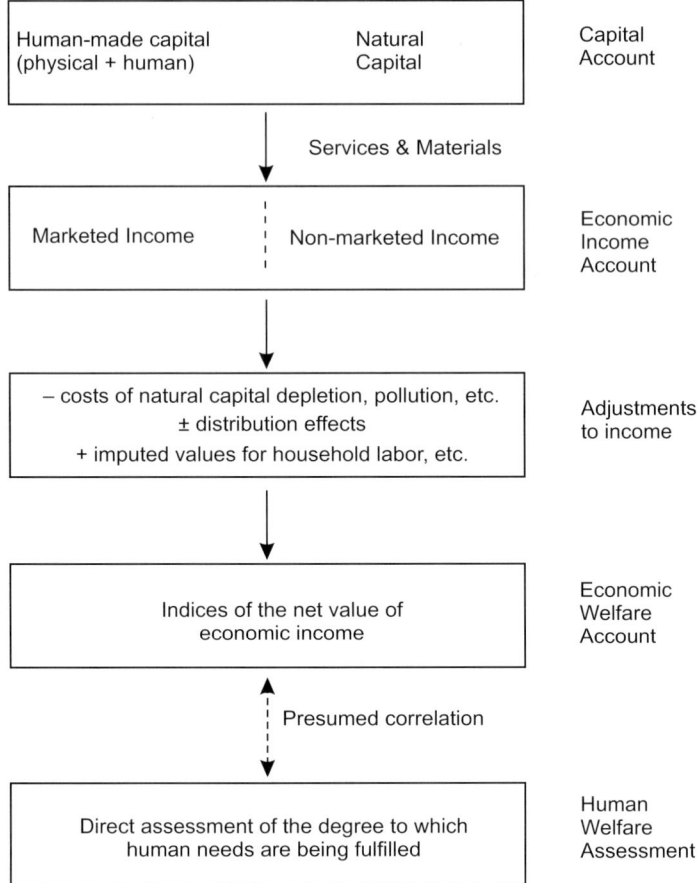

Fig. 2. Distinctions between economic income, economic welfare, and human welfare.

incorporation into further goods and services). As an accounting procedure, if production activity is fully compensated by monetary payments, *either* aggregate incomes obtained from production activity or aggregate expenditures can be used as an indicator. These two ways of measuring the total (income or expenditure) should be equal. This measure is referred to as Gross National Product (GNP; table 1, column 1), and is by far the most widely used of the measures presented.

There are a number of serious flaws with GNP as an indicator of QOL. Economic income is a measure of the production and use of goods and services, with variations (i.e., between columns 1–3 in table 1) concerning the treatment of environmental services, natural capital, and other non-marketed products. Clearly, total economic income is ultimately generated from the stocks of both human-

Table 1
A range of goals for national accounting and their corresponding frameworks and measures

Goal		Economic income		Economic welfare	Human welfare
	Marketed	Weak sustainability	Strong sustainability		
	1	2	3	4	5
Basic framework	value of marketed goods and services produced and consumed in an economy	1 + non-marketed goods and services consumption	2 + preserve essential natural capital	value of the welfare effects of income and other factors (including distribution, household work, loss of natural capital, etc.)	assessment of the degree to which human needs are fulfilled
Non-environmentally adjusted measures	**GNP** (Gross National Product) **GDP** (Gross Domestic Product) **NNP** (Net National Product)			**MEW** (Measure of Economic Welfare)	**HDI** (Human Development Index)
Environmentally adjusted measures	**NNP'** (Net National Product including non-produced assets)	**ENNP** (Environmental Net National Product) **SEEA** (System of Environmental Economic Accounts)	**SNI** (Sustainable National Income) **SEEA** (System of Environmental Economic Accounts)	**ISEW** (Index of Sustainable Economic Welfare)	**HNA** (Human Needs Assessment)

made and natural capital (the "wealth" accounts) and includes both marketed and non-marketed items. But conventional measures of marketed economic income and expenditure (i.e., GNP) do not adequately pick this up. Measures of *sustainable* economic income attempt to incorporate non-marketed natural capital changes. If it is assumed that natural and human-made capital are substitutable, then the goal is to measure *weakly sustainable income* (column 2). If it is assumed that natural and human-made capital are not substitutable in all cases, then the goal is to measure *strongly sustainable income* (column 3). But increases in economic *income* may not correlate with increases in economic *welfare* (loosely speaking, welfare is a synonym for QOL), especially if the income measures do not adequately distinguish "costs" from "benefits".

Economic welfare (column 4, table 1) attempts to look at not just how much income is generated, but also at how much economic welfare is produced. As shown in fig. 2, these measures generally adjust income to better reflect which items in the income measures are costs and benefits. They do this by subtracting costs (such as natural capital depletion and pollution), imputing values to missing services (such as household labor), and adjusting for income distribution effects using indices of income distribution. Finally, economic welfare measured as the production of net benefits may still not correlate with overall human welfare, since many human needs are not related to consumption of economic products or services (Max-Neef, 1992). Human welfare (column 5, table 1) looks directly at the degree to which human needs are being met, economic production being only one of many possible means to these ends. These distinctions and the specifics of sustainable economic income, economic welfare, and human welfare are further elaborated below.

2.3. Sustainable economic income

"Accounting Income", as measured by GNP, is simply the sum of monetary payments to owners of the inputs used in production during an accounting period. "Hicksian Income", however, subtracts from accounting income the costs of maintaining the productive capacity of capital stock (Hicks, 1946). These costs may include a variety of defensive actions that maintain effective capital stock, such as replacement, repair, and maintenance. It also includes avoidance costs that are designed to avoid losses in capital productivity. Weitzman (1976), Atkinson et al. (1997), and others have shown that Net National Product as measured by national income statistics – consumption plus net investment, or GNP – is theoretically equivalent to sustainable income, or at least would be under some 'heroic' assumptions, such as the inclusion of all forms of capital, investment, and consumption in national accounts.

How could we extend NNP to include other forms of capital, for example, marketed natural capital? While the principle is the same as that outlined above, the actual adjustment is more complicated for natural capital stocks. We can distinguish between two types of marketed natural capital, renewable and non-renewable[1], and the adjustment method differs somewhat for each. In either case, the method of adjustment (net cost or user cost) depends on the cheapest manner in which the lost productive capacity of the natural capital stock can be replaced (El Serafy, 1989; Costanza et al., 2001).

2.3.1. Green accounting
An effective measure of sustainable income must also account for non-marketed goods and services, in particular, those produced by healthy ecosystems. Efforts to incorporate these goods and services into national accounts are referred to as 'green accounting' (e.g., Nordhaus and Kokkelenberg, 1999). In addition to deducting the loss of natural capital from GNP, green accounting adjustments for full income require valuations of income flows from natural capital. A variety of methods can be used to estimate these flows, depending on the type of income received. Table 2 outlines some of these valuation methods. The previous section described a number of methods to account for sustainable income from natural capital. Without too much oversimplification, green accounting changes these methods primarily by redefining productive capacity to account for both marketed and non-marketed benefits of this capital. For further details on specific methodologies, see Costanza et al. (2001) or Nordhaus and Kokkelenberg (1999).

Clearly then, green accounting demands that cost adjustments be made for losses in natural capital as well as human-made capital. As human-made capital is, by definition, replicable at a cost, cost adjustments to account for capital losses are relatively simple; but for natural capital it is far more difficult to calculate future incomes lost or costs necessary to replace or avoid the loss or degradation in capital productivity. This is true for several reasons.
1. There are no well-functioning markets for measuring prices of all natural capital forms.
2. There are no well-functioning markets that would equate the price of natural capital with its replacement cost, and, of course, non-substitutable, non-reconstitutable natural capital, as defined above, cannot be replaced.
3. Natural capital productivity is more complex and less amenable to measurement than human-made capital.
4. The productive state, or health, of natural capital is more difficult to measure than that of human-made capital.
5. We are profoundly ignorant of how human impacts change ecosystem health, how ecosystem health affects natural capital productivity, and when deterioration in ecosystem health leads to irreversible impacts on ecosystem productivity.

---

[1] It is worth noting that non-renewables, such as oil, are difficult to exhaust, since eventually it becomes more expensive to discover and extract than it is worth. Renewables, such as trees, can be exhausted.

Table 2
Valuation techniques for some environmental functions [a]

Functions	Valuation technique

System value
  Erosion control
  Local flood reduction
  Regulation of streamflows
	Change in productivity, preventive expenditure, trade-off games, cost effective analysis, replacement cost

Ecological values
  Fixing and cycling nutrients
  Soil formation
  Cleansing air and water
	Change in productivity, loss of earnings, opportunity cost, trade-off games, cost effective analysis, replacement cost

Biodiversity
  Gene resource
  Species protection
	Opportunity cost, cost effective analysis, replacement cost, shadow project, relocation cost

Aesthetic	Property value, wage differential
Recreation	Travel cost
Cultural	Travel cost

[a] Modified from Dixon and Sherman (1990).

These difficulties are most severe for ecosystems, which provide a variety of goods and services in a complex manner. Forests and wetlands are good examples, both representing complex ecosystems where measurements of health and productivity are complicated, where some goods and service flows are marketable and others are not, and where private market values of the capital forms do not necessarily reflect their value or cost of replacement (Daily, 1997; Costanza et al., 1997).

2.3.2. Weak vs. strong sustainability

The sustainability of income requires replacement, or avoidance of loss, of some forms of capital sufficient to maintain consumption opportunities. This means that substitutability plays a crucial role in implementing any sustainable income adjustments to GNP. "Weak" sustainability requires maintenance of the total capital stock. It assumes a high degree of substitutability between all forms of capital. "Strong" sustainability presumes limited substitutability between natural capital and other capital forms; therefore, strong sustainability requires the maintenance of some natural capital separately from other capital forms (Costanza and Daly, 1992; Pearce, 1993; El Serafy, 1996). Ultimately, of course, we cannot make something from nothing, which means that natural capital is an essential input into any other form of capital. However, the fact that it is also impossible to make nothing from something means that some natural capital will always be available, even if only in states of very high entropy, and the argument for weak sustainability implies that improvements in human capital could allow us to use even the highest entropy natural capital.

A wide range of replacement cost options are available under the weak sustainability case, including the forms of capital lost or degraded as well as substitute forms of capital. In the case of strong sustainability, degraded natural capital must be replaced in comparable form. There is no well-defined line dividing the two cases of weak and strong sustainability. The essence of the distinction relies on the ability of various capital forms to provide a flow of income; i.e., the degree of substitutability between capital forms. There are no reasonable substitute capital forms for those types of natural capital which provide basic life-support functions at large spatial and temporal scales (such as availability of the proper mix of ambient gases, hydrologic flows, protection from ultraviolet rays, etc.), although there may be substitutability at small scales.

Risk and uncertainty are critical elements in the debate between strong and weak sustainability. We do not understand ecosystems sufficiently to predict the impacts of human action on their ability to reconstitute themselves, nor can we predict what technologies will evolve in the future to substitute for ecosystem functions. Therefore, we could define strong sustainability as a social notion. A society may need to identify a level of natural capital beyond which it will not substitute for fear of approaching an irreversible ecological threshold. For example, society could define levels of biodiversity below a certain threshold as unacceptable. Accounting adjustments for degradation beyond these points require estimates of costs of repair to acceptable levels. That is, if biodiversity in a wetland falls below the threshold level, adjustments to income must account for the cost of repair. This cost of repair may be the engineering costs of wetland restoration and include restoration of extirpated species. If engineering methods will not successfully repair the damages, and natural processes will, accounting adjustments must be made for income losses attributable to not using those wetlands during their natural regeneration period.

The Environmental Net National Product (ENNP) (Mäler, 1991; Hamilton and Lutz, 1996) and the UN's System of Environmental Economic Accounts (SEEA; Bartelmus, 1994) are both measures that account for weak sustainability. Accounting for strong sustainability requires adjusting for the cost to return specific forms of degraded natural capital to their "acceptable" conditions (Hueting, 1989). The Sustainable National Income (SNI; Hueting, 1995) and some versions of the SEEA incorporate this perspective.

2.4. Measuring economic welfare

So far, we have been discussing various measures of economic income, with various adjustments for the sustainability of that income. Column 4 in table 1 moves from the goal of Economic Income assessment to the goal of Economic

Welfare assessment. The latter goal is more complex and requires clearly distinguishing between costs and benefits. While this distinction between costs and benefits is absolutely essential if one wants to talk about welfare rather than income, it is inherently a difficult and somewhat subjective and arbitrary distinction.

2.4.1. Adjusting for defensive expenditures

Both sustainable income measures and green accounting as outlined here do not necessarily adjust GNP for expenditures designed to reduce or mitigate the impacts of environmental degradation and pollution. Such costs, referred to as 'defensive expenditures', contribute to GNP when undertaken by households or the government (when undertaken by businesses, they show up as intermediate costs) (Markandya and Perrings, 1993). A problem in adjusting for defensive expenditures is distinguishing between "incurred" and "defensive" expenditures. For example, medical expenses may be purely to offset adverse consequences of economic activity and permit the maintenance of original welfare levels (i.e., fully repair or avoid degrading human capital). On the other hand, some medical expenses may truly result in improvements in welfare above original levels, and thus should be considered net investment in human capital stock. In practice, distinguishing between these two types of expenditures, one welfare-enhancing and the other welfare-maintaining, is difficult. Furthermore, there may be costs associated with economic activities that are not mitigated by defensive expenditures. For example, untreated health costs from pollution or work days lost from pollution are not explicit defensive expenditures. Increased time costs necessary to catch recreational fish or diminished recreational enjoyment attributable to a degraded watershed do not have explicit defensive expenditures that can be observed and used for adjusting GNP. These costs would have to be deducted from income to obtain a net income measure since they do not reflect positive utility-creating consumables. Explicitly defensive expenditures, such as the increased travel time costs necessary to access an adequate recreational facility when a former one becomes too degraded for use, should be deducted from income as costs. However, traditional accounting would add them into income, erroneously suggesting welfare improvements.

Nordhaus and Tobin (1972) produced an early version of this kind of indicator in their Measure of Economic Welfare (MEW). MEW starts with GNP and makes three types of adjustments: "Reclassification of GNP expenditures as consumption, investment, and intermediate; imputation for the services of consumer capital, for leisure, and for the product of household work; and a correction for some of the disamenities of urbanization" (Nordhaus and Tobin, 1972, p. 5).

MEW focuses on the aggregation of individual welfare; it is "atomistic". MEW does not include any adjustments for distributional effects or for environmental costs. Daly and Cobb (1989) developed an Index of Sustainable Economic Welfare (ISEW) that takes consumption as a starting point, but incorporates some

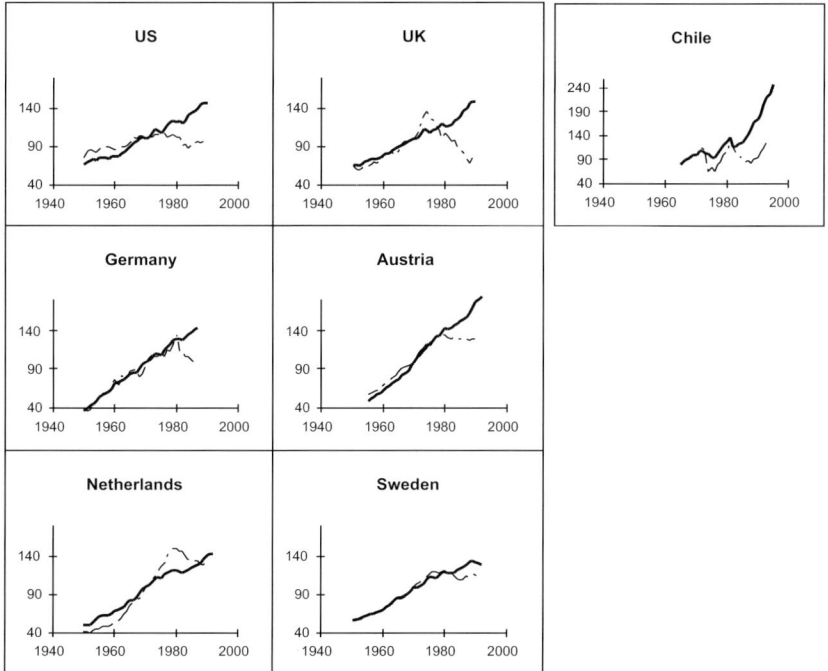

Fig. 3. Indices of GNP (solid) and ISEW (dashed) for several countries. 1970 = 100 in all cases.

of the environmental and distributional issues ignored by MEW. To summarize, the ISEW
- Allows for an income distribution adjustment;
- Includes changes in the stock of fixed reproducible capital, but excludes land and human capital in this calculation;
- Includes estimates for the costs of air, water, and noise pollution;
- Includes estimates of costs of the loss of wetlands and farmlands, depletion of non-renewable resources, commuting, urbanization, auto accidents, advertising, and long-term environmental damage;
- Includes imputed values for the value of unpaid household labor; and
- Omits any imputation of the value of leisure.

Daly and Cobb (1989) and Cobb and Cobb (1994) calculated ISEW for the US economy for the period 1950 to 1993. Other researchers estimated ISEW for several other countries and the indices of ISEW for several cases are shown in fig. 3 along with GNP indices for the same countries. In most of these cases, ISEW and GNP per capita run parallel for some initial period, but separate during the 1970s and 1980s. Max-Neef (1995) has postulated that this separation is evidence for a "threshold hypothesis" in which growth of economic income increases welfare only until a threshold is reached where the costs of additional growth (which are counted as benefits in GNP) begin to outweigh the real benefits. Nordhaus and Tobin (1972) calculated their MEW in 1972 before the threshold was reached, concluding that GNP was an adequate proxy for economic welfare.

The ISEW is certainly far from a perfect measure of Economic Welfare or QOL, but it is decidedly better than GNP for this purpose. This is because, as we have pointed out, GNP is not a welfare measure at all, but only an income measure.

2.5. Assessing human welfare directly

While the ISEW provides a measure of environmentally adjusted economic welfare, it is still based on measuring how much is being consumed, with the tacit assumption that more consumption leads to more welfare. A completely different approach would be to look directly at actual well-being or QOL achieved. This would separate the means (consumption) from the ends (QOL) without assuming one is correlated with the other (fig. 2). The UN's Human Development Index (HDI) is a crude attempt to assess human well-being by using an index comprised of generally available data on four basic needs variables at the country level: (1) life expectancy at birth; (2) literacy; (3) average number of years of schooling; and (4) GDP per capita (converted at purchasing power parity). Although it includes more than economic income by adding the other three elements, it is still based on "means" assessment and excludes any measures of environmental degradation. Max-Neef (1992), in contrast, has developed a matrix of human needs and has begun to address well-being more directly from the "ends" perspective by involving people in interactive dialogues to perform a Human Needs Assessment (HNA; table 3). The key idea is that humans do not have primary needs for the products of the economy; the economy is only a means to an end. The end is the satisfaction of primary human needs. Food and shelter are ways of satisfying the need for subsistence. Insurance systems are ways to meet the need for protection. Religion is a means to meet the need for identity. Max-Neef suggests,

> Having established a difference between the concepts of needs and satisfiers, it is possible to state two postulates: first, fundamental human needs are finite, few and classifiable; second, fundamental human needs (such as those contained in the system proposed) are the same in all cultures and in all historical periods. What changes, both over time and through cultures, is the way or the means by which the needs are satisfied.   *Max-Neef (1992, pp. 199–200)*

This is a very different conceptual framework from the others in table 1, which assume that human desires are infinite and that, all else being equal, more consumption is always better. According to the alternative conceptual well-being framework, we should be directly measuring how well basic human needs are being satisfied since overall human well-being and consumption are not necessarily correlated and may, in fact, be going in opposite directions. Quantifying HNA, however, is even more difficult than HDI or ISEW or other

Table 3
Matrix of human needs

Axiological categories	Existential categories			
	Being [b]	Having [c]	Doing [d]	Interacting [e]
Subsistence	Physical health, mental health, equilibrium, sense of humor, adaptability	Food, shelter, work	Feed, procreate, rest, work	Living environment, social setting
Protection	Care, adaptability, autonomy, equilibrium, solidarity	Insurance systems, savings, social security, health systems, rights, family, work	Cooperate, prevent, plan, take care of, cure, help	Living space, social environment, dwelling
Affection	Self-esteem, solidarity, respect, tolerance, generosity, receptiveness, passion, determination, sensuality, sense of humor	Friendships, family, partnerships, relationships with nature	Make love, caress, express emotions, share, take care of, cultivate, appreciate	Privacy, intimacy, home, space of togetherness
Understanding	Critical conscience, receptiveness, curiosity, astonishment, discipline, intuition, rationality	Literature, teachers, method, educational policies, communication policies	Investigate, study, experiment, educate, analyze, meditate	Settings of formative interaction, schools, universities, academies, groups, communities, family
Participation	Adaptability, receptiveness, solidarity, willingness, determination, dedication, respect, passion, sense of humor	Rights, responsibilities, duties, privileges, work	Become affiliated, cooperate, propose, share, dissent, obey, interact, agree on, express opinions	Setting of participative interaction, parties, associations, churches, communities, neighborhoods, family
Idleness	Curiosity, receptiveness, imagination, recklessness, sense of humor, tranquility, sensuality	Games, spectacles, clubs, parties, peace of mind	Daydream, brood, dream, recall old times, give way to fantasies, remember, relax, have fun, play	Privacy, intimacy, space of closeness, free time, surroundings, landscapes

continued on next page

Table 3, *continued*

Axiological categories	Existential categories			
	Being[b]	Having[c]	Doing[d]	Interacting[e]
Creation	Passion, determination, intuition, imagination, boldness, rationality, autonomy, inventiveness, curiosity	Abilities, skills, method, work	Work, invent, build, design, interpret	Productive and feedback settings, workshops, cultural groups, audiences, spaces for expressions, temporal freedom
Identity	Sense of belonging, consistency, differentiation, self-esteem, assertiveness	Symbols, language, religion, habits, customs, reference groups, sexuality, values, norms, historical, memory, work	Commit oneself, integrate oneself, confront, decide on, get to know oneself, recognize oneself, actualize oneself, grow	Social rhythms, everyday settings, settings in which one belongs, maturation stages
Freedom	Autonomy, self-esteem, determination, passion, assertiveness, openmindedness, boldness, rebelliousness, tolerance	Equal rights	Dissent, choose, be different from, run risks, develop awareness, commit oneself, disobey	Temporal/spatial plasticity

[a] From Max-Neef (1992).
[b] The column of **Being** registers attributes, personal or collective, that are expressed as nouns.
[c] The column of **Having** registers institutions, norms, mechanisms, tools (not in material sense), laws, etc., that can be expressed in one or more words.
[d] The column of **Doing** registers locations and milieus (as time and spaces). It stands for the Spanish **Estar** or the German **Befinden**, in the sense of time and space.
[e] Since there is no corresponding word in English, **Interacting** was chosen à faut de mieux.

"means"-based measures, especially across time and between different countries – those conditions for which we would most like to have the quantification for comparisons. This is obviously an area in need of much further research.

In summary, we have presented several alternative frameworks for measuring national income and well-being, all of which are intended to some degree to measure QOL. As one moves to the right in table 1, the suggested changes to simple national income accounts become more controversial and difficult, but also many would argue, more closely reflective of QOL. The correlation between QOL and the accumulation and consumption of wealth and resources also grows increasingly tenuous as we move to the right. Establishing that the connection between wealth and resources and QOL is not necessarily a one-to-one monotonic relationship has important implications for the following discussion on fairness and distribution.

3. A comparison of two approaches to fairness in the distribution of wealth and resources

Any concern for distribution implicitly assumes some normative goal. In this discussion of the distribution of wealth and resources, the explicit normative goal is fairness, both within the current generation and between generations. We will lay out two approaches to fairness, first the procedural approach of market economics, and second an explicitly outcome-based approach derived from theories of justice. While market economics is not traditionally viewed as laying out a 'theory of fairness', it is the dominant system determining the distribution of wealth and resources on the planet and, therefore, its implications for fairness must be assessed.

3.1. Fairness across individuals in space

Free market economists claim that economics is a positive science with no normative judgements involved. However, if forced to, they would probably define fairness as an inevitable outcome of the freedom of choice allowed by a free market economy; markets are fair because they (theoretically) offer the "justice of earned deserts" (Lane, 1986, p. 1). All wages are fair since the individual is free not to work for that wage if he/she believes it to be unfair. All prices are fair since no one is physically coerced into buying or selling anything. It may not be fair that some individuals are born more intelligent, beautiful, or talented than others and therefore do better than others within a free market; however, this disparity in genetic assets is simply a fact of nature and there is no reason it should be dealt with differently in human society than it is in nature. In addition, because the free market provides abundant incentives for production and innovation, it leads to continuous increases in consumption and invention. The resulting new discoveries and declining prices make people with poorer genetic endowments better off than they would have been under a different system. It is not only economists who might pursue such a line of reasoning. Polls have shown that at least as recently as 1977, the majority of Americans believed that the free enterprise system is "fair and wise" (82%), "gives everyone a fair chance" (65%), and is a "fair and efficient system" (63%) (McClosky and Zaller, 1985, cited in Lane, 1986). This theory of justice will hereafter be referred to as the 'just deserts' theory.

Ultimately, the test of whether or not the market system is fair should be the fairness of wealth and resource distribution as determined by that system. Industrial countries with only 25% of the planet's population consume 40–86% of the earth's various natural resources (Durning, 1992, p. 50). The USA alone, with 4% of the global population, consumes 25% of global resources and generates a similar percentage of global waste. Bill Gates controls the same wealth as 45% of the poorest households in the USA, and the three most well-to-do individuals in the world control wealth greater than the combined GDP of the 48 least-developed countries. Both within and between nations, the concentration of wealth is becoming ever more pronounced (Gates, 1999). Given these facts, it becomes increasingly difficult to argue that the market system produces fair outcomes, even if we believe it to be procedurally fair. However, economics is largely a prescriptive 'science', not a descriptive one. When the empirical facts contradict economic theory, the standard response of economists is not to change their theory, but rather to suggest policies that will bring the world more in line with that theory. Thus, to prove to many economists that the market system is unfair we must show that the theory is unfair on its own terms.

A variety of market failures such as externalities, public goods, and missing markets do make economic theory unfair on its own terms, especially when this theory drives the allocation of natural capital. We will define and discuss the impact of externalities in section 4 (*Can we measure fairness?*), and define and discuss public goods and missing markets in chapter 12. In the meantime, we will look at fairness from a theory of justice point of view.

Based largely on market outcomes, many justice theorists (and average citizens) find the lack of fairness to be one of the weak points of the market system (Lane, 1986). Many justice theorists argue that fairness is an essential condition for justice, and in some senses the two are virtually indistinguishable. As Barry states, "the central issue in any theory of justice is the defensibility of unequal relations between people" (Barry, 1989, p. 3).

One theory of justice, introduced in Plato's *Republic*, is that justice arises from mutual advantage. Succinctly, "[j]ustice is the name we give to the constraints on themselves that rational self-interested people would agree to as the minimum price that has to be paid in order to obtain the cooperation of others" (Barry, 1989, p. 7). Glaucon in Plato's *Republic* says that someone capable of inflicting injustice without suffering it would be 'insane' to make a pact prohibiting him from doing so. In this view, there is no such thing as intergenerational justice, since we are capable of inflicting any injustice on future generations without fear of reprisals. If we accept that we have ethical duties to future generations, then we must believe in an alternative theory of justice.

Perhaps the most widely accepted alternative theory is commonly referred to as 'justice as impartiality', or 'justice as fairness'. Under this theory, it is just to give to others that which they cannot demand, but which would be fair (Shue, 1992). It is this understanding of justice that motivated John Rawls's *A Theory of Justice* (1971), in which he lays out precisely what is 'fair'. Given the extensive influence of this work, it is worth briefly laying out Rawls's approach.

Rawls used the thought experiment of a 'veil of ignorance' to arrive at the criteria for a just society. The idea was that if rational people could decide on the type of society they would want to live in without knowing anything about what role they would play in that society or what their personal attributes would be (i.e., from behind a 'veil of ignorance'), the resulting society must be just. Rawls first assumed that a person's wealth and position in society arose as a result of three morally arbitrary lotteries: one for parentage and natal social position, a second for luck, and a third for genetic potential. Since all inequalities in wealth and power arose from morally arbitrary lotteries, the only initially justifiable position must be one of equal distribution. However, equal distributions remove incentives. By providing incentives, allowing inequality would lead to greater wealth, and could improve the welfare of everyone over the initial equal distribution. As long as increasing inequalities improved the welfare of everyone (Pareto improvements[2]), everyone would agree to it. However, beyond a certain point, increasing inequality might make some groups better off while making others worse off. If we would arrive at one of these positions starting from equality, we are going beyond mere Pareto improvements.

What Rawls concludes is that inequality should be increased until it maximizes the welfare of the worst-off group. While those at the bottom should favor this since it maximizes their welfare, it is possible that some groups would have higher welfare with lower inequality, and would therefore favor a stop to increasing inequality where they maximized their welfare. Rawls asserts roughly that as differences arise from morally arbitrary advantages in the first place, what was fair to the worst-off group must also be fair to any group better off than they are. Basically then, the worst-off group has veto power over the degree of inequality. Rawls refers to this concept as 'the difference principle'. However, Rawls does not venture to explicitly outline what sort of system meets these criteria. The Rawlsian theory outlined here will hereafter be referred to as 'justice theory'.

---

[2] A Pareto improvement is any change in distribution or allocation that leaves at least one person better off without leaving anyone worse off. Theoretically, all market transactions lead to Pareto improvements, since market transactions are voluntary, and no one would undertake a voluntary transaction if it left her worse off.

Many people would argue that although they arrive there from different starting points, believers in 'just deserts' and 'justice theory' might reach the same practical conclusion. The argument is that a free market economy is indeed the institution that, by providing adequate incentives, leads to the greatest QOL for the worst-off individual. Economists have explicitly argued that the statistical data reveal an "inescapable relationship", that is, "the greater the concentration of wealth, i.e., income, in the hands of the few, the greater the capital supply, and therefore the greater the gain in national well-being" (Snyder, 1936). Supply-side economics, popularly known as 'trickle down theory', makes the same basic assumptions. If it is true that the market system is just in the Rawlsian sense, it must mean either that remaining poverty is a result of market imperfections that can be eliminated, or else society has insufficient resources to end poverty. While it may seem absurd to think that the western nations lack the resources to end poverty, some people do believe that steps taken to redistribute wealth in order to eliminate poverty would inevitably destroy the incentive structure to such an extent that in the long run poverty would actually increase[3]. However, a market economy is unlikely to generate conditions more fair than those from which it begins. That is, a market economy might be able to allocate goods fairly, but only if the initial distribution of wealth and resources is fair. Further, even in theory a market allocation is only fair (in the sense of 'just deserts') if market conditions prevail for all resources, which we will show later is clearly not the case.

3.2. Fairness across individuals in time

While the above discussion applies to fairness across individuals in space, fairness through time is considerably more complex. There are three important ways the current generation can have an impact on future generations: population growth, the quality and quantity of man-made capital created, and the quality and quantity of natural capital preserved. Man-made capital can be sub-divided into capital goods (infrastructure, machinery, etc.) and knowledge (technology, culture, etc.). Natural capital can be subdivided into three categories with distinct attributes: non-renewable natural resources, renewable resources (defined to include the waste absorption capacity of the environment), and environmental services (which

---

[3] For example, Milton Friedman implicitly makes this argument when he states that "few trends could so thoroughly undermine the very foundation of our free society as the acceptance by corporate officials of a social responsibility other than to make as much money for their stockholders as possible" (Friedman, 1963). The same is true for followers of Ayn Rand, who explicitly presents the thesis in her fiction and non-fiction works that a more equal re-distribution of wealth will lead to increasing misery for all (e.g., Rand and Peikoff, 1996; Rand, 1964). It is interesting to note that Alan Greenspan, the current chairman of the Federal Reserve, is a professed admirer of Ayn Rand's philosophy, and has contributed to one of her books (Ramo, 1999).

include life-support functions of ecosystems, amenity values, genetic resources, climate stability, etc.). Further, natural capital can be essential, life supporting, non-substitutable and/or non-reconstitutable, as defined in the first section of this chapter.

Economists rarely hesitate in supporting the free market economy as the 'fairest' system across space and time. Up through the mid-20th century, continuous advances in technology and standards of living led most intergenerational economic analysis to focus on the optimal savings rate issue. How much capital should one generation accumulate for the next, at the cost of current consumption? One of the most influential results of this analysis was Phelps's Golden Rule of Capital Accumulation (Phelps, 1961, 1965). Basically, the idea was that current generations should sacrifice for future generations until the maximum sustainable per capita consumption was achieved. Since successive generations were living better owing to the contributions of past generations, it was virtually taken for granted that we should sacrifice for the greater well-being of future generations.

As human populations and per capita resource consumption continued to grow, it became clear that humans had the potential to exhaust certain resources in the foreseeable future. The intergenerational distribution debate in economics shifted to the question of non-renewable resource use and non-sustainable use of renewable resources. Hotelling (1931) offered perhaps the earliest important contribution, that a non-renewable resource should be exhausted at such a rate that the price increases at the same rate as returns on capital[4]. With regards to renewable resources, economists suggested that harvesting rates which maintained the maximum sustainable yield were unlikely to be economically optimal, while in some cases, resource extinction could be (e.g., Clark, 1990). However, these conclusions were reached by maximizing the net present value of the resource in question. This implicitly assumes that all rights to that resource belong to the present, and none to the future. In fact, this assumption is virtually essential for economic analysis[5]. If instead economists assumed that future generations did have rights to resources, the market could no longer be used to allocate resources, for clearly it would be impossible for generations not yet born to participate in the market for resources. The assumption that the future has no rights to wealth and resources would hardly meet anyone's definition of fair.

---

[4] While theoretically compelling, there is little if any empirical evidence that this ever holds true.

[5] To quote from Arrow et al. (2000, p. 1402): "That the utility discount rate is positive is, according to Koopmans, a mathematical necessity. Otherwise, we would not be able to define preferences for the consequences from now and into the infinite future ... The discount rate ... is perhaps the most influential parameter governing cost–benefit analysis and management choices, yet its proper choice is extremely difficult". Thus, economists appear to recognize that discounting is driven by methodology (mathematics) and not by the problem, which in this case is the explicitly ethical issue of intergenerational equity.

Solow (1974) and others explored the implications of resource exhaustion for intergenerational equity. The emerging consensus appeared to be that imminent resource exhaustion led to higher prices and, hence, incentives to innovate substitute resources or use existing backstop resources. Self-regulating market mechanisms made non-market intervention unnecessary. Therefore, in a free market economy, resources through time are infinite for all practical purposes, and we really need not concern ourselves with fairness towards future generations. While many scientists were alarmed by the imminence of resource exhaustion (e.g., Meadows et al., 1972), economists generally tended towards greater complacency.

The 1960s and 1970s also brought increasing attention to environmental degradation. Pollution was becoming a serious problem, which many economists recognized as a market failure. The field of environmental economics evolved quickly and developed mechanisms for valuing environmental goods and internalizing externalities to the production/consumption process. Economists recognized that uncertainty and irreversibility were critical aspects of environmental issues and incorporated these into their models. Still, the general belief has been that if we resolve the problems of externalities and public goods (i.e., extend the free market to cover all goods and resources) and include option values to compensate for uncertainty, market prices will determine the optimal usage of resources and environmental goods for the present and ensure the invention of substitute resources for future generations. The entire ethical argument from free market economists then is finessed through faith in technological advance induced by market forces.

Rawls's analysis cannot easily be extended to deal with intergenerational fairness either. It is clear in Rawls's work that the dominant intergenerational question in his mind was that of capital accumulation rather than resource depletion, and applying Rawls's difference principle to intergenerational justice would lead to us doing nothing to make future generations better off than we are. Rawls himself realized this problem, and did not attempt to apply the difference principle to issues of intergenerational justice. In fact, Rawls states that

> ... the question of justice between generations ... subjects any ethical theory to severe if not impossible tests ... I believe that it is not possible, at present anyway, to define precise limits on what the rate of savings should be. How the burden of capital accumulation and of raising the standard of civilization is to be shared between generations seems to admit of no definite answer. *(quoted in Solow, 1974)*

Instead, Rawls offers a

> deliberately vaguer principle, given by the balance between what a typical person feels it is reasonable to ask of his parents and what this same person is prepared to do for his children. *(quoted in Solow, 1974)*

This 'vaguer principle' is similar to the frequently heard claim that as long as we care about our children and grandchildren, the free market will lead to the appropriate amount of resources for future generations. This might be true if determining an optimal savings rate were the only problem, but it ignores the potential for very long-term environmental problems, such as global warming, nuclear wastes, etc. Our actions today can affect generations far enough in the future that kinship is scarcely felt.

Rawls's problem with applying his difference principle to intergenerational issues was apparently his belief that 'raising the standard of civilization' should be a goal. The difference principle fails because it does not lead to the increased well-being of future generations. Since Rawls wrote, however, the intergenerational distribution debate has increasingly shifted towards sustainability, i.e., making sure future generations have as much as we have, or at least enough to comfortably survive. We are no longer at all certain that future generations will be better off than our own, and the worry is that they may be worse off. What implications does this have for extending Rawls's analysis to intergenerational issues?

Certainly, the generation into which someone is born is morally arbitrary. This would then imply that an equitable division of resources and capital between generations would be just. Renewable resources could only be used at the rate which they can replace themselves, a constant capital stock would be maintained and passed on to the next generation, and exhaustible resources would be divided equally among generations. There are two absurdities inherent in this result. First, human-made capital accumulates as we make advances in science and technology; in the absence of catastrophe or fundamental changes in human society, future generations will inherit more knowledge. Second, equitable division of exhaustible resources over infinite generations leads to each generation receiving an infinitely small amount. Any other division which awards finite quantities to a finite number of generations is a Pareto improvement; those generations receiving the resource will be better off, while those not receiving it will be no worse off than before.

The difference principle would not hold one generation responsible for making the next generation better off, but at a minimum it would seem to forbid one generation from causing subsequent ones to be worse off than itself *and* worse off than they would have been with equal division of resources. Since equal division of resources implies zero use of non-renewable resources, there would appear to be no special obligation for one generation to share non-renewable resources with future generations, as long as the following conditions are met:
- Future generations are not left dependent for survival upon non-renewable resources in danger of exhaustion. This implies that human populations cannot rely on non-renewable resources to exceed the carrying capacity of the earth, or at least must cease to do so before the necessary resources are exhausted.
- If in the absence of the non-renewable resource some generations would be worse off than others, then these generations have the right to use those

resources. While lack of knowledge of the future makes this impossible to know for certain, it implies that when the future is likely to be better off than the present (which was the case for much of history) then the present has the right to non-renewable resource use. If the future is likely to be worse off (as increasingly appears to be the case), then those resources should be saved for the future, especially if the future is being made worse off by current use of those resources in excess of the waste absorption capacity of the environment.
- When one generation extracts and uses non-renewable resources at a rate that generates sufficient waste flows and degradation to destroy renewable resources, the resulting destruction of renewable resources meets the criteria for the fair use of renewable resources outlined below.

The difference principle has different implications for different types of renewable natural capital (RNC). Essential RNC must be maintained above the amount required to make sufficient essential human capital to meet future needs. We must maintain sufficient life supporting natural capital to guarantee adequate provision of all forms of natural capital. Yields from non-substitutable RNC must be maintained intact, or at least in sufficient quantities that the marginal contribution to QOL is zero, because by definition, it is not possible to compensate the future for its loss. If non-essential, substitutable RNC is harvested at greater than maximum sustainable yield (where yield is taken to include both market and non-market goods and services), the future must be compensated. If yields exceed this maximum yet remain sustainable (i.e., the same yield may be taken every year, without further degradation of the resource), compensation (in the form of increased quantities of other forms of capital) need only make up for the QOL lost by a future generation decreasing its harvest of the renewable resource until stocks increase sufficiently to again support the maximum sustainable yield. If non-essential, substitutable, and non-reconstitutable RNC is harvested beyond its capacity to recuperate, this is a finite loss for infinite time, and compensation must be an equivalent resource flow from an alternative capital stock for infinite time.

Almost certainly, in the real world some resources are being depleted at such a rate that future generations may suffer the negative impacts of the resource use without having the benefits of the resource itself. Fossil fuel use leading to global warming is a good example. Future generations are also being left dependent for survival upon exhaustible resources. Without oil for energy, fertilizers, pesticides, and transportation, it might be very difficult to maintain sufficient agricultural output to feed Earth's population. Many renewable resource stocks are also threatened by irreversible exhaustion. Any sense of fairness to future generations demands we leave intact the life-support functions of natural capital, and the non-substitutable natural capital.

The question is, does our accumulation of man-made capital compensate future generations for the loss of substitutable, non-reconstituable capital? The answer depends on the unknown costs of environmental damage to future generations, the unknown benefits of man-made capital accumulation, and the unknown ability of human-made capital (particularly future inventions) to substitute for natural capital (where we are not entirely sure what natural capital is non-substitutable) and the unknown thresholds below which natural capital becomes non-reconstitutable. Uncertainty therefore is a crucial factor that must be considered.

The treatment of uncertainty in ethical analysis is difficult, but as a rule of thumb, we might say a gamble is appropriate if the gains from winning the gamble are approximately equal to the losses from losing[6]. In this case, are the gains from gambling that man-made capital is a substitute for natural capital approximately equal to the losses if it is not? Summarizing the differences between human-made and natural capital can help us answer the question. Perhaps the most important distinction is that natural capital can be irreversibly damaged or destroyed, but we typically do not know when irreversible change occurs in ecosystems or renewable resource dynamics, and in the ecosystem case we rarely understand the full implications. Second, the first law of thermodynamics tells us that man-made capital must always rely on natural capital – the two are ultimately complements, and man-made capital can never completely substitute for natural capital. Third, there is as yet little evidence that man-made capital can substitute for the life-support functions of ecosystems, and we are not certain which ecosystem resources are critical to generating these life-support functions. Fourth, capital goods depreciate, and if they are left to the future in compensation for an exhausted renewable resource, the resource flow necessary to maintain them must be left as well. Fifth, man-made capital tends towards obsolescence[7]. Sixth, technology is not always beneficial, and even beneficial technologies may have seriously negative side effects, many of which are not immediately apparent. Seventh, probably the most important man-made resource we leave for future generations is accumulated knowledge, which will lead to new technologies. However, prior to the invention of a new technology, nothing can be said with certainty about what it will be[8]. Thus,

---

[6] Theoretically, a risk-neutral individual is indifferent towards a gamble with payoffs of A or B if the probability of outcome A times the value of outcome A is equal to the probability of outcome B times the value of outcome B. However, in the type of uncertainty we are discussing, we do not know the probabilities of each outcome.

[7] A well-maintained road system left for the future to compensate for global warming will be of limited use if the future no longer uses cars. If global warming becomes a serious problem, many of our fossil-fuel consuming technologies may become obsolete even before fossil fuels are exhausted. Any technology that relies on exhaustible resources will eventually become obsolete.

[8] Clearly, many inventions are predicted. Jules Verne, for example, predicted travel to the moon and submarines. He could not have assigned any realistic probabilities to when these inventions might occur. This only became possible when knowledge and technology had reached a more advanced level. Verne could not have predicted the mechanics of jet engines, nor the computer technology required for space flight, nor the transistor, which set off the computer revolution. Application of known principles is less invention than innovation (Proops and Faber, 1990).

we cannot say with certainty if future technologies will compensate for natural resource depletion. The benefit of winning the substitution gamble is presumably higher levels of consumption now and in the future, with most of the people benefiting from this being those in the highest consumption class in human history. The potential loss is damaging non-reconstitutable, life-supporting RNC, which would have potentially catastrophic consequences for the QOL of all humans and even for human survival.

Even if weak sustainability were to hold true, and production in a world with virtually no resources were possible, still this would not mean that we could ignore resource scarcity. First of all, while scarcity does induce innovation, technological advance is a function of accumulated knowledge. Accumulation of knowledge requires both time and effort. Natural resource prices respond to political turmoil, imperfect markets, and imperfect knowledge, so price increases may thus be far faster than expected (Reynolds, 1999), greatly decreasing the time available for developing substitutes and more efficient technologies. Also, the quicker a resource is exhausted, or the more sudden the price surge as a crucial resource nears exhaustion, the more likely is economic disruption. Severe disruption can slow down the creation of substitute resources. For example, during the great depression, or the recent breakup of the Soviet Union, economic chaos led to the unemployment of numerous scientists and decreased investments in research and development. Yet, the faster resources are depleted, the more rapidly technology will have to advance to compensate. Thus, there is no guarantee that efficiency-increasing and resource-substituting technologies will develop at the same rate as depletion of resources. This implies that justice demands we slow resource use while substitutes are being developed rather than await imminent resource exhaustion to trigger research into alternatives[9]. Regulations slowing resource use create artificial scarcity, inducing research into substitutes more quickly, and avoid the economic disruption which may accompany more sudden resource exhaustion.

Nor does the fact that Malthus has so far been wrong prove the assumption of infinite substitutability of resources. We are living in a constantly evolving world. Every day sees a greater increase in the world population and a greater depletion of renewable and non-renewable resources than any previous day. Population growth

---

[9] Modern agriculture provides a good example. We currently depend on petroleum for producing and running farm equipment, for transporting goods produced, for manufacture of pesticides, herbicides and fertilizers, and indirectly for almost every other facet of modern agricultural production. There are feasible substitutes for all of these petroleum-based inputs, but to implement all substitutes as rapidly as petroleum supplies run down will be costly. Food supplies might suffer, and the consequences could be severe. Given the instability of oil prices, and the paramount importance of maintaining constant food supplies, relying on as-yet-undiscovered technologies may be a very risky strategy.

has accelerated far beyond what it was in Malthus's time. It is difficult to base predictions on past experience when such unprecedented changes are occurring. Blind faith in the ability of technology in a free market system to overcome resource and environmental constraints as they occur, and thus compensate for depleted natural resources, seriously jeopardizes the QOL of this and future generations [10].

With respect to uncertainty, then, fairness demands that we assume strong sustainability until proven otherwise. That is, we should only deplete resources when substitutes have been proven, and not before. Only when the survival of the current generation is threatened is it free to use resources to meet current needs even when it risks the survival of future generations, as those generations living at or below the subsistence level arguably have no obligations to future generations. This last point suggests that some level of intragenerational fairness is a prerequisite for intergenerational fairness.

3.3. Fairness across countries in space and time

In theory, there is nothing in the economic or in the justice theory 'definition' of fairness that is affected by geography. Market economics, however, is in the midst of a historically unprecedented rate of expansion, unprecedented not just for an economic system but for any type of human institution. Yet the claims of fairness for market economics are only valid when the actual market system is an accurate depiction of the theoretical system [11]. Unfortunately, the global market place is far less 'perfect' than most national markets. In the first place, market fairness across countries would require capital immobility (Daly, 1996), while market fairness for individuals within a global economy would require labor mobility. Instead, capital is free to move across international boundaries, yet labor is not. Second, market fairness requires a very large number of nearly identical firms, not a handful of grain exporters, oil companies and car companies that dominate the world market. Third, market fairness requires that firms make profits by competing with each other, not by 'rent seeking' activities such as bribing politicians for lucrative contracts. It is widely recognized that international companies routinely bribe government officials in less-developed countries, and that businessmen in

---

[10] A good example is nuclear power. When nuclear generators were first built, the full dangers of radiation were not understood. The assumption in the 1950s was that technology would develop a means for safely disposing of nuclear wastes when they became a problem. The half-life of plutonium is 24 300 years, and we are no closer to solving the disposal problem then we were in 1950.
[11] While most scientists seek theories that are an accurate depiction of reality, economists have developed a theory that does not accurately depict reality, but is widely used as guide for policies designed to bring reality closer to that theory.

Europe can actually write the bribes off as tax deductions (Trade Compliance Center, 2001). Fourth, in Adam Smith's classic, *The Wealth of Nations*, he points out that trade secrets (read 'patents' in today's world) are essentially a form of monopoly and "the monopolists by keeping the market constantly understocked, by never fully supplying the effectual demand, sell their commodities much above the natural price ... The price of monopoly is upon every occasion the highest which can be got" (Smith, 1970, p. 164). Citizens of developed countries currently hold 97% of global patents, and even within less-developed countries, they hold 80%. Empirically, in countries undergoing rapid trade liberalization, wage inequality has increased and unskilled workers have suffered often dramatic drops in real wages (Wallach and Sforza, 1999). In Latin America over the last five years, the number of people living below poverty level has soared in the countries most avidly pursuing trade liberalization (Faiola, 1999), and globally over the same time period the number of people living in absolute poverty has increased by 20%. By any measures of QOL presented above, the worst off have suffered a decline with increasing integration into the world market system.

4. Can we measure fairness?

Rawls's theory of justice fails to provide an empirical measure of fairness. It does, however, suggest a means to measure progress towards fairness; if the worst off are becoming better off, then society is becoming fairer. If the worst off are becoming even worse off, then society is becoming less fair. Economic theory is supposedly value-neutral, but many people do believe that economics is fair in terms of awarding just deserts, and further contend that the free market system provides the appropriate incentives to maximize the QOL of the worst off. This suggests that from the just deserts viewpoint, an appropriate measure of *intra*generational fairness might be society's proximity to a perfect free market system. This is a highly contentious assumption, skirts the issue of the fairness of the initial distribution of resources before market transactions take place, and is still quite difficult to measure. However, it would allow us to assume that goods and services not distributed in accordance with free market principles are unlikely to be fairly distributed, even procedurally, and focus our attention on these. In terms of intergenerational fairness, the main difficulty lies with ensuring the provision of sufficient natural services to ensure an acceptable QOL for future generations. Any consideration of intergenerational fairness from market economics hinges on the ability of human-made capital to replace natural services, which is taken on faith. It is perhaps more reasonable, therefore, to measure intergenerational fairness as the extent to which our society ensures the provision of natural services in accordance with Rawls's theory modified for intergenerational issues as laid out above. However, when we recognize that the *intra*generational allocation of natural capital is plagued by market failures, it becomes clear that the allocation of natural capital is a reasonable focus for assessing fairness both within and among generations.

Our emphasis on natural capital makes even more sense when we remind ourselves that environmental services are essential and indispensable to human society. Without them, society has no material or energy resources with which to produce goods, and perishes in its own waste. A degraded or diminished environmental base reduces the long-term QOL because it contributes less service to individual humans, society, and the economy [12]. Conserving sources and sinks means that more will be available for future generations, so intergenerational equity is served, and there is a greater chance for a sustainable society.

The environment can be viewed as infrastructure that contributes to QOL and the economy in the same fashion as the more traditional infrastructure of water delivery systems, solid waste and sewerage disposal, and other public services. The public welfare suffers if water and waste disposal systems are abused; environmental abuse will also result in lowered public welfare. If resource use and waste releases exceed the environment's capacity to provide these services, then natural capital is diminished and pollution rises. Those benefiting from the activity are externalizing some of their costs to the public domain and are, in effect, appropriating public property rights to the open access sources and sinks for their own use without compensation – unfair by any standards. For example, polluters can gain an internal subsidy by spending less than their peers on pollution control, thus externalizing their pollution costs by consuming more public natural capital. Reappropriation of these public property rights by the reduction of pollution and the associated subsidies would improve environmental quality, and as we shall see, reduce poverty and inequality. Society normally penalizes those who steal or embezzle public *financial* capital, but it is unclear why we allow the unauthorized appropriation of public *natural* capital for private gain. Perhaps it is related to the fact that society has not yet recognized natural capital as a valuable form of capital. Since economists' models and theories rarely recognize the value of natural capital in creating wealth, perhaps we should not expect the public to act differently.

The effect of externalities on markets and the need for prices to reflect all production costs has been of concern to economists since Adam Smith. A theme running through ecological economics is that laissez-faire free market activities

---

[12] For example, industrial discharges that exceed the assimilation capacity of the environment can accumulate in fish and wildlife and inhibit commercial and sport fishing. Acid rain and other forms of air pollution slow crop and tree growth and reduce economic returns. Cities and regions that maintain clean air and an appealing aesthetic environment are more pleasant places to live and have a better chance of attracting businesses and tourists and their dollars. Clean, unpolluted surface or ground water available for human consumption without much treatment means better health and less public expense. In addition, pollution often results in the loss of more jobs than are created by the economic activity that produced the discharge.

using common resources inevitably lead to market failure, ecological abuse and inequity due, primarily, to externalities or spillover costs (Hardin, 1968; Perrings, 1987; Tisdell, 1991; Ophuls and Boyan, 1992). If prices do not reflect all costs, then market failure results with all of its attendant inefficiencies and inequities. However, while many recognize that externalities are pervasive and growing (Bromley, 1986; Baumol, 1967), few economists seem overly concerned with their effects on public welfare. In addition, it is apparent that many, if not most, externalities create subsidies for those externalizing their costs (Templet, 1995a) and subsidies contribute to distributional inequality.

Most observers agree that when a negative externality occurs, a cost is transferred from the initiator of the externality to the receiver. For example, pollution from a facility causes nearby residents breathing the fumes to bear a cost which may range from merely irritating to life threatening. External effects are generally accepted as pervasive (Baumol and Oates, 1979, p. 77; Goodland and Daly, 1993) in market systems, and many economists have discussed them (Mishan, 1971; Cowen, 1988), but there is significant debate about whether they are a major problem, and if so, what to do about them. For example, Coase (1960) argues that external costs can be bargained away by the affected parties without government intervention although transaction costs, which accompany the bargaining transaction, will arise. Oates (1986) points out that Coase's bargaining approach will not work in the large group case due to excessive transaction costs.

However, there is little discussion of the fact that the initiator also secures an implicit subsidy by having caused the externality. For example, a firm releasing pollution to the environment, rather than internalizing the cost, is enjoying an internal subsidy of the retained dollars which would have been spent on pollution control if the firm's pollution spending equaled the US average. The result is higher pollution levels and corporate profits but increased costs, including health costs, to those burdened with the pollution. The polluter has appropriated natural capital from the public and public welfare declines as pollution rises (Templet and Farber, 1994). The pollution subsidy is a useful measure of appropriation of the sink side of natural capital. Subsidy creation also occurs when tax structures and energy pricing, and possibly other costs of production, are manipulated by government, generally at the urging of the firms which benefit, to reduce costs to one sector. In earlier analyses, Templet developed indicators of inequality in pollution, energy pricing and taxes, which were then used to calculate an equity index (Templet, 1995b) and subsidies per capita by state (Templet, 1995a). These measures were statistically compared to a number of state socioeconomic indicators, including poverty. Poverty was lower when equity was higher and when subsidies were low. The study found that as subsidies increase, economic health and sustainability both decline. In the case of taxes and energy pricing, the public directly pays the external cost and functions as a type of socioeconomic commons. Externalized energy costs are a useful measure of the extent of appropriation of the resource side of natural capital while externalized taxes affect both sources and sinks. Tax measures were included because realignment of taxes offers a means of reappropriating natural capital by forcing the internalization of costs not currently captured in market prices or disposal costs.

Generally, if costs can be passed on to others, then potential expenditures can be foregone and internal corporate dollars go unspent. In the pollution example, the costs that are externalized include impacts on people, property, and ecosystems. If these costs are greater than the internal subsidy gained by the firm, optimality, potential Pareto or otherwise, is not achieved. The effect of the externalization is, generally, a net loss to public welfare with private interests benefiting while public interests lose considerably more; public costs exceed private benefits and distributional inequities escalate.

In this way, externalities lead to subsidies and inequalities that are among the most important driving forces leading to poverty. The subsidies increase wealth for some, which is then used to maintain and increase existing subsidies through campaign contributions and political action. While externalities increase wealth for a few, they diminish wealth for many others because there are costs involved. Externalized costs diminish natural capital through resource depletion and pollution and negatively affect human, social, and built capital. The impacts can be grouped in three ways: 1) direct impacts that diminish health and QOL and reduce disposable incomes through lost productivity and other costs; 2) fiscal impacts that occur because subsidies deprive government of revenues, some of which would have helped the poor, e.g., better educational and health care systems (fiscal impacts have not been examined in this chapter); 3) power distribution impacts that heighten inequalities in political power and wealth and allow the wealthy to more effectively manipulate government and markets to their advantage. These results are summarized in fig. 4 (overleaf).

5. What is the relationship between fairness and QOL?

When private entities appropriate public assets, as discussed in the previous section, this reduces the QOL of the general public directly and through a number of indirect feedbacks. However, there is a secondary effect as well. There is substantial evidence that individuals assess their QOL relative to a reference group or an ideal state (Schuessler and Fisher, 1985; Frank, 1999; Galbraith, 1969). The greater the difference between one's own circumstances and this ideal state, the lower one's subjective QOL. The greater the inequality in society, presumably the more likely one's reference group or ideal state will be considerably better off than oneself. Thus, if unfair appropriation of wealth and resources by one group makes them better off than another, the losers in this transaction have a doubly negative impact on their QOL.

This possibility has very serious implications for government policy. Currently, most governments in the world are seeking to increase QOL through continuous

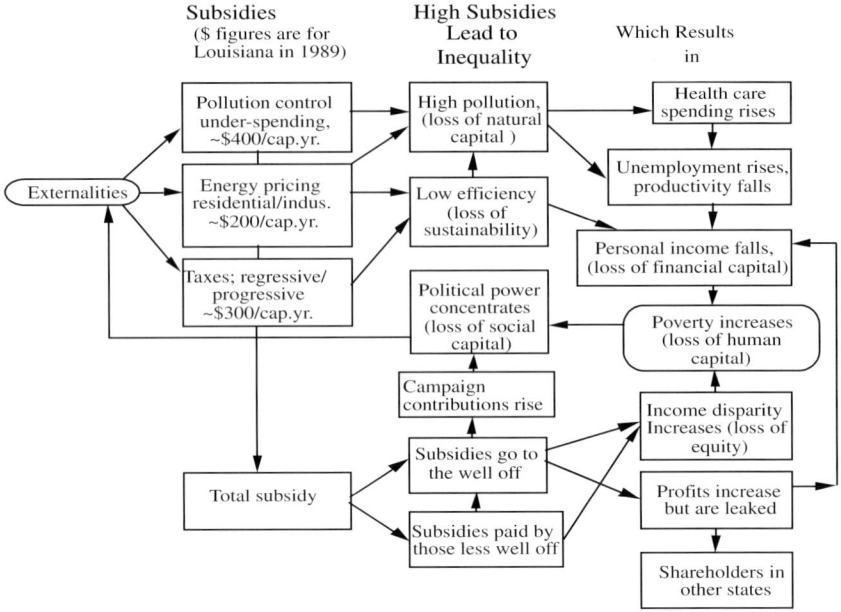

Fig. 4. The relationship among externalities, natural capital, and Quality of Life.

economic growth. While this approach will clearly increase QOL if it brings people from poverty to subsistence level or above (for it is obvious that below the subsistence level, QOL is probably not a relative measure), it may not be a fruitful policy for populations that have already met their basic needs. For example, economic growth may allow me to purchase a fancy car, which I desire because my neighbor has one. However, if my neighbor also benefits from the economic growth, she may now purchase an even nicer car. As I rate my QOL relative to hers, I am no better off than before in terms of human-made capital, and the production and use of the two cars has degraded natural capital and the QOL-enhancing services it provides. What is more, those who do not benefit from the economic growth are worse off, even if their absolute incomes are unchanged. We cannot simply grow our way to a better QOL if this is the case, and growth accompanied by increasing disparity in income distribution, as is the case in much of the world right now, will actually make some poor people worse off even if their absolute incomes increase.

It is also fairly clear that inequality in the distribution of material wealth and resources leads to an unequal distribution of political power, which can then be used to accumulate even greater wealth. A typical case is when corporations contribute to politicians in exchange for support on legislation that allows the private sector to appropriate public wealth. In an empirical study,

Templet (1995a) finds that the size of campaign contributions per candidate for congressional offices across states relates significantly and negatively to congressional environmental voting records (League of Conservation Voters, 1990). The higher the contributions, the poorer the voting scores on environmental issues. In addition, state-level subsidies per capita are significantly and negatively related to congressional environmental voting records. Finally, the size of campaign contributions is positively related to the size of the manufacturing industry within a state, indicating that the manufacturing sector is a significant contributor to federal campaigns. While these relationships do not establish a conclusive case between elected representatives' voting patterns and campaign contributions, they do suggest a role for campaign contributions in influencing voting, and probably other political prerogatives, to promote subsidies for vested interests. In support of this view, Boyce et al. (1999) have found that as political power concentrates, pollution increases and public health and welfare decline. To restate in the terms used above, as political power concentrates in vested interests, larger externalities and subsidies are created through political action that then lead to higher pollution and other inequalities. With these linkages it is apparent that vicious feedback cycles exist among subsidies, political power, and poverty, which make the subsidy–poverty cycle very difficult to change. Simply put, great disparities in wealth undermine democracy. If it is true that democratic systems are most likely to provide for the needs of their citizens, then this erosion of democracy associated with unequal distribution will further reduce QOL.

Finally, empirical studies show that people have strong negative reactions to unfairness, which may directly impact their QOL. Studies suggest that these negative reactions can be so intense that people perceiving inequitable inequalities frequently experience cognitive tensions sufficient to cause them to change their values (Walster et al., 1976, cited in Alwin, 1987).

6. Principles for achieving a sustainable, fair, and high-QOL society

Obviously, a basic requirement for creating a fair and sustainable society that offers high QOL to current and future generations is the willingness of its members to pursue such a goal. One of the major obstacles to generating the necessary will has been the perception that fairness requires a reallocation of access to resources from the well-to-do to the poor, and sustainability requires a reallocation of access to resources from the present to the future. Both circumstances would force some groups to consume less so that others might consume more. The dominant economic paradigm on the planet today measures QOL by consumption. Under this paradigm, consuming less implies a lower QOL for the well-to-do and the present generation. As shown above, in modern societies wealth equals power. Clearly, the present generation also has power over the resources left to future generations. Thus, those who have the power to impose change are those who would stand to lose by it, and change is unlikely to happen. One of the more important implications of the discussion in this chapter is that additional consumption may not be closely associated with increased QOL, and conversely, the well-to-do could reduce their consumption without reducing their QOL. To the extent people accept this argument, it dramatically increases the potential for creating a fair, sustainable, and high QOL society.

What else would be required to create a sustainable society? Costanza et al. (1998) have outlined six core principles, an indivisible collection "that embody the essential criteria for sustainable government" (p. 198). To quote at length, these are:

Principle 1: Responsibility. Access to environmental resources carries attendant responsibilities to use them in an ecologically sustainable, economically efficient, and socially fair manner. Individual and corporate responsibilities and incentives should be aligned with each other and with broad social and ecological goals.

Principle 2: Scale matching. Ecological problems are rarely confined to a single scale. Decisionmaking on environmental resources should (i) be assigned to institutional levels that maximize ecological input, (ii) ensure the flow of ecological information between institutional levels, (iii) take ownership and actors into account, and (iv) internalize costs and benefits. Appropriate scales of governance will have to be those that have the most relevant information, can respond quickly and efficiently, and are able to integrate across scale boundaries.

Principle 3: Precaution. In the face of uncertainty about potentially irreversible environmental impacts, decisions concerning their use should err on the side of caution. The burden of proof should shift to those whose activities potentially damage the environment.

Principle 4: Adaptive management. Given that some level of uncertainty always exists in environmental resource management, decisionmakers should continuously gather and integrate appropriate ecological, social and economic information with the goal of adaptive improvement.

Principle 5: Full cost allocation. All of the international and external costs and benefits, including social and ecological, of alternative decisions concerning the use of environmental resources should be identified and allocated. When appropriate, markets should be adjusted to reflect full costs.

Principle 6: Participation. All stakeholders should be engaged in the formulation and implementation of decisions concerning environmental resources. Full stakeholder awareness and participation contributes to credible, accepted rules that identify and assign the corresponding responsibilities appropriately.

What is required to create a fair society? Economic theorists would argue just deserts, and Rawlsian justice would call for making the poor as well off as possible. An additional prerequisite suggested by material presented here is adequate access to environmental resources for all. We have already shown that the market failure of negative externalities can deprive society's members of these resources, and the following chapter will show how other market failures do the same. Thus, unregulated market allocation of such resources is unfair, and nonmarket institutions will be required to ensure adequate access for current and future generations. Gross concentrations of wealth, unfair in their own regard, also convert to political power that introduces even greater inequities in the distribution of natural capital.

How do we create a society with higher QOL? We focus on the satisfaction of human needs rather than ever-greater production. To create a society that provides a high QOL and is sustainable and fair, we will need to alter society's preferences towards satisfiers that are non-consumptive, so that meeting the needs of one group or generation does not impose on those of another.

The following chapter will flesh out these ideas in greater detail and suggest concrete policies to achieve our objectives.

References

Alwin, D.F., 1987, Distributive justice and satisfaction with material well-being. Am. Sociol. Rev. 52:83–95.
American Heritage Dictionary of the English Language, 1992, 3rd edition (Houghton Mifflin, Boston, MA) 2140 pp.
Arrow, K., Daily, G.C., Dasgupta, P., Levin, S., Mäler, K.-G., Maskin, E., Starrett, D., Sterner, T. and Tietenberg, T., 2000, Managing ecosystem resources. Environ. Sci. Technol. 34:1401–1406.
Atkinson, G., Dubourg, W.R., Hamilton, K., Pearce, D.W., Munasinghe, M. and Young, C., 1997, Measuring Sustainable Development: Macroeconomics and Environment (Edward Elgar Publications, Aldershot) 252 pp.
Barry, B., 1989, Theories of Justice, Vol. I (University of California Press, Berkeley, CA) 428 pp.
Bartelmus, P., 1994, Towards a Framework for Indicators of Sustainable Development. Department for Economic and Social Information and Policy Analysis, Working Paper Series No. 7 (United Nations, New York).
Baumol, W.J., 1967, Macroeconomics of unbalanced growth; the anatomy of urban crisis. Am. Econ. Rev. 57(June):415–426.
Baumol, W.J. and Oates, W.E., 1979, Economics, Environmental Policy and the Quality of Life (Prentice Hall, Englewood Cliffs, NJ.) 377 pp.
Bloom, D.E., Canning, D., Graham, B. and Sevilla, J., 2000, Out of Poverty: On the Feasibility of Halving Global Poverty by 2015 (Discussion Paper No. 52). Consulting Assistance on Economic Reform (CAER II). Available on-line: http://www.cid.harvard.edu/caer2/htm/content/papers/paper52/paper52.htm.
Boyce, J.K., Klemer, A.R. and Templet, P.H., 1999, Power distribution, the environment, and public health: a state level analysis. Ecol. Econ. 29:127–140.
Bromley, D.W., 1986, Markets and externalities. In: D.W. Bromley (Editor), Natural Resource Economics (Kluwer–Nijhoff Publishing, Boston, MA) ch. 2.

Clark, C., 1990, Mathematical Bioeconomics: the Optimal Management of Renewable Resources (Wiley, New York) 386 pp.
Coase, R.H., 1960, The problem of social cost. J. Law Econ. 3(October):1–44.
Cobb, C.W. and Cobb, J.B., 1994, The Green National Product: A Proposed Index of Sustainable Economic Welfare (University Press of America, New York) 343 pp.
Collados, C. and Duane, T.P., 1999, Natural capital and quality of life: a model for evaluating the sustainability of alternative regional development paths. Ecol. Econ. 30:441–460.
Costanza, R. and Daly, H.E., 1992, Natural capital and sustainable development. Conserv. Biol. 6:37–46.
Costanza, R., d'Arge, R., de Groot, R., Farber, S., Grasso, M., Hannon, B.M., Limburg, K., Naeem, S., Paruelo, J., O'Neill, R.V., Raskin, R.G., Sutton, P. and van den Belt, M.J., 1997, The value of the world's ecosystem services and natural capital. Nature 387:253–260. See http://www.floriplants.com/news/article.htm.
Costanza, R., Andrade, F., Antunes, P., van den Belt, M.J., Boersma, D., Boesch, D.F., Catarino, F., Hanna, S., Limburg, K., Low, B., Molitor, M., Pereira, J.G., Rayner, S., Santos, R., Wilson, J. and Young, M., 1998, Principles for sustainable governance of the oceans. Science 281:198–199.
Costanza, R., Farber, S., ñeda, B. Casta and Grasso, M., 2001, Green national accounting: goals and methods. In: C.J. Cleveland, D.I. Stern and R. Costanza (Editors), The Nature of Economics and the Economics of Nature (Edward Elgar Publishing, Cheltenham, England).
Cowen, T. (Editor), 1988, The Theory of Market Failure; A Critical Examination (George Mason University Press, Fairfax, VA) 384 pp.
Daily, G.C. (Editor), 1997, Nature's Services – Societal Dependence on Natural Ecosystems (Island Press, Washington, DC) 392 pp.
Daly, H.E., 1996, Beyond Growth: The Economics of Sustainable Development (Beacon Press, Boston, MA) 253 pp.
Daly, H.E. and Cobb, J.B., 1989, For the Common Good: Redirecting the Economy Toward Community, the Environment, and a Sustainable Future (Beacon Press, Boston, MA) 482 pp.
Dixon, J. and Sherman, P., 1990, Economics of Protected Areas: A New Look at Benefits and Costs (Island Press, Washington, DC) 234 pp.
Durning, A.T., 1992, How Much is Enough? The Consumer Society and the Fate of the Earth, 1st edition (Norton & Company, New York) 200 pp.
Ekins, P., 1991, The sustainable consumer society: a contradiction in terms? Int. Environ. Affairs 3:243–258.
Ekins, P. and Max-Neef, M. (Editors), 1992, Real-life Economics: Understanding Wealth Creation (Routledge, London) 460 pp.
El Serafy, S., 1989, The proper calculation of income from depletable natural resources. In: Y.J. Ahmad, S. El Serafy and E. Lutz (Editors), Environmental Accounting for Sustainable Development (World Bank, Washington, DC) pp. 10–18.
El Serafy, S., 1996, In defense of weak sustainability: a response to Beckerman. Environ. Value 5:75–81.
Faiola, A., 1999, Argentina's lost world: rush into the new global economy leaves the working class behind. Washington Post, December 8, 1999.
Farquhar, M., 1995, Elderly people's definitions of quality of life. Soc. Sci. Med. 41:1439–1446.
Frank, R., 1999, Luxury Fever: Why Money Fails to Satisfy in an Era of Excess (Free Press, New York) 326 pp.
Friedman, M., 1963, Capitalism and Freedom (University of Chicago Press, Chicago, IL) 202 pp.
Galbraith, J.K., 1969, The Affluent Society, 2nd, revised edition (Houghton Mifflin, Boston, MA) 333 pp.
Gates, J., 1999, Statistics on Poverty and Inequality. Available on-line: http://www.globalpolicy.org/socecon/inequal/gates99.htm (Global Policy Forum).

Goodland, R. and Daly, H.E., 1993, Why northern income growth is not the solution to southern poverty. Ecol. Econ. 8(2):85–102.
Haas, B.K., 1999, A multidisciplinary concept analysis of quality of life. West. J. Nurs. Res. 21(6):728–743.
Hamilton, K. and Lutz, E., 1996, Green national accounts: policy uses and empirical experience, Environment Department Paper no. 039 (The World Bank, Washington, DC) 47 pp.
Hardin, G., 1968, The tragedy of the commons. Science 162:1243–1248.
Hicks, J.R., 1946, Value and Capital (Oxford University Press, Oxford) 340 pp.
Hotelling, H., 1931, The economics of exhaustible resources. J. Polit. Econ. 2:137–175.
Hueting, R., 1989, Correcting national income for environmental losses: toward a practical solution. In: Y.J. Ahman, S. El Serafy and E. Lutz (Editors), Environmental Accounting for Sustainable Development (World Bank, Washington, DC) pp. 32–39.
Hueting, R., 1995, Estimating sustainable national income. In: W. van Dieren (Editor), Taking Nature into Account: Toward a Sustainable National Income (Springer, New York) pp. 206–230.
Lane, R.E., 1986, Market justice, political justice. Am. Polit. Sci. Rev. 80(2):383–402.
League of Conservation Voters, 1990, The 1990 National Environmental Scorecard (League of Conservation Voters, Washington, DC).
Mäler, K.-G., 1991, National accounts and environmental resources. Environ. Resour. Econ. 1: 1–15.
Markandya, A. and Perrings, C., 1993, Accounting for an ecologically sustainable development: a summary. In: A. Markandya and C. Costanza (Editors), Environmental Accounting: A Review of the Current Debate (Harvard Institute for International Development, Cambridge, MA) ch. 1.
Max-Neef, M., 1992, Development and human needs. In: P. Ekins and M. Max-Neef (Editors), Real-life Economics: Understanding Wealth Creation (Routledge, London) pp. 97–213.
Max-Neef, M., 1995, Economic growth, carrying capacity and the environment: a response. Ecol. Econ. 15:115–118.
McClosky, H. and Zaller, J., 1985, The American Ethos: Public Attitudes Toward Capitalism and Democracy (Harvard University Press, Cambridge, MA) 342 pp.
Meadows, Donella, Meadows, Dennis, Randers, J. and Behrens III, W., 1972, The Limits to Growth (Universe Books, New York) 205 pp.
Mishan, E.J., 1967, The Costs of Economic Growth (Frederick Praeger, New York) 190 pp.
Mishan, E.J., 1971, The postwar literature on externalities: an interpretive essay. J. Econ. Lit. 9:1–28.
Nordhaus, W.D. and Kokkelenberg, E.C. (Editors), 1999, Nature's Numbers: Expanding the National Economic Accounts to Include the Environment. Panel on Integrated Environmental and Economic Accounting, National Research Council (National Academy Press, Washington, DC) 250 pp.
Nordhaus, W.D. and Tobin, J., 1972, Is growth obsolete? In: Economic Growth, Fiftieth Anniversary Colloquium V (National Bureau of Economic Research) pp. 509–532. Available on-line: http://cowles.econ.yale.edu/P/cp/p03b/p0398a.pdf.
Oates, W.E., 1986, Comment to 'markets and externalities' by David Bromley. In: D.W. Bromley (Editor), Natural Resource Economics, (Kluwer–Nijhoff Publishing, Boston, MA).
Ophuls, W. and Boyan Jr, S.A., 1992, Ecology and the Politics of Scarcity Revisited; The Unraveling of the American Dream (Freeman, New York) pp. 195–216.
Pearce, D.W., 1993, Economic Values and the Natural World (Earthscan, London) 129 pp.
Perrings, C., 1987, Economy and Environment; A Theoretical Essay on the Interdependence of Economic and Environmental Systems (Cambridge University Press, Cambridge) 179 pp.
Phelps, E., 1961, Communications: the golden rule of accumulation: a fable for growthmen. Am. Econ. Rev. 51:638–643.
Phelps, E., 1965, Second essay on the golden rule of accumulation. Am. Econ. Rev. 55:793–814.

Proops, J. and Faber, M., 1990, Evolution, Time, Production and the Environment (Springer, New York) 240 pp.
Ramo, J.C., 1999, The three marketeers. Time 153(6):34–42.
Rand, A., 1964, The Virtue of Selfishness: A New Concept of Egoism (Signet Books, New York) 144 pp.
Rand, A. and Peikoff, L.I., 1996, Atlas Shrugged (Signet Mass Market Paperback, New York) 1088 pp.
Rawls, J., 1971, A Theory of Justice (Harvard University Press, Cambridge, MA) 607 pp.
Reynolds, D.B., 1999, The mineral economy: how prices and costs can falsely signal decreasing scarcity. Ecol. Econ. 31:155–166.
Schuessler, K. and Fisher, G., 1985, Quality of life research and sociology. Annu. Rev. Sociol. 11:129–149.
Shue, H., 1992, The unavoidability of justice. In: A. Hurrell and B. Kingsbury (Editors), The International Politics of the Environment (Clarendon Press, Oxford) pp. 373–397.
Smith, A., 1970, The Wealth of Nations: Books I–III (with an introduction by Andrew Skinner) (Penguin Books, Harmondsworth, Middlesex, UK) 535 pp.
Snyder, C., 1936, Capital supply and national well-being. Am. Econ. Rev. 26:224.
Solow, R., 1974, Intergenerational equity and exhaustible resources. Rev. Econ. Stud., Symp.: 29–45.
Templet, P.H., 1995a, Grazing the commons; externalities, subsidies and economic development. Ecol. Econ. 12:141–159.
Templet, P.H., 1995b, Equity and sustainability; an empirical analysis. Soc. Nat. Resour. 8:509–523.
Templet, P.H. and Farber, S., 1994, The complementarity between environmental and economic risk: an empirical analysis. Ecol. Econ. 9:153–165.
Tisdell, C.A., 1991, The environment and economic welfare. In: D.L. McKee (Editor), Energy, the Environment, and Public Policy; Issues for the 1990s (Praeger Publishers, New York) pp. 6–18.
Trade Compliance Center, 2001, Addressing the Challenges of International Bribery and Fair Competition: July 2001. Available on-line: http: // www.tcc.mac.doc.gov / cgi-bin / doit.cgi?204:71:5700519:1.
Wallach, L. and Sforza, M., 1999, Whose Trade Organization?: Corporate Globalization and the Erosion of Democracy (Public Citizen, Washington, DC) 229 pp.
Walster, E., Berscheid, E. and Walster, G.W., 1976, New directions in equity research. In: L. Berkowitz and E. Walster (Editors), Advances in Experimental Social Psychology, Vol. 9 (Academic Press, New York) pp. 1–42.
Weitzman, M., 1976, Welfare significance of national product in a dynamic economy. Q.J. Econ. 90(1):156–163.

Consensus

Chapter 12

Quality of Life and the Distribution of Wealth and Resources

J. Farley, R. Costanza, P. Templet, with M. Corson, Ph. Crabbé, R. Esquivel, K. Furusawa, W. Fyfe, O. Loucks, K. MacDonald, L. MacPhee, L. McArthur, C. Miller, P. O'Brien, G. Patterson, J. Ribemboim and S.J. Wilson

Abstract

All anthropocentric definitions of sustainability, at least implicitly, place a central focus on sustaining an acceptable level of human quality of life (QOL). Within the dominant ideology of free market capitalism, it is believed that reducing wealth and resource consumption also reduces QOL within a generation, yet it appears that excessive resource consumption on the part of the current generation threatens dramatic reductions to the QOL of future generations. Continued economic growth substantially increases this threat. If current levels of QOL do indeed depend on current consumption levels, this would mean that ensuring sustainability for future generations requires a reduction in QOL for at least some of the people alive today. We show in this chapter that in reality, above a certain level, greater wealth and resource consumption are not tightly linked to QOL. Thus, a more fair distribution of resources and wealth within and between generations need not require a sacrifice in QOL for the current generation, increasing the feasibility of policies directed towards this outcome.

1. How do we define Quality of Life (QOL)?

Philosophers have been discussing the issue of QOL at least since the time of Aristotle, and have yet to reach any kind of consensus on what it means. In chapter 11, we presented the following definition of QOL, "a multidimensional evaluation of an individual's current life circumstances in the context of the culture in which they live and the values they hold. QOL is primarily a subjective sense of well-being encompassing physical, psychological, social, and spiritual dimensions. In some circumstances, objective indicators may supplement or, in the case of individuals unable to subjectively perceive, serve as a proxy assessment of QOL" (Haas, 1999).

Understanding and Solving Environmental Problems in the 21st Century
Edited by R. Costanza and S.E. Jørgensen
© *2002 Elsevier Science Ltd. All rights reserved*

We also drew upon the work of Max-Neef to present a discussion of human needs. Integrating human needs with the above definition suggests a concise working definition of the determinants of QOL with practical policy implications: Quality of life is determined by our ability to satisfy our needs and wants.

1.1. What are human needs?

This definition requires that we clearly define what we mean by needs. First, we define absolute needs as those required for survival, which are biologically determined. Some 1.2 billion individuals globally and 28% of the population in the third world currently live in extreme poverty (World Bank, 2000; Bloom et al., 2000), and have difficulty meeting even these absolute needs. For this group, greater consumption is probably very closely correlated to greater QOL. Once absolute needs have been met, as is the case for about 80% of the human race, then QOL is determined by the satisfaction of a whole suite of primary human needs that have evolved with us as a species. Numerous researchers have proposed a variety of human needs, typically claiming that they are pursued in hierarchical order – Maslow's hierarchy (Maslow, 1954) being only the most famous. The hierarchical ordering, though generally not seen as rigid by these researchers, still leaves something to be desired. Even the 1.2 billion people living in absolute poverty seek to fulfill needs other than mere subsistence. For example, malnourished children have not met their basic physiological needs, but will still seek love and protection. And as Maslow recognized, numerous people have gone on hunger strikes or risked life and limb to pursue higher needs for esteem and self-actualization (the highest levels in the Maslow hierarchy). Max-Neef (1992), in contrast, has summarized and organized human needs into non-hierarchical axiological and existential categories (table 3 of chapter 11). In this non-hierarchical framework, needs are interrelated and interactive, many needs are complementary, and different needs can be pursued simultaneously. In our opinion, this reflects reality better than a hierarchy in which we only pursue higher needs after lower ones have been fulfilled. Another important point to make is that in Max-Neef's conception, needs are both few and finite. This stands in stark contrast to the dominant belief across countries and ideologies that unending economic growth is the best way to meet human needs.

1.2. Satisfiers and wants

We are not concerned solely with the needs themselves, but also with the means we use to satisfy our needs, which we shall call satisfiers (table 3 of chapter 11). While needs remain consistent across time and across cultures, satisfiers differ. In general, different satisfiers may be required by different people to meet a given need, and the same satisfiers can meet given needs to a different extent for different people. Further, and in contrast to neo-classical economic theory, people do not always make optimal choices among satisfiers to meet their needs. In fact, many apparent satisfiers are not satisfiers at all. Max-Neef defines 'violators and destructors' as supposed satisfiers intended to satisfy a need, but which in fact "annihilate the possibility of its satisfaction, [and] also render the adequate satisfaction of other needs impossible" (Max-Neef, 1992, p. 208). He provides the example of an arms race intended to provide protection but which actually makes us less safe, while at the same time depriving us of resources useful in meeting other needs. At the national level, an example would be the increasing private ownership of weapons in the USA. He next defines 'pseudo-satisfiers' as "elements that stimulate a false sensation of satisfying a given need" (Max-Neef, 1992, p. 208). Visiting a prostitute may be a pseudo-satisfier for someone's need for affection. Finally, 'inhibiting satisfiers' are those that satisfy (or over-satisfy) one need, but simultaneously inhibit the satisfaction of others. For example, commercial television satisfies our need for leisure, but inhibits understanding, identity, and creation. We define the desire for violators and destructors, pseudo-satisfiers, and (to a lesser extent) inhibiting satisfiers as 'wants' which are quite distinct from needs.

Additional examples may be helpful. First, recall the definition of consumerism offered in chapter 11 as the cultural orientation that holds that "the possession and use of an increasing number and variety of goods and services is the principal cultural aspiration and the surest perceived route to personal happiness, social status and national success" (Ekins, 1991). By this definition, consumption should satisfy our needs for happiness, status, and success, clearly seen as elements of a good QOL. However, though we consume more than twice as much as our grandparents' generation, it is not readily apparent that we enjoy a higher QOL. Increasingly, studies find the opposite: there is a pronounced trend towards greater rates of depression and suicide in the market democracies, and especially in America where the number of people who declare themselves 'very happy' in studies of subjective well-being is declining[1] (Lane, 2000). Empirical studies find that regardless of income, people believe they would be happier if only they earned twice as much (Lapham, 1988, in Durning, 1992). Income and consumption in this context are thus pseudo-satisfiers; many pursue them without fulfilling their needs. If carried to the extreme of damaging ecological services, as we increasingly risk doing, consumption becomes a violator and destructor. Similarly, sufferers of anorexia nervosa believe they would be more attractive and thus better able to fulfill their need for affection if only they could lose a few more pounds. Many

---

[1] For individual domains of life, the same trend is found. Between 1972 and 1994, studies found a decreasing percentage of Americans declared themselves 'very happy' with their marriage, 'very satisfied' with their jobs, 'pretty well-satisfied' with their financial situation, or very satisfied with their place of residence (Lane, 2000).

weight lifters believe they are small and would be attractive if only they could add bit more muscle mass. When taken to the extremes of starvation and steroid abuse, thinness and muscularity as measures of beauty also become destructors and violators. Thus, demand for the wrong types of satisfiers may be infinite precisely because they fail to satisfy our finite needs.

1.3. Implications of our definition for improving QOL

Now that we have defined needs and wants, of what use is our new definition, in particular with respect to the distribution of wealth and resources? Concisely put, it provides us with three general policy paths towards greater QOL for all. Most obviously, we can attempt to increase people's ability to satisfy a given set of needs or wants. This can be done by providing greater access to the necessary satisfiers or by using satisfiers more efficiently. The latter approach is particularly appropriate when the satisfiers in question consume finite physical resources, and thus use by one person reduces the amount available for others. For example, we mentioned several studies in chapter 11 suggesting that relative amounts of wealth and resources affected QOL more than absolute amounts. Thus, if some people meet their need for identity by consuming more than others to enhance their self-esteem, we could reduce everyone's material consumption above and beyond absolute needs by half without affecting relative consumption nor anyone's ability to fulfill the need for identity. We would need to work less to meet our consumption demands and would have more time to devote to satisfying other needs. A second option is to change society's preferences[2]. One approach would be to intentionally alter a society's cultural preferences for satisfiers in such a way that fewer resources allow us to better meet our needs. Decreasing our dependence on single occupancy vehicles for leisure and participation needs comes readily to mind. Similarly, society could work to reduce or eliminate the individual's wants, where wants are defined as the demand for satisfiers that in some way

---

[2] Undoubtedly, any suggestions for manipulating wants, needs, and cultural preferences will be viewed with concern by those who fear it impinges on personal freedoms, and rightfully so. Needs and wants can be manipulated towards different ends, many of which would not be morally acceptable to the majority of us. But we should not let a valid concern over appropriate ends obfuscate the fact that our wants and needs are already constantly being manipulated. As Rawls (1971) points out, "an economic system is not only an institutional device for satisfying existing wants and needs but a way of creating and fashioning wants in the future. How men work together now to satisfy their present desires affects the desires they will have later on, the kind of person they will be. These matters are of course, perfectly obvious and have always been recognized. They were stressed by economists as different as Marshall and Marx" (pp. 259–260; quoted in Goodwin, 1997). And advertising, of course, is an enormous industry that does little else than manipulate wants. We must simply ensure that any efforts to manipulate wants and needs involve public discussion, are transparent, and are subject to the principle of adaptive management.

diminish our ability to satisfy our needs, as described above. This is a particularly promising approach, because unlike needs, wants can be infinite, and many wants are for wealth and resources. As wealth and resources are the only physical components of satisfiers and hence QOL, they are the only ones that can be depleted, and thus the ones most relevant to the questions of distribution, fairness, and sustainability. Third, society should avoid anything that would increase wants or needs without simultaneously increasing the ability to satisfy them, since that creates the conditions for lowering QOL.

1.4. QOL and the four capitals

Recent research in the social sciences can provide us with useful insights into the nature of potential satisfiers for human needs. While it is clear from table 3 of chapter 11 that economic production only provides satisfiers for some human needs, a focus on economic production can still provide insights into what is required to satisfy our needs. Economic production is not only the result of man-made (built) capital; it also requires inputs from natural capital, human capital, and social capital. For example, all built capital requires inputs of some sort, which are ultimately derived from natural capital. The technology and knowledge inherent in the production process is the product of human knowledge, or human capital. Social capital refers to the institutions, relationships, and norms that shape the quality and quantity of a society's social interactions. Social capital is not just the sum of the institutions that underpin a society; it is the glue that holds them together (World Bank, 2001). Social capital reduces transaction costs via co-operation and lubricates social interactions. It is thus essential to the production process in society. Hence, economic production requires inputs from all four of these capitals.

In an analogous manner, all four capitals are required to satisfy human needs and generate QOL. Natural capital supplies not only the basic raw materials essential for our survival, but also recycles our wastes, regulates our climate, and provides us with clean air and water. According to the 'biophilia' hypothesis (Wilson, 1986; Kellert and Wilson, 1993), humans have an innate affection for nature, which may be as important to our psychological well-being as forming personal attachments with other humans. Studies have shown that people experience lower levels of stress-related illness, lower blood pressure, faster postoperative recovery, greater levels of happiness, and reduced fear when exposed to nature scenes rather than urban scenes (Ulrich et al., 1991). Immersion in nature can generate self-reported feelings of 'wholeness' and comfort (Kaplan and Kaplan, 1989). Nature also fulfills spiritual, cultural, and aesthetic needs, and has intrinsic values unrelated to consumption of its material bounty. In fact, we must emphasize the primacy that natural capital holds in determining QOL, both in history and actuality. Long before we evolved into thinking, social, tool users, most of our needs were met directly by nature, and even today nature contributes substantially to the continued satisfaction of all of our human needs.

Humans are also innately social creatures, and human relationships, trust, and community are essential components of our well-being. Just as the biophilia hypothesis asserts a genetic basis for our love of nature, eons of evolution as a social creature have no doubt engendered a similar need for social capital. Human capital in the form of acquired knowledge and skills and physical health further contributes to our QOL. An education, it has been said, makes your mind a better place to spend your leisure time. Skills and knowledge instill pride and status and offer greater opportunities for less dangerous, more fulfilling employment. And few would deny that health plays an important role in QOL. In historical terms, it is built capital that is the most recent arrival, and the basic needs of the human psyche were no doubt largely established before the first tools were invented. While built capital also contributes to fulfilling many human needs, it has shown continuous growth for several centuries, has the greatest negative impact on natural capital, and is becoming increasingly abundant relative to the other forms of capital (Daly, 1993). Thus, increasing built capital, so long emphasized as the critical element in achieving a high QOL, and in the past perhaps justifiably so, may now play a relatively minor role. Built capital continues to play a major role, however, in the depletion of resources, and ownership of built capital strongly influences the distribution of wealth in the current economic system.

2. How can we measure QOL?

We must recognize that existing national accounts focus primarily on built capital. To the extent this is true, it would appear that these national accounts may be better measures of our ability to pursue wants rather than needs. If we are to know if our policies for maintaining and increasing QOL both now and in the future are successful, then we will need to develop measurable indicators that serve as suitable proxies for needs fulfillment and QOL.

To state the obvious, we cannot precisely measure QOL. In the words of Clifford Cobb (2000, p. 5) "[t]he most important fact to understand about QOL indicators is that all measures of quality are proxies – indirect measures of the true condition we are seeking to judge. If quality could be quantified, it would cease to be quality. Instead, it would be quantity. Quantitative measures should not be judged as true or false, but only in terms of their adequacy in bringing us closer to an unattainable goal. They can never directly ascertain quality."

2.1. Are objective measures suitable?

In chapter 11, we reviewed several different approaches to objectively measuring the generation of wealth, both natural and human-made, on a national scale. All of the approaches that have been operationalized appear inadequate as measures of QOL. The problem is that numerous studies have found only weak relationships between objective measures of QOL and the subjective assessments of the same by the subjects concerned (Haas, 1999). However, both these studies and the various types of national accounts seem to include a relatively narrow range of objective indicators, and often place what we consider to be an excessive emphasis on consumption. Quite possibly the problem is that QOL is too rich a gumbo to allow us to recapture its flavor with so few ingredients. We propose then, as a research agenda, a serious effort to measure access to satisfiers for Max-Neef's axiological and existential categories of human needs, for use as indicators of QOL.

Using Max-Neef's human needs as the basis of a QOL measure is a dramatic departure from existing national accounts as well as from most of the proposed alternatives reviewed in chapter 11, differing even in its theoretical underpinnings. Neo-classical economics and GNP are explicitly utilitarian. Within utilitarian philosophy, individual QOL is determined by the degree to which individuals can satisfy their desires, and it is generally accepted that the goal of society is to provide the maximum amount of 'utility' for its citizens. As utilitarian philosophy has been operationalized by neoclassical economics, citizens are best able to determine what provides utility. As it is extremely difficult to measure 'utility' directly, economists have taken to using revealed preferences as a proxy. Preferences are revealed by people's objectively measurable choices in the market. In the market economy, preferences are revealed through market decisions. Market decisions can only be made with money, and even Jeremy Bentham (one of the founding fathers of utilitarianism) believed that "[m]oney is the most accurate measure of the quantity of pain or pleasure a man can be made to receive" (Bentham, 1830). Under this conception of utilitarianism, the philosophy only values end states and requires only 'having' such things as possessions and experiences. Sustainable income accounting, green accounting and measurements of economic welfare are basically just extensions of this philosophy and similarly value only 'having' (Cobb, 2000). In Max-Neef's framework, having things is important, but is only one of the elements required to meet our needs. Thus, a benevolent dictator with the resources to provide us with all the physical things we require for happiness would fail to meet our existential needs for being, doing, and interacting, as well as our axiological needs for creation, participation, and freedom. Also, within Max-Neef's conception, people are not always best able to determine what contributes to their QOL, as discussed above when we distinguished between 'needs' and 'wants'.

The approach we propose, which values human actions independently of their outcomes, has been dubbed the "human development" approach to QOL. Its main proponents include Nobel Prize winning economist Amartya Sen and Martha Nussbaum. In a similar tone to Max-Neef, they argue that 'capabilities' and 'functionings' are critical to QOL (Cobb, 2000; Sugden, 1993; Nussbaum, 1990).

Roughly speaking, functionings correspond to human needs, while capabilities include both states of being and opportunities for doing. In utilitarian theory, we might have several different options, of which we choose one. If all options but that one were eliminated, it would not affect our QOL. In the human development approach, losing options restricts our capabilities and would therefore affect our QOL. In a stark illustration, there is a fundamental difference between someone fasting out of choice or fasting because he or she does not have the option of eating (Kiron, 1997). The human development approach is less concerned with the actual choices that people make than with the options they are free to choose from, and the marketplace is only one of many spheres in which choice is important.

2.2. Operationalizing human needs assessment as a measure of QOL

Measuring the extent to which human needs are satisfied is, of course, an exceptionally difficult task and a highly subjective one. Following the lead of Sen and Nussbaum, it would be most useful to measure capabilities, that is, the extent to which individuals have access to satisfiers. However, as noted in chapter 11 and above, specific satisfiers may vary by culture, and the difference in satisfiers required to meet a human need may indeed be one of the key elements that defines a culture. This means that objective 'QOL accounts' must be very culture specific. Second, as discussed earlier, some satisfiers might help fulfill several human needs, while other needs require several satisfiers. Further complicating matters, satisfiers may change through time. And humans are social creatures who inhabit a complex environment; needs are not satisfied only in regards to the individual, but also in regards to the social group and the environment in which individuals find themselves (Max-Neef, 1992). Finally, while needs are interactive and may complement each other, they are nonetheless different and distinct, and therefore not additive. Abundant access to satisfiers for one set of needs does not compensate for a lack of satisfiers for another set of needs. This suggests that separate 'accounts' should be kept for access to satisfiers of different needs.

In developing QOL accounts based on Human Needs Assessment (HNA), it would be useful to test measurements of satisfiers empirically in studies comparing these objective measures against subjective assessments of QOL to determine their effectiveness. These empirical tests as well as efforts to operationalize HNA accounts must involve people in interactive dialogues which will confirm or refute the validity of the needs Max-Neef specifies, as well as the validity of the satisfiers we use to assess the degree to which needs are met. Such dialogues would almost certainly elicit additions and alternatives to the satisfiers shown in table 3 of chapter 11. While the average person may not always know exactly what satisfiers will best meet his or her needs, interactive discussion with people is nonetheless essential to select and test appropriate indicators. We would also need to develop group-based methodologies to determine the effectiveness of our indicators in a social setting.

2.3. Ecosystem services: indicators to integrate with QOL

Finally, when measuring QOL, we must account for its relationship with ecosystem services generated by natural capital. In some way or another, all of the human needs listed by Max-Neef depend on natural capital. However, we are tremendously ignorant concerning how ecosystem structure generates ecosystem function, how ecosystem function generates services valuable to humans, how human impacts affect ecosystem functions, and where the thresholds lie beyond which natural capital fails to reconstitute itself. Hence, it is virtually impossible to say precisely how specific ecosystem functions affect specific human needs. Nonetheless, we recognize that the relationship between ecosystem services and human needs is absolutely fundamental. Given the unacceptable risks of overestimating ecosystem resilience or underestimating human dependence on the ecosystem, we assert that a healthy ecosystem is essential to human well-being[3]. A healthy ecosystem is defined as well-functioning, and well-functioning means an ecosystem's ability to supply services. Hence, ecosystem health is a prerequisite to fulfilling Max-Neef's human needs matrix, and any accounting system designed to measure human QOL through time must account for ecosystem health.

2.4. The implications of using HNA as a measure of QOL

It is clear that Max-Neef's approach is very difficult to operationalize, even if theoretically more compelling than the alternatives presented. The debate over which approach to take to national accounting – theoretically sound measures or ease of accounting – is old. As Irving Fisher argued back in 1906, the appropriate measure even of income is one that captures the psychic flux of service (i.e., satisfaction of needs and wants) and not simply the final costs of goods and services (Daly and Cobb, 1989). And at the time Fisher wrote, the absence of suitable data for calculating either psychic flux of service or final costs no doubt led many to ignore the debate as entirely academic, as no doubt some will regard the arguments we are putting forth here. The widespread use of GNP indicates that in practice Fisher lost this earlier debate. However, measures such as the ISEW (preceding chapter; Daly and Cobb, 1989) suggest that the GNP is becoming

---

[3] Assessing ecosystem health will require another set of indicators and measurements. While we lack space here to discuss the nature of appropriate indicators, Costanza (1992) suggests that indicators must cover at least 3 aspects of ecosystem health, including (1) vigor, which is a measure of system activity, metabolism, or productivity; (2) organization, referring to the number and diversity of interactions between system components; and (3) resilience, referring to a system's ability to maintain its structure and pattern of behavior in the presence of stress.

increasingly less capable of measuring economic welfare, much less QOL. Even if we can never quantify access to satisfiers as accurately as we currently quantify GNP, as Amartya Sen suggests, perhaps it is better to be vaguely right than precisely wrong (Crocker, 1995).

Accepting Max-Neef's human needs matrix as a framework for the specific elements of human QOL, and access to satisfiers as potentially the best objective indicator of QOL, has profound implications with respect to the distribution of wealth and resources and our capacity to sustain human QOL. First, most of the possible indicators suggested by Max-Neef require few, if any, material resources, and hence are not subject to physical exhaustion. Thus, for most elements of human QOL use by one person or generation does not leave less for others. Second, by explicitly accepting that there is a limit to needs, we can limit consumption without sacrificing QOL. This result is critical, because the laws of thermodynamics make it impossible to delink physical consumption from resource use and waste production. As abundant evidence suggests, current levels of consumption could not be sustainably met with renewable resources alone, and therefore, we must limit consumption or else threaten the supply for future generations of life-supporting, non-substitutable, and essential natural capital.

Yet within the current dominant ideology of neoclassical economics with its belief in insatiable wants (which are not distinguished in any way from needs) and the use of GNP as a proxy for QOL, it is unlikely that the current generation will voluntarily limit its consumption for the sake of the future. People are extremely reluctant to sacrifice their own well-being for others, and if wealthy individuals and nations refuse to make sacrifices for the poor alive today, how much less likely are they to do it for those yet to be born? Since in reality wealth translates to power and the powerful make the rules, rules that 'punish' the powerful rarely evolve. In addition, the dominant institution for distributing wealth and resources in use today is the market system; yet it is absolutely impossible for future generations to participate in this system. Only if people accept that limiting current consumption of material resources beyond a certain threshold has little negative impact on the QOL of people alive today, are we likely to create a more sustainable society for the future.

From this perspective, the difficulty of operationalizing Max-Neef's framework may actually be a point in its favor. Why is it that we want to measure QOL in the first place? It is not just to track the rise or fall of QOL, but also to help us create policies to improve it. Simply providing statistical data on QOL is insufficient to achieve this end. It is also necessary to relate those data to theories that show not only why the data are relevant, but also how change can be achieved. Theories concerning QOL and its appropriate indicators are little more than ideologies, and the ideology behind HNA as the basis for QOL accounts provides an important alternative to the ideology behind GNP. To attain a more just distribution of goods and services that generates a greater QOL for all, we must change people's perceptions about what actually contributes to our QOL. This requires a compelling story supported by statistical measures of QOL, and the story we present is based on the ideological assumptions inherent in the human-development approach to QOL. The very effort to operationalize HNA-based QOL accounts and the extensive dialogue it requires will expose people to the theory behind it. Exposure to a theory is the first step towards acceptance. Once people accept this theory, leaving vital resources for future generations will not be viewed as much of a sacrifice by the current generation. This perception is a vital step towards meeting our goals (Cobb, 2000).

3. Development of indicators of fairness in the distribution of wealth and resources

In chapter 11 we presented the argument that the market system was potentially fair within a generation, since it awarded people their 'just deserts'. However, many of the outcomes we actually see from this system are clearly not fair in most people's eyes. Two possible explanations of this unfairness include the fact that the economic system is only fair if the starting point of all the players is fair, and the fact that there are market failures for many resources, in particular those provided by natural capital. Turning now to justice theory, Rawls defines a fair society as one in which the worst-off individuals are as well off as possible, but does not state what that society looks like. For practical purposes we are left only with the notion that a society is becoming increasingly fair if the worst-off are improving their lot and less fair if the converse is true.

In terms of intergenerational justice, market economics confronts more serious difficulties; future generations cannot participate in today's markets, hence the market system no longer functions. There is no guarantee that these future generations will receive their 'just deserts'. Still, many supporters of the system are reluctant to admit defeat. Instead, they argue that as resources become scarce, prices increase, inducing innovation of substitutes. Thus, future generations will always be provided for [4]. However, if the market fails to place the appropriate price on a resource to begin with, then the price will not respond correctly to scarcity, and there will be no incentive for the market to develop substitutes. By definition, goods and services characterized by market failures are not appropriately priced

---

[4] However, substantial evidence suggests that previous civilizations have perished from over-exploitation of resources. If we believe the market system is to avoid this fate, we must assume that the profit motive is more powerful than the survival motive, or else that technology has reached a point where infinite substitution is possible. Either assumption is based on faith and inductive reasoning, not science, and cannot be ethically justified if we accept that we have obligations to future generations.

by the market. Justice theory, as we have presented it, would demand three things for intergenerational justice: do not leave the future worse off than it would have been with an equal intergenerational distribution of resources, assume strong sustainability until proven otherwise, and maintain the yields from non-substitutable natural capital.

Rather than attempt the perhaps impossible task of developing a detailed theory of fairness acceptable to 'just deserts' and justice theorists alike, we will seek instead to draw forth a limited number of specific indicators of unfairness and requisites to fairness on which both approaches should agree. These can then form the basis for objective measures of fairness in the following section.

3.1. Natural capital and market failures

Natural resources, and ecosystem services in particular, are plagued by market failures. As we have argued above, natural capital plays a critical role in meeting human needs and in providing a satisfactory QOL. We can assess how market failures relate to fairness through a close examination of two specific market failures: public goods and externalities (for details, see chapter 11).

3.1.1. Excludability and 'rivalness'

Virtually any good or service (or at least specific properties of any good or service) can be classified according to two characteristics: excludability and 'rivalness'. Excludability is essentially a question of enforceable property rights. An excludable good is one that an individual or an institution can keep others from using, and a non-excludable good is one where this is not possible. Since a person can use non-excludable goods whether she pays for them or not, few individuals will pay, and the market will not provide them. A rival good is one where use by one person leaves less for use by someone else, and a non-rival good is one where use by one person does not affect the quantity or quality of the good remaining for another user. Essentially, the cost of an additional person using a non-rival good is zero. Since economic efficiency demands that the price of a good be equal to its marginal cost, market provision of non-rival goods will be inefficient. In other words, if there is a price on a non-rival good, a person will use less than if it were free, potentially resulting in a lower QOL for that person; yet additional use would not incur additional costs for society.

Any goods that are not both excludable and rival are therefore not efficiently provided by the market[5]. This is a market failure. Goods such as oceanic fisheries

---

[5] Note that if people are not the rational maximizers of self-interest depicted by neoclassical economic theory, a market economy could supply public goods and minimize externalities. However, if we accept this supposition to argue that market failures are not a problem, we also undermine the assumptions on which the optimality of market allocation is based.

that are non-excludable and rival are 'open access' resources subject to the 'tragedy of the commons', and will be overexploited by market forces. Goods such as information (for example, the information stored in biodiversity) that are non-rival, but can be made excludable through appropriate institutions can be provided by the market, but the resulting price will not be efficient. Goods such as the ozone layer or global climate regulation that are both non-rival and non-excludable are pure public goods, and will only be efficiently provided (or preserved) by extra-market institutions. Many types of natural capital are complex mixtures of these different categories of goods. For example, trees in the Amazon when seen simply as timber are market goods, but when in areas too vast to monitor, they are open access resources. Genetic information contained within those trees could be made excludable by the Convention on Biodiversity, but the information is not depleted no matter how many people use it. As contributors to rainforest function, these trees provide the ecosystem services of climate regulation, gas regulation, disturbance regulation, habitat, and a host of other pure public goods. It is worth noting that most life-supporting services of natural capital are pure public goods.

The relationship between excludability, rivalness, and fair distribution can now be drawn out. *Open access resources* in a market system are subject to first-come first-served treatment, and lacking proper institutions, those who arrive too late receive nothing. Few disagree that this outcome is both unfair and inefficient. *Non-rival excludable goods* will not be efficiently distributed according to economic theory, but it can be difficult to assess what is fair in this case. If someone invents something, it is probably fair that she receives some payment for it, yet she would not receive payment if it were made non-excludable. If the inventor receives payment from individuals using the invention, then it is likely to be used less than is socially optimal (at least assuming that it is an invention that makes a positive contribution to QOL). If we accept the economists' contention that the free market is fair, then the distribution of *market goods* will also be fair, but only if we assume a fair initial distribution of resources. However, once a *pure public good* is made available, fair distribution is automatic. Whoever wants to use it can do so, and to the extent they desire without leaving less for anyone else. It follows then that destruction of public goods for private gain is clearly unfair.

The next issue we must examine then is the relationship between natural capital, market goods, and public goods. We can distinguish two types of natural capital: goods and services. Goods are simply the raw material inputs from nature, such as timber, fish, and minerals. All natural capital goods are rival, in that if one person removes a tree from the forest or a fish from the ocean, it is no longer there for someone else to remove. Whether or not natural capital goods are excludable depends on property rights and how well they are enforced. For example, oceanic fisheries are mostly non-excludable, while forests on private land are theoretically excludable. On private land in the middle of the Amazon, of course, it may not be possible to enforce property rights, and the trees become non-excludable.

Once a natural capital good is harvested, however, it is almost always excludable. Hence, natural capital goods are essentially market goods. Natural capital services, on the other hand, include such things as climate regulation, gas regulation, water regulation, etc., which for the most part cannot be owned, and use does not lead directly to depletion. These services are public goods.

What is the relationship between natural capital goods and services? Natural capital goods as described here can be thought of as components of ecosystem structure – that is, they are the mineral resources, organic matter, and individuals and communities of plants and animals of which an ecosystem is composed. When all the structural elements of an ecosystem are in place, they create a whole that is greater than the sum of the parts, and generate ecosystem functions as an emergent phenomenon from the complexity of ecosystem structure. An ecosystem function that has value to human beings is called an ecosystem service. As all market goods must be produced from the structural elements of natural capital, and depletion of structure diminishes function, production of market goods in general must reduce the ability of the ecosystem to generate public goods (Farley, 1999).

How does this relate to the fairness question? Market goods specifically benefit individuals and public goods benefit everyone, hence the production of market goods implies the destruction of public goods for individual benefit. Thus, there is built-in unfairness in the production of market goods. 'Just deserts' would demand that whoever produces or consumes a market good compensate all those who suffer from its loss. Justice theory would tolerate the increasing unfairness inherent in ever-greater conversion of natural capital to market goods only as long as it continues to make the worse off better off. Eventually, excessive production of market goods may undermine ecosystem health and the ability of global ecosystems to generate critical life-support functions, making everyone worse off. The outcome in this case would be extreme unfairness, particularly towards future generations.

3.1.2. Externalities

Another market failure closely related to distribution and fairness is that of externalities. Externalities occur when one actor's activity causes unintended impacts on another actor, and no compensation occurs. Because no compensation occurs, externalities do not enter into market decisions. Many negative externalities are in the form of destruction of public goods provided by natural services. In fact, this is exactly what was described in the discussion of public goods; one actor harvests ecosystem structure, which has an uncompensated negative impact on other individuals who previously benefited from the ecosystem services generated by that structure. Similarly, all negative externalities are likely to contribute to unfair distributions of wealth and resources, as some individuals benefit while others pay the costs. Templet (1995a) and chapter 11 provide many empirical examples of this.

Hence, both justice theory and 'just deserts' should agree that to the extent society allocates resources (and particularly essential ones) characterized by market failures via the market system, society is unfair to both the present and the future.

3.2. The elimination of poverty

A second point of agreement should be that poverty – broadly defined as the lack of access to the satisfiers required to fulfill human needs – in a society with sufficient resources to prevent it is unfair. This is very clear in the case of Rawlsian analysis. The poorest individuals are the worst off, and if an alternative society would make them better off, then the society in which they exist is unfair. Neoclassical welfare economics, whose foundations are utilitarian philosophy and diminishing marginal utility, certainly should call for elimination of poverty. If the goal of society is to maximize utility summed over individuals, and wealth and income offer diminishing marginal utility, then clearly an additional unit of wealth for a poor person provides more utility than the same unit would provide for a wealthy person. Economists reluctant to accept this conclusion have asserted that different people have immeasurably different capacities to enjoy, and hence we cannot make interpersonal comparisons of utility. Thus, economists have focused on maximizing production rather than utility, which effectively skirts the distribution issue (Robinson, 1964). However, can anyone be foolish enough to believe that on average a unit of additional income would not benefit someone living in absolute poverty more than the same amount would benefit a millionaire? People may have different capacities to enjoy at some level, but our biological needs are the same, and the additional utility when one moves from below these needs to above them is obviously immense.

It is far less obvious why the 'just deserts' principle should call for alleviation of poverty. Solow (1993) has pointed out that the whole discussion of sustainability generally assumes that some sacrifices may be required by this generation to make future generations better off. If we are concerned about the potential poverty of people not yet born, what ethical system will allow us to ignore the actual poverty of those alive today? The 'just deserts' theorists might claim that the market is fair within a generation, but not between them, because future generations cannot participate in today's market. Therefore, 'just deserts' could justify concern for providing sufficient resources for potential future generations while essentially ignoring poverty today, strange as this may sound. Further, most Americans profess to believe that the current distribution of income in the USA is unjust, yet they remain reluctant to provide income to those who have not 'earned' it. However, the 'just deserts' argument basically claims that people are paid according to their contribution to society. Yet the last two centuries have seen a fairly steady upward trend in real incomes. This is not so much because people make more substantial contributions to society on their own, but because they benefit from past contributions to productivity. That is, many people are awarded more than their just deserts already, and if anybody is to be awarded more than they deserve, shouldn't it be the worst-off? Further, if a lack of opportunity is the cause of poverty, then the fairness criterion of 'just deserts' is not met. It would appear then that the 'just deserts' argument should at a minimum favor equal opportunity. Perhaps direct transfer payments to the poor are inappropriate under this ethical system, but at a minimum, guaranteed jobs at a living wage and equal access to education and job advancement could be defended (Lane, 1986).

3.3. Maximum income level

A third point of agreement should be that unlimited income and accumulation of material wealth on a planet with finite resources is unfair. Justice theorists could argue that allowing unlimited accumulation of wealth creates incentives that increase total production and make the worst-off better off than before. 'Just deserts' theorists could argue that the wealthy are wealthy solely because they have earned it, and society has no right to take away someone's just deserts. However, on a finite planet subject to the laws of thermodynamics, if too many people consume too much, they will reduce the resources available to future generations. This means that in the future, society may be worse off than it is today, or individuals in the future will have to work harder than individuals today to consume as much. Thus, the 'just deserts' principle would not apply between generations. 'Just deserts' would demand that society today cannot consume so much that future generations lack the same opportunities to be rewarded for their work as we enjoy. We have already argued that society is consuming too much by these standards. However, to demand that society as a whole must reduce consumption and yet not demand that those in society who have the most also restrict consumption simply cannot be defended in terms of 'fairness'. Some people might go on to argue that the wealthiest are not necessarily the largest consumers. If this is so, then there is even greater reason not to allow unlimited accumulation of wealth, as we shall explain.

Why would anyone accumulate wealth if they do not intend to consume it? The only reasonable answer is to amass power and status. Certainly, no one can rationally argue that wealth does not bring power in existing political systems. While many people argue that inequitable distribution of wealth is acceptable, far fewer accept that inequitable distribution of power is (Lane, 1986), at least in those countries that profess to be democratic. What's more, once people have accumulated power, they then use that power to accumulate even more wealth and power. For example, it is painfully clear that corporate donations to political parties in most countries are not made to strengthen democracy, but rather to promote legislation that provides greater economic advantage for the contributors. Great wealth allows people to get more than their 'just deserts' in the political arena, and then use that power to take unfair advantage in the economic arena as well. Examples of this were provided in chapter 11, and also in Templet (1995a,b). Strangely enough, however, Americans are far more opposed to limiting maximum income than they are to ensuring a minimum income (Lane, 1986). Americans seem to have two completely incompatible core beliefs: we live in a democratic society, and anyone is entitled to become filthy rich. However, as Supreme Court Justice Louis Brandeis said, "We can have a democratic society, or we can have the concentration of great wealth in the hands of the few. We cannot have both."[6]

These last two shared principles of a fair society outlined here are hardly modern. Perhaps the earliest known western philosopher, Thales of Miletus, wrote in 1600 BC: "If there is neither excessive wealth nor immoderate poverty in a nation, then justice may be said to prevail" (Quoted in Durning, 1992, p. 143).

3.4. Geographical fairness

Notions of fairness should not depend on geographical proximity. Historically, there may have been a genetic justification for greater fairness towards one's neighbors, since they were more likely to share one's genes. In some countries this may still hold. In others, immigration mixes the gene pool, and ease of travel continues to do so. In any case, we have argued that we have ethical obligations to the future, including far distant generations that are as little related to us as anyone in the remotest corner of the earth. Thus, rather than searching for specific nuances of fairness that apply across space, we will instead focus on two particularly egregious examples of unfairness.

3.4.1. Third world debt

Total third world and Eastern European debt is now in the neighborhood of $2.6 trillion dollars, and in some countries up to 40% of government expenditures go towards servicing the debt. Currently there is net flow of debt-related financial capital from the poor countries to the rich, and this has been the case for at least

---

[6] One school of philosophy argues that simply ensuring a more equal distribution of wealth will do little good. There are numerous spheres of justice, each of which pertains to a different social arena. In western capitalist society, monetary wealth is dominant. Distributing wealth more equally would require a powerful political apparatus, and politics would replace wealth as the arena of dominance. If political power were divided more equally, than the dominance of monetary wealth would return. Justice is only achieved if we sever the links between the numerous spheres of justice so that inequality in one sphere cannot translate into inequality in another (e.g., Walzer, 1990). While the argument is compelling and autonomy of spheres of justice should be pursued to the extent possible, it seems that relying solely on this approach to justice would require far more radical changes to society than those we will propose.

10 of the last 20 years. Many of these poor countries are forced to spend more on debt service than on health and education combined (Roodman, 2001). Debt crises have caused considerable hardship, and most recently high loads of short-term debt were linked to currency crashes and severe depressions, which began in South East Asia. The unfair nature of this debt is obvious in the terms of 'justice theory'. Nonetheless, the 'just deserts' school claims that these countries entered into these agreements of their own free will and are therefore obliged to honor them. This argument holds little weight. First, despotic dictators acquired much of this debt. Marcos of the Philippines, Mobutu of Zaire, Suharto of Indonesia and the Duvaliers of Haiti are some of the most infamous, but there are dozens of examples. Some of the loans they acquired went to corrupt cronies, some went into bank accounts in Switzerland and other financial havens. Worse, much of the money was used to maintain illegitimate power. Now that these dictators have been thrown from power, western banks claim that the very people this money was used to subjugate must repay this debt. Even if the lenders were ignorant of how their money was used, and it is clear that they were not, they would not be morally entitled to collect this debt. Nor are they according to established precedence in international law. In 1898, after the USA essentially seized Cuba from Spain in the Spanish American war, the USA declared all Cuban debt to Spain null and void, because it was 'odious debt'. The argument was that the money had been loaned to dictators that did not represent the people, and therefore the people had no obligation to repay it (Chomsky, 1998). As John Maynard Keynes (1919, p. 210) maintained, "nations are not authorized, by religion or by natural morals, to visit on the children of their enemies the misdoings of parents or of rulers." If we cannot visit them on our enemies, we certainly cannot visit them on anyone else. Demanding repayment cannot be considered a case of 'just deserts'[7]. The numerous other arguments for canceling the debt typically accept the false premise that we are demanding payment from the actual debtors, and need not be reviewed here.

3.4.2. Ecological debt

If there is any moral obligation to repay a debt, it is the obligation of the overdeveloped countries (ODCs)[8] to pay the less developed countries (LDCs) for centuries of accumulated ecological damage. The ODCs are responsible for

---

[7] The fact that the USA and other western nations now insist (with minor concessions) on repayment of many similarly odious debts is based on a different but far more ancient concept, might makes right.

[8] We define overdeveloped countries as those where the net marginal benefits to aggregate QOL for the country from consumption and economic growth are less than or equal to zero, or alternatively, where the marginal external cost of this consumption imposed on other countries and future generations is greater than aggregate marginal benefits.

the vast majority of natural resource use and waste output. Even though much resource extraction takes place in the LDCs, it is the consumers in the ODCs who are ultimately responsible. Toxic chemicals produced in the ODCs are now found even in Antarctica (McGinn, 2000). Public outcry over pollution in the ODCs has forced many factories to shut down and relocate to the LDCs where environmental laws are weaker or enforcement is lax. Over-consumption of potentially renewable natural resources not only threatens to leave less for future generations but for the present as well. For example, European nations have purchased fishing rights from some West African countries, and the fishermen in those countries find the resulting depleted stocks are adversely affecting their livelihood (Brown, 1998). Oil production by western companies in the Nigerian delta region has severely damaged one of the world's largest mangrove ecosystems, with seriously adverse affects on the health of the local communities (Constitutional Rights Project, 1999). Worse, excessive burning of fossil fuels now threatens to induce (if it has not already) global climate change. Resulting sea-level rise will literally inundate low-lying island countries such as Mauritius and the Seychelles, and threaten coastal zones of numerous others. Hypocritically, the ODCs clamor that Brazil's destruction of the Amazon threatens biodiversity and will contribute to greenhouse gases, yet the clearing of forests over past centuries in OECD countries has contributed more CO_2 to the atmosphere than is contained in the entire Amazon (Bueno and Marcondes, 1991). The LDCs have far fewer resources with which to cope with global warming, are more dependent upon agriculture, which is the sector most affected, and hence will likely suffer more from the impacts. Now that ODC-caused problems such as ozone depletion and global warming have reached crisis proportions, all countries must cooperate to minimize damage. In many cases this might mean slower economic growth for those countries with the highest proportion of citizens in absolute poverty, who could still benefit from greater production and consumption. There is little serious talk of compensation for ecological damages caused, and most ODCs are arguing that technologies which reduce greenhouse gas emissions and replace ozone depleting substances should be sold to the LDCs, not given. Some 'just deserts' theorists such as Lawrence Summers (1991) argue that we should ship toxic wastes to the LDCs since (1) they are 'under-polluted', (2) they value safe environments less, and (3) the lives of people in LDCs are worth less. However, one cannot credibly argue that the poor countries receive their 'just deserts' when no compensation occurs for the harm they suffer at the hands of the ODCs.

4. Approaches to measuring fairness

Measuring an ethically based notion such as fairness is perhaps even more difficult than measuring QOL. In this section we will not lay out measures of fairness in detail, but rather suggest possibilities that would capture elements of fairness too often ignored. Many of these suggestions would require substantial amounts of research and modification to be made practical. This does not mean that they are 'naïve'. Bear in mind that when GNP-style national accounts were first suggested, we did not have the data available to calculate them, and it took decades from first discussion to practical implementation. As suggested above, a good starting point for measuring fairness should focus on objective indicators of unfairness and requisites to fairness on which both the 'just deserts' and 'justice theory' schools agree. We will therefore look at ecosystem health and market failures affecting the environment as a measure of fairness, as well as income distribution and the ability of wealth to provide political power.

4.1. Ecosystem health and functioning markets

We concluded above that both damaging public goods for private gain and negative externalities are by nature unfair. Damage to public goods and negative externalities result from normally functioning markets. Extra market institutions, such as the government, must be responsible for supplying and preserving public goods. Thus, the extent to which a society supplies and preserves public goods and eliminates negative externalities (especially those which affect public goods) is probably a reasonable indicator of its fairness. Alternatively, if society subsidizes market goods or market-good production that do not generate positive externalities, and particularly if the market-good production in question degrades public goods, the subsidies are indicators of unfairness. Templet (1995a) has used various types of government subsidies as an indicator of unfairness, verifying their validity through statistical analysis (see chapter 11).

To reiterate, most environmental services are pure public goods. All market goods require raw material inputs and generate waste outputs, and raw materials are extracted from ecosystem structure that would otherwise generate ecosystem function. Thus, production of market goods in general creates negative externalities in the form of damage to environmental services. We defined ecosystem health above as the well-functioning of an ecosystem, where well-functioning is the ability of an ecosystem to generate services. Obviously, life-support functions – by which natural capital reconstitutes itself – are the most important of these services. Thus, a healthy ecosystem generates public goods, and is not too severely affected by the negative externalities of market-good production. Further, we have argued that ecosystem health plays an important role in the satisfaction of all human needs, some directly, some indirectly. Particularly in rural and coastal areas, many people depend directly on ecosystem goods and services for their livelihood, and the poorest often depend on healthy ecosystems for their survival. Some of the endless examples of this include mangrove ecosystems that provide building materials and food sources and act as a 'nursery' to many fish species upon which local populations depend (e.g., Nickerson, 1999); extractive reserves in the Amazon that sustain a number of the region's poor (Schwartzman, 1989); or the forest services in Thailand and Ivory Coast (and no doubt worldwide) shown to significantly improve local crop yields (Panayatou and Parasuk, 1990; Ehui et al., 1990). Thus, it would appear that ecosystem health could serve as an important indicator of fairness both within and between generations.

However, accepting that ecosystem health is a reasonable indicator of fairness still provides little insight into how we could use it as an indicator. Some ecosystem services accrue to people at the local level, as described in the previous paragraph. Others are regional, such as the impact of deforestation on rainfall, regional climate, and agricultural yields hundreds or even thousands of miles away. Yet others are international, such as global climate regulation and planetary life-support systems. And just because someone lives far from unpolluted air and water, that does not necessarily imply unfairness. For example, Donald Trump at home in Trump towers with its carefully controlled climate is not exactly surrounded by direct and tangible ecosystem services, but he does have the capacity to substitute for them on a small scale, and he has access to them if he so desires. It would appear then that the appropriate indicator of fairness would be access to the services provided by healthy ecosystems. If someone lives in a degraded ecosystem because it is the only place they can afford to live, that is unfair. Considerable research is required to operationalize ecosystem health as an indicator of unfairness (see Costanza, 1992), but the concept does show promise.

4.2. Poverties and pathologies

If poverty is unfair as we argued above, then one measure of fairness should be the degree to which a society has eliminated poverty, defined as the inability to satisfy any one of the human needs. In this context, Max-Neef refers to 'poverties' and not just poverty. The problem with poverty is that it generates pathologies in the systems in which it is found. Max-Neef (1992, p. 200) provides the following examples: "... persistent economic pathologies are unemployment, external debt and hyperinflation. Common political pathologies are fear, violence, marginalization, and exile." This notion of system-wide pathology also has counterparts on the level of the individual. For example, subsistence poverty creates the pathology of malnutrition, protection poverty creates the pathology of preventable disease, affection poverty creates the pathologies of violence and intolerance. One could use the presence of such pathologies as indicators of poverties and hence as a measure of the fairness or unfairness of a given society.

4.3. Wealth and power

We have also argued that the concentration of material wealth and power is an indicator of an unfair society, through both space and time. The simplest measures of fairness include the percentage of the wealth owned by the top 1% of the population and the top 20% relative to lower deciles, both within and between countries, and the trend in fairness can be determined by how these statistics change over time. In the USA in 1995, the Federal Reserve estimated that wealth of the top 1% was greater than that of the bottom 95%, up from the bottom 90% only three years earlier. In 1998, the people in well-to-do countries were 82 times better off than people in countries where the poorest 20% of the world's people live. Three decades ago, they were 'only' 30 times better off (Gates, 1999). Since wealth implies excessive consumption and power in modern society, concentration of the wealth is probably the best single indicator of its unfairness within a generation. In contrast, total wealth, independent of distribution, may be the best indicator of unfairness towards future generations. Thus, in terms of national measurements, we could consider societies such as the OECD countries the least fair through time, while the Latin American countries with their notoriously unequal distributions of wealth show greater domestic unfairness in the current period. By international measures, the OECD countries both benefit the most from current unfairness and impose the greatest costs on future generations.

We should also attempt to measure to what extent wealth buys political power. In the USA in the year 2000 election campaign, less than 1% of the population donated 71% of Bush's campaign donations, and 61% of Gore's. Not surprisingly, polls find that policies espoused by Bush and Gore were far more closely aligned with their big donors' views than with the views of average Americans. For example, Gore wanted to use the government surplus to pay down the national debt, and Bush proposed tax cuts. Almost two-thirds of voters preferred investment in health care and education, with the remaining one-third divided between debt reduction and tax cuts. In contrast, 52% of major donors favored tax cuts or debt reduction, with Republicans the most in favor of tax cuts (Lake and Borosage, 2000).

The simplest indicator of the influence of wealth on political power in nominally democratic societies would be to calculate the share of donations provide by the top 1%, 5%, and 10% of a society, as well as the percentage of the population that donates nothing. More difficult but more interesting would be to estimate the correlation between a politician's votes and the preferences of his largest donors vs. the preferences of his constituents. More difficult still but also interesting would be to calculate a Gini coefficient of political donations and lobbying expenditures by both eligible voters and corporations. Commonly used to compare income distribution between nations, the Gini coefficient (GC) is simply a measure of the area between the Lorenz Curve and the 45-degree equality line. The Lorenz curve is a diagram showing the cumulative percentage of national income (or in this case political donations) received by a certain percentage of individuals or households (or in this case donated by a certain percentage of individuals and corporations). A GC of zero refers to a perfectly equal distribution of voter donations (or income) and a coefficient of one to the case where one person makes all the donations (or earns all the income).

Corporations must be included in these calculations because their dollars have just as much influence as the dollars of citizens. Non-voting but eligible voters must also be included in a democratic society. This measure could be used to compare politicians within a country with each other and also to compare countries. Of course, this measure is only applicable in the nominally democratic societies on the higher end of the income scale, where individuals have sufficient resources to donate to politicians. Other measures must be developed for the bulk of the world's countries. The disadvantage with the GC measurement is that it requires explanation to understand what it measures, and therefore would be primarily useful for comparative purposes when the user only needs to understand that a higher GC indicates a less equal distribution than a lower one.

It would also be worthwhile to examine the relationship between gross political donations and voting records or political donation GCs and voting records on issues that affect the environment. As discussed in chapter 11, Templet (1995a,b) found that candidates with larger campaign donations have statistically significant worse environmental voting records as measured by the League of Conservation voters. More generally, Boyce et al. (1999) found that as political power concentrates, pollution increases and public health and welfare decline.

4.4. A Quality of Life Gini Coefficient?

While Gini Coefficients (GC) are used to calculate fairness in income distribution, our concern with fairness is not limited to the distribution of income, but also to the distribution of all the factors that contribute to a high QOL. This raises the question as to whether a GC based on the Human-Needs approach to QOL accounts proposed above – a Quality of Life Gini Coefficient (QOLGC) – might be a more appropriate measure of fairness. While quite an abstract concept and currently beyond our means to calculate, the QOLGC would capture many aspects of fairness not captured by the standard GC. However, there are some serious problems with this approach. First, we would need to assign a specific number to people's QOL derived from objective measures of people's access to satisfiers of human needs, or at the very least a cardinal measure of the level of satisfaction of each specific human need. Second, not all satisfiers depend on the consumption of physical resources. Those that do not consume physical resources then may not impinge on others' ability to enhance their own QOL, and hence it is not 'unfair' if one group has more than another. In addition, excessive consumption of physical resources is unfair, but beyond a certain level it probably fails to contribute substantially to QOL, and therefore would not be captured in objective measures of QOL. This point was discussed above in relation to 'violators and destructors', 'pseudo-satisfiers', and 'inhibiting satisfiers'. While these false satisfiers may ultimately be destructive of QOL, people may use considerable resources to gain access to them, and this access should be included in any measure of fairness. That is, rather than a QOLGC, a more broad-spectrum GC designed to measure fairness should be based on access to satisfiers, violators and destructors, pseudo-satisfiers, and inhibiting satisfiers.

Further complications arise if we attempt a broad-spectrum GC-like measure of fairness across nations. Satisfiers are culturally specific, so it is very difficult to judge fairness in terms of access to satisfiers across culture. What's more, some countries emphasize satisfiers that are by nature less fair. Specifically, many national cultures emphasize consumption as a satisfier and consumption depletes the world of resources that could otherwise be used by other individuals and other generations. As noted earlier, consumption is often an inhibiting satisfier or for many human needs a pseudo-satisfier, and, in excess, a violator and destructor. Thus, attention in these cultures to consumption has probably led to reduced access to family, community, nature, etc., and reduced satisfaction of human needs. However, one cannot claim that American society, for example, has been treated unfairly because we build strip malls and sit through traffic jams that reduce our QOL.

For international measures of fairness then, perhaps the best approach is to calculate a simple income-based global Gini-coefficient. Income is probably the best measure of consumption of physical resources, which due to the laws of thermodynamics deprives others of access to those resources and spews waste into the environment, and hence may be the best indicator of fairness. To our knowledge, the GC has never been used to calculate trends in concentration of wealth on an international level. It would be possible to calculate the GC of all the nations by using per capita income or of the entire global population ignoring national boundaries and using individual incomes. In either case, it would be best to adjust for purchasing power parity. Both measures would convey useful information and statistics are readily available [9]. These measures could be tracked through time to indicate whether global fairness in income distribution is improving or declining.

5. Implications of the relationship between fairness and QOL

It is implicit in the definition of unfairness that those who experience it suffer as a result and enjoy a lower QOL than they would if treated fairly. However, unfairness that is attributable to the actions of others presumably would not occur unless someone else benefited from it or at least perceived a benefit from it. Certainly the common perception is that reducing unfairness must also reduce the QOL of those who benefit from it. The fear on the part of the affluent and powerful that a fairer allocation of resources will inevitably reduce their QOL is a major obstacle

---

[9] There is reasonably good data available on per capita income in different nations, but data on income distribution within nations is likely to be less accurate.

to greater fairness nationally, internationally, and intergenerationally. Since the affluent and powerful have the greatest ability to change the current distribution, this is a serious obstacle to greater fairness. However, significant evidence suggests that a fairer distribution of wealth and resources may actually improve the QOL, not only for those who are currently impoverished, but for the affluent as well.

5.1. Positional wealth

First, we return to the fact that above a certain level, resource consumption and wealth may be 'positional', that is, we derive QOL from comparing our position with that of others. It appears that we are currently engaged in a never-ending wealth and consumption race, where greater consumption by our reference group demands greater consumption on our part simply to maintain the same relative position. With current economic growth patterns leading to greater concentration of the wealth in the hands of the few, the majority of the population is falling behind in this race. The wealthy obviously compare themselves with each other and not with the poor, and therefore they are not achieving greater QOL either. To the contrary, the blind pursuit of positional wealth and consumption places substantial demands on our time and resources, and leaves us with ever less ability to meet our other human needs (Frank, 1999; Broome, 1991). Further, as all market consumables must be produced from natural capital, we inevitably diminish the ability of natural capital to generate public goods. Hence, the more resources we consume in this positional race, the more natural capital is depleted and the fewer ecosystem services we enjoy. Eventually, we risk the destruction of life-supporting natural capital, threatening our very subsistence. Basic subsistence is certainly not a positional good and the loss of life-supporting natural capital will have an unacceptable, negative impact on global QOL. In Max-Neef's terms, excessive consumption or accumulation of natural capital is a pseudo-satisfier, and if carried to extremes becomes a violator and destructor.

If above a certain level, positional wealth and consumption matters more than absolute wealth and consumption, then if we could somehow reduce all consumption above that level by 90%, for example, people might suffer little direct change in their QOL. Indirectly, lower consumption needs would require less work, leaving more time to pursue satisfaction of other human needs. Ecosystem services would be more abundant, contributing to the fulfillment of all of our needs. We would move farther from ecological thresholds, be relieved of the stress of worrying about ecosystem degradation, and better fulfill our need for protection. Since ecosystem services are public goods, this would also be fairer to both current and future generations.

5.2. Income inequality as a detriment to QOL

As mentioned earlier, QOL was first introduced as a concept to address issues such as increasing crime rates in a society experiencing ever-greater economic production. Thus, almost by definition, crime – and in particular violent crime – reduces QOL. In terms of human needs assessment, violent crime reduces society's ability to satisfy the need for security. It is fairly obvious that absolute poverty provides an incentive to commit crime. However, numerous studies have found significant correlations not only between poverty and violent crime but also between income inequality and violent crime, even when controlling for poverty (Kennedy et al., 1998; Hsieh and Pugh, 1993; Fajnzylber et al., 1998). QOL is also an important concept in the field of medicine, and ceteris paribus, most people would agree that ill health reduces QOL. Again, numerous studies have found a significant correlation between poor health and income inequality (Lynch et al., 1998; Kawachi et al., 1997[10]). For example, Wilkinson (1996) found that among developed countries, it is not the richest societies that have the best health, but those that have the smallest income inequality between rich and poor. Both inequality and relative poverty translate into increased death rates. Many of these studies of both violence and health find that it is the lack of social cohesion, or social capital, resulting from income inequality that contributes to these undesirable outcomes. It is likely that social capital contributes to QOL in many other ways not captured by these studies and offers yet another reason that fairness contributes to QOL.

5.3. Do we still need incentives to produce?

As a final thought on the relationship between fairness and QOL, Rawls (1971) initially justified some inequality because it provided incentives for greater production and hence increased the QOL of the worst-off. However, ever-greater production on a finite planet is impossible. Beyond some point, the costs economic growth imposes in terms of diminished ecosystem services outweigh the benefits of greater consumption. If we have not yet reached the point where this occurs, we are probably nearing it. Thus, it is increasingly likely that we would all be better off if there were fewer incentives to produce, not more. To the extent that this is the case, justice theory should call for greater equality.

6. How do we achieve sustainable, fair, and high QOL?

The discussion so far has addressed the definition of QOL and of fairness, suggested indicators to serve as proxies for the two, and examined their relationship to each other. This discussion is only of use, however, to the extent that it can suggest policies that will lead to a fair distribution of wealth and resources, a prerequisite for ensuring the best possible QOL for this and future generations. What would such policies look like?

---

[10] See http://www.worldbank.org/poverty/inequal/abstracts/health/read.htm for other examples.

It is quite likely that current consumption levels are unsustainable and threaten the QOL of future generations, and continued economic growth is sure to make them so. We believe that to achieve sustainability at the local, national, and global levels, we must respect the 6 Lisbon Principles as outlined in chapter 11: responsibility, scale-matching, precaution, adaptive management, full cost allocation and participation (see chapter 11 or Costanza et al., 1998). Fairness requires (at a minimum) healthy ecosystems, an end to poverties, and limits on wealth and consumption. It is further enhanced by the provision of public goods and diminished by over-consumption of market goods. QOL is enhanced by increasing our ability to satisfy our human needs or by reducing our wants. Perhaps the most important conclusion of analysis up to this point is that QOL, fairness, and sustainability are intimately linked and predominately complementary. The question is, what general policies will help us achieve a sustainable future with a high QOL for all?

The issue that most directly links sustainability, fairness, and QOL is the accumulation and consumption of wealth and resources. At the risk of ad nauseam repetition, consumption of physical resources deprives others of access to those resources, degrades the environment, threatens our planetary life-support functions, and diminishes other environmental services that benefit all. While there is no fixed link between consumption above a certain level and QOL, there is the widespread and growing perception that we would all be happier if we could just consume a bit more, and governments measure their success in terms of how well they achieve this goal. This is an ideological position that is not well supported by existing evidence. If excessive consumption is not necessary for QOL (and in fact may reduce it) and is unfair and threatens sustainability, why is increasing our production and consumption not just a national but a global obsession? More important, how can this be changed?

We will present two important answers to the first question. Detailed answers to the second question do not yet exist and those that do exist are subject to intense debate and would require innumerable volumes to elucidate. We will, however, suggest some general policies for achieving this goal.

6.1. Current world setting

6.1.1. The changing world

As the first part of our answer to the first question, we must remember that existing social, economic, and political institutions, as well as academic disciplines, evolved at a time when natural resources and ecological services were vast relative to the human presence, tightly bonded communities were essential to survival, and human impacts were relatively small and local. Scarcity of human-made and market goods were the binding constraints on improving QOL. Economics has been called the science of scarcity, dedicated to the allocation of scarce resources among alternative ends, and the market system historically was remarkably good at producing consumer goods and improving the QOL (at least as measured by longevity and health) from generation to generation. The effectiveness of the market system in meeting our needs in a world of plenty influenced our value system, promoting those values of individualism, competition, and materialism, which helped the market economy to function. Now, however, natural resources and ecosystem services have become the scarce goods, but we are slow to adapt to this change. We must develop a system in which an economic equilibrium will be compatible with an ecological equilibrium, an issue neglected by traditional economics. That is, we must fit the scale of our economic system within the scale of the ecosystem that sustains it. Also, resource exhaustion and environmental degradation now threaten to make future generations worse off than the present, so the issue of distribution both within and between generations must become a central focus (Daly, 1991; Costanza et al., 1991).

The problem is that values that helped us achieve desirable ends under one set of circumstances seem to lead us towards undesirable ends under another, and cultural values can be slow to change.

Fortunately, human economic systems are dynamic; they evolve and adapt in response to changes in the human environment. For example, the development of agriculture required the innovation of property rights to land, with radical implications for existing economic systems. Now, a growing body of scientific literature suggests that human activities threaten resources such as the ozone layer and climate stability, whose efficient allocation is not amenable to the type of property rights and associated values underlying our current economic system. Hence, we require a fundamentally different way of looking at economic development taking place within the earth's life-support system. Sustainability demands that we extend our social goals to address the issues of scale and distribution in addition to efficient allocation. We have sacrificed other human needs on the altar of production and we must now attend to these if we hope to increase our QOL. However, social evolution is slow, and the changes we are discussing have arrived very quickly. People are slow to accept new ideas, and institutions and individuals in power are reluctant to alter the society that confers that power. Thus, many continue to act as if increasing consumption is the best path towards a high QOL.

6.1.2. "The Good Life at a Great Price, Guaranteed[11]"

The second answer to why we have a global obsession with economic growth and consumption is that the market system as it currently exists provides a serious

---

[11] The Sears advertising slogan, which the Sears CEO says is "built around our core value proposition" (Martinez, 1999).

obstacle to the diffusion of ideas concerning the growing need for environmental services and non-marketed satisfiers of human needs. Most people get information and ideas through profit-driven media that depend on advertising for survival. In contrast to 70 years ago, when most words a person heard were spoken to them or to someone nearby, today most words we hear are direct sales pitches and the programs sponsored by them (Durning, 1992). Insidiously, advertising is only profitable if it convinces us to buy. Therefore, virtually all advertising is designed to stimulate our demand for market goods, and businesses are betting an estimated $652 billion per year that the strategy is effective [12] (International Advertising Association, 2000). Virtually no money is spent convincing us to prefer public goods or other non-marketed satisfiers of human needs, and such advertising would not automatically generate the revenue to be self-supporting. Since we have limited time and income to spend on satisfying our needs, if we spend more on one thing, we must spend less on another. Economists argue that the consumer is sovereign and is best able to determine what activities most increase his/her QOL, so the impact of advertising on relative preferences need not be a problem. Advertising will make people spend more on market goods than non-market goods, but only because it has altered their psyche to make those goods have a higher impact on their QOL. Unfortunately, stimulating demand for consumer goods means greater depletion of natural resources and expulsion of waste into the environment. Essentially, advertising convinces us to damage or destroy public goods for individual gain. Sovereignty over preferences for market goods for some consumers denies other consumers sovereignty over their preferences for public goods.

Further, the existence of social traps means there is serious reason to doubt that people make the best decisions regarding their QOL. Costanza (1987) defines "[a] social trap [as] any situation in which the short-run, local reinforcements guiding individual behavior are inconsistent with the long-run, global best interest of the individual and society." At least five types of social traps have been identified. First is time delay, where the reward is immediate and the negative impacts delayed. Second is ignorance, where we simply are not aware that long-run pay-offs are negative. Third is the sliding reinforcer, where the rewards change (diminish) over time. Fourth is the problem of externality discussed previously. Fifth is the collective trap, where an action is good for the individual, but when everyone engages in it, it is bad for society. Social traps may also be hybrid, combining two or more of these other traps. Thus, for a number of reasons we may make decisions that are not the best for our long-term QOL. From the examples offered above and numerous others, it would appear that

---

[12] To place this figure in context, only 7 countries in the world had GNP's higher than $600 billion in 1997.

nature's services might be particularly prone to social traps. Hence, if advertising changes our preferences from public goods to private goods, it may be leading us into a hybrid social trap by persuading us to pursue activities that actually reduce our QOL. Thus, to the extent that consumption induced by advertising threatens life-supporting natural capital and sustainability and reduces the supply of public goods, advertising is unfair.

More needs to be said about how advertising affects the QOL. As stated earlier, our QOL improves if we are better able to meet our needs and wants, and diminishes if we are less able to meet our needs and wants. Advertising creates wants by making us believe we need some product or another, yet gives us no greater ability to satisfy that want. In this sense, advertising directly diminishes our QOL. In the words of the advertisers themselves, B. Earl Puckett, former head of Allied Stores Corporation, "it is our job to make women unhappy with what they have" (Quoted in Durning, 1992, pp. 119–120). Anthony Reilly, CEO of food conglomerate H.J. Heinz, claims that "[o]nce television is there, people of whatever shade, culture, or origin want roughly the same things" (Quoted in Durning, 1992, p. 126). Unfortunately, while even third world slum dwellers increasingly have access to TV, they do not have access to the resources necessary to satisfy the wants that TV creates. Advertisers are keenly aware of the wide variety of human needs and try to make us believe that consumption will meet those needs. In the words of Alan Durning (1992), "they cultivate needs by hitching their wares to the infinite existential yearnings of the human soul." Experts in consumer behavior claim that consumers identify with brands as a means to differentiate themselves from one another (Durning, 1992); that is, advertising makes us believe that a particular brand will satisfy our need for identity. Other human needs especially targeted by advertising include affection, participation, and freedom, though none are left out. In fact, advertisers often attempt to make us believe that consumption of a particular good is a 'synergistic satisfier', meeting several needs at once, when in reality it is at best a pseudo-satisfier or an inhibiting satisfier, and through excessive consumption it becomes a violator and destructor.

Max-Neef's (1992) work can shed even more light on the relationship between advertising and QOL. He points out that needs have a two-fold character, encompassing both deprivation and potential. When we lack something, we feel deprived, but we also are engaged, mobilized, and motivated to fulfill that need. Hence, the need for participation or the need for affection is potential for participation and affection. In this sense, needs are a resource. However, if we are led to believe that consumption will fill our need for affection or participation, we do not seek to fulfill it elsewhere and the potential inherent in the need is lost. In addition, while needs may be finite, and hence demand for satisfiers finite, if we attempt to fulfill our needs with a pseudo-satisfier, we are unable to do so. Demand for pseudo-satisfiers cannot be satiated. Thus, people in consumer cultures, stimulated by advertising, continue to believe that if we only consumed a bit more or had twice our current income, we would attain the QOL we seek. In reality, this will not happen because consumption does not actually fulfill our needs.

6.2. Policy suggestions

6.2.1. Curbing the impact of advertising

We do not deny that advertising plays a useful role in providing us with information about the products that we consume. However, in most cases, the information content of advertising is quite low and often misleading. Most of the effort is designed instead to convince us that consumption is the best means to satisfy our human needs, yet it appears that current levels of consumption in the overdeveloped countries are incompatible with a sustainable future and are unfair. Reducing consumption levels will be exceedingly difficult in the presence of so much advertising. Thus, advertising has many elements of a 'public bad', and consequently should be curbed. People have argued that efforts to curb advertising interfere with the right to freedom of expression and furthermore are naïve. One rebuttal is that consumption induced by advertising interferes with the even more fundamental right to survival of future generations and the belief that we can substantially reduce consumption without limiting market-based advertising is exceedingly naïve. The problem is, what are the most feasible and effective means for controlling advertising for consumer goods? This is a very contentious issue but we present several possibilities here.

6.2.1.1. Charging for airwaves and removing tax exempt status for advertising. Currently, advertising over the airwaves in many countries is essentially subsidized. The airwaves are public property, but are typically given free of charge to communications corporations. Since airwaves have properties of public goods in that they are non-excludable and non-rival, there is a solid rationale for making them free. However, if the government charged corporations for the use of airwaves for advertising, it would target only that portion of the airwaves devoted to private profit.

Also, advertising is currently considered a business cost and is tax exempt. For the reasons listed above, however, it would be more appropriate to tax advertising. We do confront a problem with a tax on advertising, in that advertising can provide information, which is also a public good. Ideally, a tax should be targeted only at that portion of advertising that does not convey information. Unfortunately, it is extremely difficult to decide exactly what aspects of advertising do convey information (e.g., Coke tastes great!!). Such a tax would require a non-biased, non-government (due to the influence of money on politicians) institute, such as the non-profit Consumer Guide, to make these decisions. Such an institute could be funded from sales of airwaves devoted to advertising.

6.2.1.2. Full disclosure advertising and altering preferences. While taxes would presumably reduce the quantity of ads, it would not help to generate concern for non-market satisfiers of human needs. There are several alternatives for helping achieve this goal. Perhaps most effective would be a law mandating 'full disclosure' advertising. Just as medicines are labeled with all their potential adverse side effects, so should advertisements list all the potential adverse side effects of the products they advertise. This would, of course, include all negative impacts on the environment and the implications of those negative impacts. While this would not directly attempt to stimulate demand for non-market goods, it would at least make people more aware of their existence and more aware of the impacts of their consumption on those goods. This would have to be accompanied by efforts to educate consumers on how to use this information, perhaps funded by the suggested tax on advertising. Another alternative would be to provide free airtime for public service announcements that specifically seek to create demand for environmental services and other non-consumptive satisfiers of human needs. The media is a phenomenally powerful tool for altering preferences for satisfiers. If we are to create a more sustainable and fair world, we must alter people's preferences toward satisfiers that do not limit the ability of others, now and in the future, to attain a high QOL.

A problem with both of these restrictions on advertising, however, is that people will complain that they infringe on the basic right of free speech. However, the right to free speech does have restrictions. For example, no one is allowed to shout 'fire!' in a crowded theater if there is no fire, because it threatens the well-being of others. Shouting 'fire!' may not be fundamentally different from encouraging people to consume when such consumption threatens the well-being of future generations. Many nations already curb advertising on alcohol and tobacco, and the Australian Consumers Association has attacked the right to advertise unhealthy foods on children's TV shows (Durning, 1992).

6.3. Natural capitalism, increased efficiency, industrial ecology, and dematerialization

Given the political and economic power of large corporations and the advertising industry, the global dominance of the market paradigm, and the near universal belief that capitalism depends on growth for survival, is anything resembling a curb on markets at all feasible? One popular alternative that strives for reduced consumption of natural capital while allowing continued increases in consumption by consumers is the "natural capitalism" approach to business, which involves reducing resource consumption through business redesign. Natural capitalism aims to achieve major increases in 'productivity of natural resources', focusing on biologically based production (e.g., closed-loop, waste-free production), solutions-based models of business, and reinvestment in natural capital (Hawken et al., 1999). Because increased energy efficiency, reduced waste, and increased product quality (e.g., fuel-cell technology for vehicles) present revenue opportunities, many argue that this can be successful business strategy.

Some questions arise, however. If natural capitalism can compete successfully with more resource- and waste-intensive industries, why is it not more widespread? Do the environmentalists extolling this approach understand more about earning profits than the corporations? In reality, it appears that under current conditions, in most cases natural capitalism is probably not more profitable than intensive resource use. However, it may be simpler to make such an approach competitive than it would be to curb advertising, and there are success stories. For example, The Natural Step has used intensive education to influence some businesses to move toward sustainable and natural capitalism, and Paul Hawken's Ecology of Commerce (1994) has introduced these concepts to business students. Educating citizens on the benefits of sustainability so that their market preferences drive businesses to provide sustainable options could further strengthen the natural-capitalism approach. Of course, obtaining the resources to carry out this educational task would be difficult, especially if it must overcome the $650 billion spent annually educating people in the opposite direction. Also, to argue that people will voluntarily pay more to purchase goods that do less harm to public goods is to argue that people are inherently altruistic. While this may certainly be true, it is curious to argue that we can only make the market system compatible with sustainability by assuming that the underlying assumption of market economics – the primacy of 'rational' self-interest – is false. Perhaps the most effective approach to encouraging natural capitalism would be green taxes, discussed below. By increasing the costs of resource- and pollution-intensive industry, such taxes would make natural capitalism more competitive.

Even if we could bring about natural capitalism, would it be sufficient? Certainly there is enormous inefficiency in economic production that could be removed. Eventually, however, any industrial process must reach a limit beyond which it cannot become significantly less resource intensive. We cannot keep reducing the raw material inputs into consumer goods indefinitely: total dematerialization of production is physically impossible. No matter how efficient our production techniques, if consumption continues to grow we will continue to degrade natural capital and eventually threaten life-support functions. We will then be confronted with the current problems but at higher levels of consumption. Given our level of ignorance about ecosystem function and existing threats to ecosystem life-support functions, as well as the inevitable difficulties we will face in reducing consumption by consumers or producers, the precautionary principle suggests we should act on both fronts at once. We must strive to reduce final consumption while making production processes as efficient as possible.

6.3.1. Green taxes and human needs accounting

Green taxes were mentioned above as a way to stimulate natural capitalism. In general, green taxes could serve as path towards high QOL and sustainability. We use green taxes here as shorthand for a suite of financial mechanisms that incorporate the full cost of market production and consumption into market prices, as required by the Lisbon Principles. The basic idea is that if we have to pay for the ecological and social damage caused by our consumption, we will consume less and/or shift our consumption towards goods that have fewer negative impacts. Price increases will also encourage us to develop substitutes for those consumables that damage the environment. Even economists agree that market allocation is only efficient if prices reflect all costs.

Many governments under-price natural resources or even subsidize their extraction with the intention of promoting economic growth. Such subsidies are a direct transfer of resources from the public sector to the private sector, and indirectly lead to reduced public goods from environmental services. A first step must be to eliminate these distortions. Some of these subsidies are mentioned in chapter 11 and are discussed in greater detail in Templet (1995a). Others include the small stumpage fees charged by so many governments for logging rights, the below-market-price grazing fees charged by the US government, and the sale of timber rights to US national forests at times for even less than the cost of preparing the bids. There are numerous types of green financial mechanisms, including emissions taxes, tradable permits, and quotas, which have been outlined in great detail elsewhere, and would help reduce and shift consumption. Space does not permit discussion here, but for greater details, we refer you to Roodman (1998), Pearce and Turner (1990), Bernow et al. (1998). One point worth emphasizing is that while economists argue that quotas and taxes are quite similar, quotas ideally are determined by ecological factors, and are not subsequently affected by economic shocks [13]. Thus, they are more compatible with the precautionary principle and sustainable scale (Daly, 1996).

We would like to provide some details about two proposals that have received perhaps less attention than they deserve. The first is a highly progressive consumption tax, proposed by Frank (1999), that is particularly appropriate for

---

[13] Both taxes and tradable quotas/permits will provide an incentive for the individual to reduce pollution. With taxes, every reduction is a direct decrease in expenditures. With permits, reductions allow excess permits to be sold, increasing revenue. Fixed taxes apply a constant pressure to reduce pollution. If there are a fixed number of polluters generating an approximately constant amount of goods that pollute (i.e., the demand for pollution is constant), new innovations to reduce pollution will eventually decrease the demand for permits, driving the price down. Under this circumstance, permits may be less effective than taxes on reducing pollution. Alternatively, if the demand for pollution increases, the price of permits will increase, leading to an increase in price. Under these circumstances, taxes may be less effective than permits.

addressing the problem of positional wealth and over-consumption. The idea is to impose a highly progressive tax only on the portion of income that is spent on consumption. Such a tax would obviously deter consumption and would do so without threatening investment. Investment itself is a problem if it stimulates excessive growth. However, with limited ability to spend returns on market investments on personal consumption, the tax would provide greater incentives for investing in the public good (e.g., environmental restoration, community centers, and education)[14]. To the extent that consumption above and beyond a certain level is mostly positional, the big consumers would not suffer significant declines in their QOL. The negative impacts of excessive wealth accumulation would be avoided and there would be no need to impose unpopular caps on income.

The second proposal is an assurance bond on activities with potentially environmentally or socially damaging outcomes. Any individuals or corporations contemplating such activities would have to post a bond or purchase insurance sufficient to cover any potential damages from their activity. After the risk of environmental damage is past, the bond would be refunded and the insurance could be cancelled. These bonds would ensure that whoever causes environmental damage would be forced to pay for it, and market forces could set fair prices on the cost of insurance for any given project without the need for additional government regulation. Essentially, this is a market mechanism for implementing the precautionary principle (Costanza and Perrings, 1990).

To know if we are achieving our goals, we must be able to measure them. In the short run this implies the implementation of green accounts, and in the longer run, of accounts that measure our ability to sustainably satisfy human needs. These topics have already been sufficiently addressed in this chapter and the preceding one.

6.3.2. Poverty alleviation and income caps

We laid out earlier the need for ending poverties (i.e., insufficient satisifers for any of our needs) in a fair society and suggested some possible approaches (debt forgiveness, payment of the ecological debt, ensuring equal opportunity to all). The orthodox solution to ending poverty, increasing the size of the economic pie so that everyone can have a larger piece of it, has not proven itself effective over decades and even centuries of rapid growth, and it cannot be sustained indefinitely on a finite planet. It many ways it has already become counterproductive. A more fair distribution of existing wealth is the alternative to growth, but it is impossible in the space allowed to examine the myriad policies available for achieving this.

---

[14] Of course, there would be considerable danger that the wealthy would spend their money on politics, with negative consequences. Such a tax would have to be accompanied by limits to political donations.

However, the common denominator in any of these alternative policies is that they require political will. Political will is an expression of cultural values, even if only the cultural values of the ruling class in most countries. Hence, we argue that the prerequisite for any policy of poverty alleviation and income caps is a change in cultural values that will provide this political will. We will make our case with respect to two types of poverty: absolute poverty, where individuals fail to adequately meet their basic survival needs; and other poverties, where individuals fail to adequately satisfy the remaining human needs.

It would certainly seem that within the poorest countries economic growth (and population control) is required to end absolute poverty. However, this is not necessarily the case. For example, Amartya Sen (1984) has documented that even during many of the world's most severe famines, the countries where those famines occurred produced sufficient food for the starving population. The problem was one of entitlements, not abundance. When the poorest countries do produce more, in the current global system most of the wealth created goes abroad or to the upper classes, so economic growth seems to offer little hope. Certainly on a global scale there are sufficient resources to end global poverty, so the problem is one of distribution (although if populations continue to increase unchecked, inevitably absolute resource scarcity will also play a role). The wealthy and powerful have the capacity to create a system that will distribute resources more fairly, but their perception is that they would suffer a decrease in QOL if they ceased to capture the lion's share of global wealth and resources. This perception stems from an ideology (value system) that says material consumption meets all our insatiable needs, and the more we consume, the better they are met.

This value system similarly limits our ability to eliminate other poverties. Our obsession with economic growth and consumption, and their nature as pseudo-satisfiers, deprives us of the resources and the potential needed to pursue real satisfiers for our various needs. Thus, in direct contrast to the prevalent view, eliminating poverties requires ending this obsession with growth and consumption, which in turn demands a change in the dominant value system.

Values are also the crux of the matter in efforts to limit wealth. People believe enormous wealth brings enormous happiness, and they want the chance to be enormously happy. These values mean that capping maximum wealth may prove even more challenging politically than ending poverty. Again, a change in values is a necessary step [15]. The question is then, how do we change cultural values in a way that is conducive to a sustainable, fair, and high-QOL society?

---

[15] In the meantime, however, a highly progressive consumption tax could obviate the need for income caps, and might be more politically feasible.

6.3.3. Education

Education is critically important in increasing QOL on its own. It directly increases our human need for understanding, and dramatically increases our access to numerous other satisfiers of human needs. More important, it may be an essential means for changing people's values. As suggested earlier, value systems evolve in response to changing institutions, changing environments, and changing cultures, but the speed with which human activity is changing our environment suggests we cannot simply sit back and passively wait. Fomenting rapid change in values will require extensive education. Part of the problem is that people are unaware of the impacts of human activity on the environment. Without broader understanding of ecological processes, people will not recognize the constraints these processes pose on our development. If people are educated to the negative impacts of our current development path (or as they become too obvious to ignore), they will become ripe to accept alternatives, but only if informed of the options. However, the dominant 'solution' currently offered (by highly educated people) to the damages caused by economic growth is more of the same [16]. Education within very narrow limits is little more than indoctrination within an ideology. At universities, education is typically delivered within the boundaries of narrow disciplines. It is easy to accept neoclassical economics if one has no understanding of ecology, and it is difficult to transform insights from ecology into practical policies if one has no understanding of the social sciences. The problems inherent in developing a sustainable society and ensuring that the human system is in equilibrium with the ecological system that sustains it demand a broadly interdisciplinary education.

However, we must recognize that most people who are aware that our levels of consumption threaten the QOL of others alive today and of future generations nonetheless fail to change their consumption levels in response. The likely reason for this is the fear that reducing consumption will lower their QOL. This message is conveyed in formal education, but only to a limited extent outside of business and economics. The more powerful educating force for this message is the media. Unfortunately, as we made clear earlier, most media are market driven. It therefore reinforces the dominant value system of consumerism and monopolizes the time and resources that could be used to educate people to alternatives. Modern media offer the most powerful means of mass education in the history of humankind, and as long as market forces control them, it will be exceedingly difficult to educate people to alternatives. Achieving our goals will require at least equal access to the

---

[16] In the developed countries, the argument goes, air and water quality are improving, empirical proof that economic growth solves environmental problems. Those who propose this solution appear oblivious to the physical laws of thermodynamics, overlook the innumerable environmental problems that are not getting better, and ignore the fact that the overdeveloped countries have simply exported their most polluting industries to the third world.

media to spread alternative ideologies. We are the first to admit that our view of the good is an ideology but we believe it far healthier for society to have several ideologies to choose from rather than one. The dominant consumerist ideology may have been appropriate in the past, and the ideology we are promoting here may no longer be appropriate in the future. Thus, broadly interdisciplinary and broadly inter-ideological education is a requirement for the principle of adaptive management necessary to achieve sustainability in a changing world.

6.3.4. Political reform

Politics implies action and the political arena is where many of the needed changes must come about. In the short run, we should also take full advantage of existing political structures to promote our agenda. With this in mind, we have drawn up a 'Sustainability Bill of Rights' reproduced in the Appendix, and challenge activists to work with their representatives to bring some version of such a bill into the political debate.

Action requires political will, be it for poverty alleviation, curbs on advertising, or education reforms. Promoting the Sustainability Bill of Rights will help, but unfortunately, under current conditions, political will is largely determined by the largest donors or simply the wealthiest individuals, depending on the country in question. In the short to medium run, to wrest control of political will from the wealthy will require campaign-finance reform in allegedly democratic nations, and other alternatives that limit the influence of the wealthy over the political agenda in other countries. The necessary political will is unlikely to spring from institutionalized parties, professional politicians, or established governments. Civil society must play a primary role not only in influencing governments, but also in providing the leadership for the development of the values and vision that must guide us.

In the longer run, a strong civil society can help create a strong participatory democracy, which is probably the form of government most conducive to creating a fair, sustainable, and high-QOL society (Prugh et al., 2000). In a participatory democracy, the people must discuss at length the issues that affect them to decide together how they should be resolved. This could directly meet people's need for participation and identity, educate people to the relevant issues and alternative ideologies, and help direct society's resources towards meeting human needs. As citizens come together in regular meetings to discuss the issues and work together to resolve them (even when substantial conflict exists), it should create strong bonds of social capital, and could play an essential role in forging a sense of community. This system will allow the people to define political will or government's purpose. These civic meetings must forge a shared vision of the future to guide their actions. This vision cannot be static but must adapt to new information and new conditions as they emerge. The importance of vision is difficult to overemphasize, and requires elaboration.

6.3.5. Vision

A fundamental missing element from the discussion of QOL and the distribution of wealth and resources at the level of society is a *coherent, relatively detailed, shared vision of what a sustainable high-quality-of-life society would look like* (Costanza, 2000), and how we could move from here to there. The default vision of continued, unlimited increases in material consumption is probably unsustainable, but no credible alternative is available for public discussion. A prerequisite to achieving a sustainable society is thus the creation of a shared vision of what we as a society want to sustain and the central shared values that express our hopes for the future. This vision must incorporate a broad diversity of perspectives and be based on principles of fairness and respect for individual human rights. To develop this shared vision of a sustainable society in a way that is credible requires the active participation of all the major stakeholder groups in society. Otherwise, the vision will be regarded as just another special interest agenda.

This vision of a desirable society must lie within the constraints imposed upon us by our finite ecosystem, but also recognize that constraints posed by our present culture and its emphasis on consumer goods as satisfiers are less rigid. Building a sustainable society almost certainly requires that we accept that consumption is not an ultimate goal, but merely a means to an end. We must recognize that consumption cannot grow without limits, but that QOL does not depend on consumption, and is not bound by such physical laws. We must redefine efficiency, not as the maximum market value we can create from a given allocation of resources, but rather as the most human needs we can satisfy with the least amount of resources. Rather than simply lament the negative outcomes of our current development path, we must affirm a positive vision of a sustainable, desirable future.

7. Conclusion

In conclusion, we have a long way to go before reaching a fair, sustainable, and high-QOL society. Developing a positive shared vision and alternative values to consumerism will be but the starting point, and we have discussed only a very few of the additional steps that we will need to take to develop this society. Some of the ideas presented may work and some may not. In presenting some of these ideas, many will accuse us of idealism and naiveté. However, we must bear in mind that prior to its implementation, there were few ideas more naïve than democracy proposed to a world of monarchies, or emancipation proposed to a world of slavery. Goddard was accused of naiveté for thinking that rockets could travel in the vacuum of space, Bell was told that telephones would never be in demand, and in 1943, the president of IBM estimated the world demand for computers at five. Such criticisms are often little more than a crisis of imagination.

True naiveté lies in believing that we can achieve the desired society without bold and radical proposals for change.

Appendix. The Sustainability Bill of Rights

- People have the right to live in natural environments, which will sustain their health and the health of future generations.
- The goal of sustainability is to improve or maintain Quality of Life over time.
- A sustainable society is one which will ensure fairness within a generation and across generations such that the natural capital one generation inherits is transferred intact or enhanced.
- Sustainability includes protection of biodiversity and respect for spiritual contact with nature.
- Social, geographical, and intergenerational fairness contribute to sustainability.
- Quality of Life depends directly and indirectly on four forms of capital:
  - Natural
  - Human
  - Social
  - Built
- Natural capital sustainability requires maintenance of natural services.
- Individuals must have an opportunity to challenge unsustainable activities through the courts and through dispute resolution via mediation in accordance with the precautionary principle.
- This bill will be reviewed through a stakeholder process at regular time intervals to allow adaptation to changes in knowledge, technology and environmental conditions.
- The government will publish on a regular basis a list of sustainability indicators to compare progress.

References

Bentham, J., 1830, The rationale of punishment. In: W. Stark (Editor), Jeremy Bentham's Economic Writings (George Allen and Unwin for Royal Economic Society, London).

Bernow, S., Costanza, R., Daly, H.E., DeGennaro, R., Erlandson, D., Ferris, D., Hawken, P., Horner, J.A., Lancelot, J., Marx, T., Norland, D., Peters, I., Roodman, D.M., Schneider, C., Shyamsundar, P. and Woodwell, J., 1998, Ecological tax reform. Bioscience 48:193–196.

Bloom, D.E., Canning, D., Graham, B. and Sevilla, J., 2000, Out of Poverty: On the Feasibility of Halving Global Poverty by 2015 (Discussion Paper No. 52). Consulting Assistance on Economic Reform (CAER II). Available on-line: http://www.cid.harvard.edu/caer2/htm/content/papers/paper52/paper52.htm.

Boyce, J.K., Klemer, A.R. and Templet, P.H., 1999, Power distribution, the environment, and public health: a state level analysis. Ecol. Econ. 29:127–140.

Broome, J., 1991, Weighing Goods: Equality, Uncertainty and Time (Basil Blackwell, Cambridge, MA) 255 pp.

Brown, P., 1998, The rich have inherited the sea. Weekly Mail and Guardian (Johannesburg), October 23. Available on-line: http://www.sn.apc.org/wmail/issues/981023/NEWS19.html.

Bueno, M. and Marcondes, H., 1991, Global deforestation and CO_2 emissions: past and present, a comprehensive review. Energy Environ. 3:235–282.

Chomsky, N., 1998, Reclaiming the remaining debts must be justified. The Guardian, May 12. Available on-line: http://www.nationalinvestor.com/noam_chomskyhtm.htm.

Cobb, C.W., 2000, Measurement Tools and the Quality of Life (Redefining Progress, Oakland, CA). Available on-line: http://www.rprogress.org/pubs/pdf/measure_qol.pdf.

Constitutional Rights Project, 1999, Land, Oil and Human Rights in Nigeria's Delta Region (CRP, Lagos, Nigeria).

Costanza, R., 1987, Social traps and environmental policy. Bioscience 37:407–412.

Costanza, R., 1992, Toward an operational definition of ecosystem health. In: R. Costanza, B.G. Norton and B.D. Haskell (Editors), Ecosystem Health: New Goals for Environmental Management (Island Press, Washington, DC) pp. 239–256.

Costanza, R., 2000, Visions of alternative (unpredictable) futures and their use in policy analysis. Conserv. Ecol. [online], 4(1):5.

Costanza, R. and Perrings, C., 1990, A flexible assurance bonding system for improved environmental management. Ecol. Econ. 2:57–76.

Costanza, R., Daly, H.E. and Bartholomew, J., 1991, Goals, agenda and policy recommendations for ecological economics. In: R. Costanza (Editor), Ecological Economics: the Science and Management of Sustainability (Columbia University Press, New York) pp. 1–21.

Costanza, R., Andrade, F., Antunes, P., van den Belt, M.J., Boersma, D., Boesch, D.F., Catarino, F., Hanna, S., Limburg, K., Low, B., Molitor, M., Pereira, J.G., Rayner, S., Santos, R., Wilson, J. and Young, M., 1998, Principles for sustainable governance of the oceans. Science 281:198–199.

Crocker, D., 1995, Functioning and capability: the foundations of Sen's and Nussbaum's development ethic, part 2. In: M. Nussbaum and J. Glober (Editors), Women, Culture, and Development: A Study in Human Capabilities (Oxford University Press, Oxford).

Daly, H.E., 1991, Steady-State Economics: Second Edition, with New Essays (Island Press, Washington, DC) 302 pp.

Daly, H.E., 1993, The steady state economy: toward a political economy of biophysical equilibrium and moral growth. In: H.E. Daly and K. Townsend (Editors), Economics, Ecology, Ethics (MIT Press, Cambridge, MA) pp. 324–356.

Daly, H.E., 1996, Beyond Growth: The Economics of Sustainable Development (Beacon Press, Boston, MA) 253 pp.

Daly, H.E. and Cobb, J.B., 1989, For the Common Good: Redirecting the Economy Toward Community, the Environment, and a Sustainable Future (Beacon Press, Boston, MA) 482 pp.

Durning, A.T., 1992, How Much is Enough? The Consumer Society and the Fate of the Earth, 1st edition (Norton & Company, New York) 200 pp.

Ehui, S., Hertel, T. and Preckel, P., 1990, Forest resource depletion, soil dynamics, and agricultural productivity in the tropics. J. Environ. Econ. Manag. 18:136–154.

Ekins, P., 1991, The sustainable consumer society: a contradiction in terms? Int. Environ. Affairs (Fall 1991), pp. 243–258.

Fajnzylber, P., Lederman, D. and Loayza, N., 1998, What Causes Violent Crime? (The World Bank, Office of the Chief Latin America and the Caribbean Region). Available on-line: http://wbln0018.worldbank.org/Networks/ESSD/icdb.nsf/d4856f112e805df4852566c9007c27a6/3bc3671fb195e1b0852567fd005338b2/$FILE/loayza.pdf.

Farley, J., 1999, 'Optimal' deforestation in the Brazilian Amazon; theory and policy: the local, national, international and intergenerational viewpoints, Ph.D. Dissertation (Cornell University, Ithaca, NY) 339 pp.

Fisher, I., 1906, The Nature of Capital and Income (MacMillan, New York) 427 pp.

Frank, R., 1999, Luxury Fever: Why Money Fails to Satisfy in an Era of Excess (Free Press, New York) 326 pp.
Gates, J., 1999, Statistics on Poverty and Inequality. Available on-line: http://www.globalpolicy.org/socecon/inequal/gates99.htm (Global Policy Forum).
Goodwin, N., 1997, Volume introduction. In: F. Ackerman, D. Kiron, N. Goodwin, J. Harris and K. Gallagher (Editors), Human Well-Being and Economic Goals (Island Press, Washington, DC) p. 427.
Haas, B.K., 1999, A multidisciplinary concept analysis of quality of life. West. J. Nurs. Res. 21(6):728–743.
Hawken, P., 1994, The Ecology of Commerce: A Declaration of Sustainability (HarperBusiness, San Francisco, CA) 250 pp.
Hawken, P., Lovins, A. and Lovins, H., 1999, Natural Capitalism: Creating the Next Industrial Revolution (Little Brown and Company, Boston, MA) 396 pp.
Hsieh, C. and Pugh, M.D., 1993, Poverty, income inequality, and violent crime: a meta-analysis of recent aggregate data studies. Crim. Justice Rev. 18(2):182–202.
International Advertising Association, 2000, Frequently Asked Questions on Advertising and Constitutional Protections (International Advertising Association). Available on line: http://www.iaaglobal.org/iaagenerator/default.asp?section_id=2&category_id=411.
Kaplan, R. and Kaplan, S., 1989, The Experience of Nature (Cambridge University Press, New York) 340 pp.
Kawachi, I., Kennedy, B.P., Lochner, K. and Prothrow-Stith, D., 1997, Social capital, income inequality and mortality. Am. J. Public Health 87(9):1491–1498.
Kellert, S.R. and Wilson, E.O. (Editors), 1993, The Biophilia Hypothesis (Island Press, Washington, DC) 484 pp.
Kennedy, B.P., Kawachi, I., Prothrow-Stith, D., Lochner, K. and Gibbs, B., 1998, Social capital, income inequality, and firearm violent crime. Soc. Sci. Med. 47(1):7–17.
Keynes, J.M., 1919, The Economic Consequences of the Peace (Macmillan, London).
Kiron, D., 1997, Summary of Amartya Sen's contributions to understanding personal welfare. In: F. Ackerman, D. Kiron, N. Goodwin, J. Harris and K. Gallagher (Editors), Human Well-Being and Economic Goals (Island Press, Washington, DC) p. 426.
Lake, C. and Borosage, R.L., 2000, Money talks and voters and donors know it. The Nation, August 21.
Lane, R.E., 1986, Market justice, political justice. Am. Polit. Sci. Rev. 80(2):383–402.
Lane, R.E., 2000, The Loss of Happiness in Market Economies (Yale University Press, New Haven, CT) 465 pp.
Lapham, L., 1988, Money and Class in America: Notes and Observations on our Civil Religion (Weidenfeld and Nicolson, New York) 244 pp.
Lynch, J.W., Kaplan, G.A., Pamuk, E.R., Cohen, R.D., Heck, K.E., Balfour, J.L. and Yen, I.H., 1998, Income inequality and mortality in metropolitan areas of the United States. Am. J. Public Health 88(7):1074–1080.
Martinez, A., 1999, Annual Report 1999: Letter to Our Shareholders. Available on-line: http://media.corporate-ir.net/media_files/NYS/S/reports/s_ar99_low.pdf.
Maslow, A., 1954, Motivation and Personality (Harper, New York) 411 pp.
Max-Neef, M., 1992, Development and human needs. In: P. Ekins and M. Max-Neef (Editors), Real-life Economics: Understanding Wealth Creation (Routledge, London) pp. 97–213.
McGinn, A.P., 2000, Why Poison Ourselves? A Precautionary Approach to Synthetic Chemicals, Worldwatch Paper 153 (World Watch, Washington, DC) 92 pp.
Nickerson, D., 1999, Trade-offs of mangrove area development in the Philippines. Ecol. Econ. 28(2):279–298.

Nussbaum, M., 1990, Aristotelian social democracy. In: R.B. Douglass, G.M. Mara and H.S. Richardson (Editors), Liberalism and the Good (Routledge, New York/London) pp. 203–252.
Panayatou, T. and Parasuk, C., 1990, Land and Forest: Projecting Demand and Managing Encroachment (TDRI, Bangkok) 85 pp.
Pearce, D.W. and Turner, K., 1990, The Economics of Natural Resources and the Environment (Johns Hopkins Press, Baltimore, MD) 378 pp.
Prugh, T., Costanza, R. and Daly, H.E., 2000, The Local Politics of Global Sustainability (Island Press, Washington, DC) 173 pp.
Rawls, J., 1971, A Theory of Justice (Harvard University Press, Cambridge, MA) 607 pp.
Robinson, J., 1964, Economic Philosophy (Doubleday, Garden City, NY) 150 pp.
Roodman, D.M., 1998, The Natural Wealth of Nations: Harnessing the Market for the Environment (Norton & Company, New York) 303 pp.
Roodman, D.M., 2001, Still Waiting for the Jubilee: Pragmatic Solutions for the Third World Debt Crisis, WorldWatch paper 155, April 2001 (Worldwatch, Washington, DC) 86 pp.
Schwartzman, S., 1989, Extractive reserves: the rubber tappers' strategy for sustainable use of the Amazon rainforest. In: J. Browder (Editor), Fragile Lands of Latin America (Westview Press, Boulder, CO) pp. 150–165.
Sen, A., 1984, Poverty and Famines: An Essay on Entitlement and Deprivation, 2nd edition (Oxford University Press, Oxford) 257 pp.
Solow, R., 1993, Sustainability: an economist's perspective. In: R. Dorfman and N.S. Dorfman (Editors), Economics of the Environment (Norton & Company, New York) pp. 179–187.
Sugden, R., 1993, Welfare, resources and capabilities: a review of 'inequality reexamined' by Amartya Sen. J. Econ. Lit. 31(December):1947–1962.
Summers, L., 1991, Internal World Bank Memo, written on December 12, 1991, and made public in February, 1992. Available on-line: http://whirledbank.org/ourwords/summers.html.
Templet, P.H., 1995a, Grazing the commons; externalities, subsidies and economic development. Ecol. Econ. 12:141–159.
Templet, P.H., 1995b, Equity and sustainability; an empirical analysis. Soc. Nat. Resour. 8:509–523.
Ulrich, R.S., Simons, R.F., Losito, B.D., Fiorito, E., Miles, M.A. and Zelson, M., 1991, Stress recovery during exposure to natural and urban environments. J. Environ. Psychol. 11:201–230.
Walzer, M., 1990, Spheres of Justice (Basic Books, New York) 368 pp.
Wilkinson, R., 1996, Unhealthy Societies: The Afflictions of Inequality (Routledge, London) 255 pp.
Wilson, E.O., 1986, Biophilia, reprint edition (Harvard University Press, Cambridge, MA) 157 pp.
World Bank, 2000, Global Economic Prospects and the Developing Countries, 2001 (World Bank, Washington, DC).
World Bank, 2001, Social Capital for Development: What is Social Capital. Available on-line: http://www.worldbank.org/poverty/scapital/whatsc.htm (World Bank, Washington, DC).

Conclusions

S.E. Jørgensen and R. Costanza

At first glance it seems difficult to draw conclusions based on the 12 preceding chapters that are the result of the EcoSummit workshop. The chapters are written in very different styles and their conclusions focus on a very wide spectrum of problems, which seem not to have much to do with each other. If we step back, however, it becomes clear that the six themes are very interrelated. Quality of life is of course dependent on the distribution of wealth and resources, but it is also dependent on human health, which again is dependent on ecosystem health. Assessment of ecosystem health requires a profound knowledge of ecosystems, which are complex adaptive, hierarchical systems. They cannot be overviewed unless we are able to develop integrated models, and an integrated model of an ecosystem cannot be developed unless we know the properties of ecosystems, i.e., CAHSystems. Quality of life is also dependent on a proper appreciation and use (not abuse) of ecosystem services. This is consistent with the definition of ecotechnology (Mitsch and Jørgensen, 1989): the design of human society with its natural environment for the benefit of both. Science is a prerequisite for our understanding of nature: How do ecosystems work? What do we understand by quality of life? Which factors influence human health? Decisions should always be taken on the basis of the best available scientific knowledge. Science, therefore, inevitably underlies all environmental decisions. So, all six themes are closely interrelated with all the other five themes directly and indirectly, forward and backward (fig. 1). We cannot look at any one of the themes separately, but need to integrate all six themes into a more comprehensive understanding of the environment, our impact on the environment and how to achieve a high quality of life in the framework of our society and our environment.

There has been an ongoing debate about which ecosystem theories are useful and are based on good science. It was proposed in 1992 (Jørgensen, 1992), that we have a pattern of *almost* consistent theories. This was reconfirmed at the EcoSummit: a pattern of something resembling an identifiable "CAHS Theory" is taking form. The various possible goal functions or orientors: exergy max-

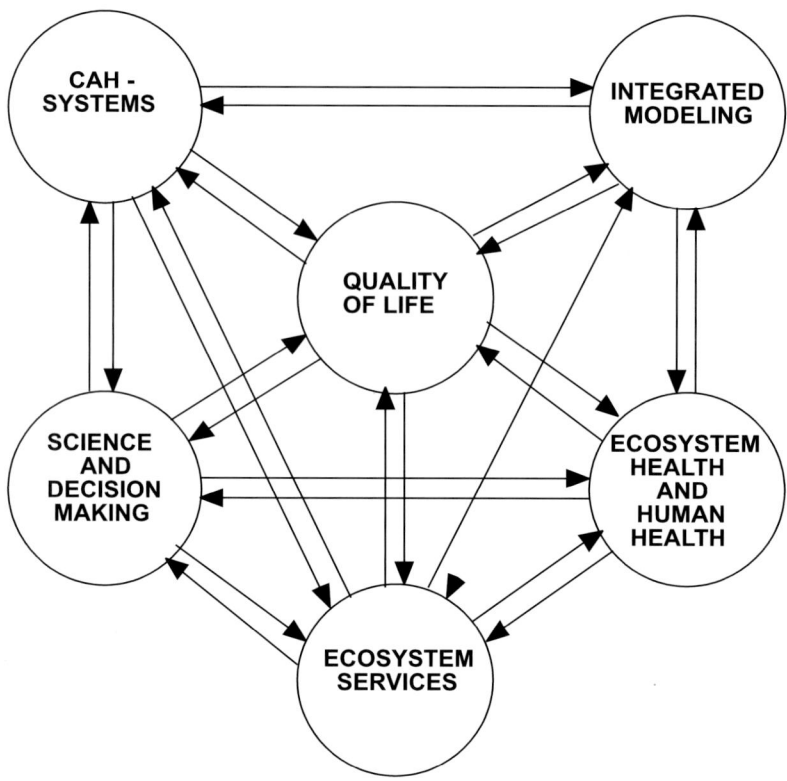

Fig. 1. All the six themes are interrelated both forward and backward.

imization, maximum power, maximum entropy production, minimum specific entropy production (to mention a few) are for the most part consistent – not necessarily in all phases of ecosystem development but in some contexts. Only exergy maximization and maximum power seem to be applicable for all forms of growth and development and in all phases, but the other orientors are important to understand the reactions of ecosystems in all detail and in all phases. A number of different complementary approaches are needed to explain all aspects of structure, organization, and dynamics of CAHSystems. This is not surprising. For example, light, which is a much simpler phenomenon than an ecosystem, requires two different explanations to cover all the observations: waves and particles.

Can we also explain our ecological observations by use of a few fundamental laws and derive rules from these fundamentals laws? Not yet, but with a pattern and spectrum of theories in hand, we are able to construct a coherent theoretical network that can be used in this context. This pattern and spectrum of theories can facilitate and inform our environmental decisions – decisions about the use of ecological services, the influence of ecosystem health on our health and quality of life. The integrated models would also be improved because they would reflect the system properties of ecosystems and social systems, which again would imply better decisions.

The application of integrated modeling has accelerated in the last decade. Particularly, integration of hydrology and ecology and ecology and economics have been reflected in modeling, but there are still extremely few models concerned with integrated ecological–economic–social systems. More modeling effort with a simultaneous focus on ecological, economic, and social problems is urgently needed because most problems of mankind today involve all three types of problems at the same time. Any decision, for example, concerning a major construction work, a dam, a bridge, or an important building, will obviously have impacts on the environment, on the economy, and on the social structure. We probably have sufficient modeling experience to develop models integrating all three systems, but it is currently difficult to establish, fund, and maintain an interdisciplinary team with sufficient expertise in all types of problems. In addition, progress is slow because there is not sufficient experience with this type of modeling yet. The first fully integrated models will probably fail, as was also the case when the first generation of more comprehensive ecological models were developed three decades ago. Many mistakes were made at first, but learning and understanding flowed from those mistakes.

A decade later, around 1980, the field of ecological modeling was maturing and reliable models could be developed, provided the experience gained was used properly (which was not always the case, of course). The conclusion is therefore, that we should get started on development of models integrating ecological, economic, and social problems, and accept that the initial models will give at best only some coarse qualitative or semi-qualitative results.

The scientific inputs to the integrated decisions humans have to make during the coming years are essential, but the open question is whether natural and social scientists should themselves play a more active role in the decisionmaking. Their role up to now has been as consultant, meaning that they have not participated in the decisionmaking process, because it was considered purely political, and most scientists want to remain "unbiased". They prefer to stay in their ivory tower. This attitude is not tenable anymore, however, because the problems are getting more and more complex. Consequently, there are thousands of wrong political decisions being made because the politicians cannot look through the fog of complexity coming from the scientific community, and rely instead purely on public opinion.

There is not a ready model for how scientists could or should influence political decisions with their expert knowledge. Clearly, there is a need for different decision procedures in the future that can adequately consider the complexity of the problems and the available scientific knowledge about the problems while embracing the democratic process. Since we do not have a clear idea how to do this, we must start to make experiments and not get stuck in the present rigid system. The structure of today's society offers many new possibilities. For example, the Internet offers the potential for contact with a large segment of the population about a focal problem very rapidly.

Ecosystem health has been an important environmental concept for the last ten years (Costanza et al., 1992). When the concept was introduced, the idea was to get a list of important ecological indicators that could be utilized to assess ecosystem health. We do not have such a list today that can be used in all situations, but we do have sufficient experience to be able to use ecological indicators to come up with a reasonably good assessment of the health of an ecosystem. The same list of indicators is not used by everybody dealing with this assessment problem, but all medical doctors also do not use exactly the same indicators for assessment of human health. There is, however, a certain consensus on the underlying information of all the proposed indicators. It has also been agreed that we need – as for human health – several indicators at the same time to get a sufficiently comprehensive image of ecosystem health.

One of the main focuses in this field of applied ecology is the interrelationship between ecosystem health and human health. There is no doubt that they are intricately interdependent, but how strong and with what implications for human health in a specific situation? These tangible questions cannot yet be answered properly, at least not quantitatively. It is therefore the hope that in the future we can develop a more complete picture of the interactions between ecosystem health and human health. The discussion about this topic at the EcoSummit meeting may enhance our effort in this direction.

Ecological engineering is a transdisciplinary field that encompasses the use of ecosystems to the benefit of nature and humans, sound ecological planning, and use of ecological restoration methods for deteriorated ecosystems. The field was advocated by H.T. Odum and M. Straskraba in the 1970s, but the advantages of ecological engineering only became clear in the early 1980s, when the debate about non-point source pollution, originating mainly from agriculture, was initiated. The use of wetlands as a filter for non-point pollution has been a core issue in ecological engineering for the last 15–20 years. The field encompasses many other possibilities to utilize our environment more prudently, considering both nature and humans. The main focus has therefore lately been on: how can we use ecosystem services in a sustainable way? This would require that we learn to appreciate these services, which we have up till now largely taken for granted. Moreover, we must understand the underlying mechanisms that create these services. This has inevitably turned ecological engineering into a discussion of basic principles and practices. We now have many good ecological engineering projects in place, and a fairly good knowledge base about what constitutes sound ecological planning, how to realize an ecotechnological project, and how to restore a contaminated ecosystem. There are also several proposals for a set of basic principles for ecological engineering. The pattern and spectrum of ecosystem theories, needs, however, to be utilized better in the development of ecological engineering principles and in setting up guidelines for practical use of ecotechnology. Moreover, a far more advanced integration of ecological economics and ecological engineering seems necessary to ensure better planning and implementation of ecological engineering projects in the future.

"Quality of life" has strong interrelationships with the other five themes, and it is perhaps the topic requiring most integration. It is also the most "political" of the six themes, which is emphasized by the heading: "Quality of Life and the Distribution of Wealth and Resources", Finally, it is probably also the most difficult of the six themes for scientists to discuss.

Not surprisingly, a major obstacle to a good quality of life for all humans on earth is the unfair present distribution of wealth and resources. The biased view of the developed world makes it very difficult to find a solution of this problem. There are obviously many initiatives that we could take to adjust this unfair distribution of wealth and resources, mainly between the developed and the developing countries, but also between the poorest and the richest in each country. The solutions are rooted in ecological economics: use of green taxes, industrial ecology, a more complex and complete national accounting system that incorporates ecological, economic, and social sustainability. All these initiatives require, however, political decisions, again linking back to the science and decisionmaking theme.

All six themes contributed to a very successful and fruitful discussion at the EcoSummit. New ideas and thoughts came up and were discussed. A good overview of the state of the art and the trends of the six themes were presented for all participants at the meeting. Discussion of all six themes, and even to a greater extent the integration of the six themes, requires a very interdisciplinary approach, which was the very basis for the EcoSummit. Without the simultaneous presence of the readers of the five journals and the members of the five societies participating in the EcoSummit, the results achieved would not have been possible. The most important outcome for the individual participants at the conference (and hopefully for the readers of this book) may be the clear vision of the very wide perspectives of our research problems in all the applied disciplines of ecology (systems ecology, ecological and environmental modeling, assessment of ecosystem health, ecological engineering, and ecological economics) and the growing importance of creating the transdisciplinary "hard problem science" necessary to address them.

References

Costanza, R., Norton, B.G. and Haskell, B.D. (Editors), 1992, Ecosystem Health: New Goals for Environmental Management (Island Press, Washington, DC) 269 pp.

Jørgensen, S.E., 1992, Integration of Ecosystem Theories: A Pattern. (Kluwer Academic, Dordrecht). Second edition: 1997, 388 pp.
Mitsch, W.J. and Jørgensen, S.E., 1989, Ecological Engineering, Introduction to Ecotechnology (Wiley, New York) 432 pp.

Author Index*

Abel, N.   12, *16*
Aber, J.D.   10, *17*
Acquay, H.   197, *218*
Acton, D.F.   197, *217*
Ahmed, M.   180, *185*
Aikenhead, G.S.   159, *165*
Alechandre, A.S.   145, *151*
Alexander, S.   109, *124*
Allen, T.F.H.   50, 82, 85, *87*, *92*
Allen-Gil, S.   196, *218*
Alwin, D.F.   253, *255*
Amann, M.   7, *16*, 162, *166*
American Heritage Dictionary of the English Language   222, *255*
Anderson, R.F.   176, *187*
Andrade, F.   254, *256*, 285, *299*
Andrews, P.K.   10, *16*
Anjema, C.M.   170, 175, *187*
Antunes, P.   99, *99*, 254, *256*, 285, *299*
Armesto, J.J.   198, *218*
Arrow, K.   241, *255*
Ashby, W.R.   44, *87*
Atkinson, G.   228, *255*

Backx, J.   162, *166*
Badii, R.   54, *87*
Bailey, P.D.   6, *16*
Bak, P.   72, *88*
Balfour, J.L.   284, *300*
Banerjee, R.D.   118, 119, *125*
Bar-Yam, Y.   54, *88*
Barnwell, T.O.   30, *38*
Barron-Tieger, B.   146–148, *152*
Barrow, C.J.   197, *217*
Barry, B.   238, *255*
Barry, R.G.   175, *187*
Bartelmus, P.   231, *255*

* Page references to text are in roman type, to bibliographical entries in italics.

Bartholomew, J.   286, *299*
Bartonova, A.   162, *166*
Baskin, C.C.   143, *151*
Baskin, J.M.   143, *151*
Bastianoni, S.   65, 69, 70, *88*, *93*
Baumler, A.J.   199, *218*
Baumol, W.J.   250, *255*
Beck, M.B.   30, 31, *38*
Beder, S.   102, *124*
Begon, M.   82, *88*, 108, *125*
Behrens III, W.   242, *257*
Bend, J.R.   170, *187*
Bendoricchio, G.   59, 69, *88*
Benjamin, C.   32, *38*
Bennett, C.H.   83, *88*
Bentham, J.   265, *298*
Berlow, E.   198, *218*
Bernow, S.   292, *298*
Berscheid, E.   253, *258*
Bhowmik, M.L.   119, *124*
Bicknell, K.B.   121, *124*
Biltonen, M.   197, *218*
Bingle, W.H.   159, *165*
Birdsey, R.A.   194, *218*
Biswas, J.K.   119, *124*
Blaauw, E.M.   162, *166*
Blake, T.   28, *38*
Bland, W.L.   10, *16*
Bloom, D.E.   222, *255*, 260, *298*
Bloomfield, J.   198, *218*
Boers, P.C.M.   162, *166*
Boersma, D.   254, *256*, 285, *299*
Boesch, D.F.   254, *256*, 285, *299*
Bohm, G.M.   175, *187*
Bongarten, B.   191, *217*
Borgstroem Hansson, C.   65, *94*
Born, S.M.   22, 26, 28, *38*, *39*
Borosage, R.L.   280, *300*
Bossel, H.   57, *88*
Bosserman, R.W.   44, *88*
Boumans, R.   28, *39*
Boyan Jr, S.A.   250, *257*

Boyce, J.K.   253, *255*, 281, *298*
Boyd, H.   32, *38*
Boyle, M.   45, 46, *90*
Brandt-Williams, S.   50, 66, 67, *88*
Brennan, T.   163, *165*
Breukelaar, A.W.   162, *166*
Bromley, D.W.   250, *255*
Brookes, A.   22, *38*
Broome, J.   283, *298*
Brower, M.   203, *217*
Brown, I.F.   145, *151*
Brown, L.R.   168, 178, 182, *185*, *186*
Brown, M.T.   65, 67, *88*, *93*
Brown, P.   277, *299*
Bruntland, G. (Chair)   112, *124*
Buczynska, A.   119, *125*
Bueno, M.   277, *299*
Butterworth, J.E.   26, *39*

Cairns, J.   105, *124*
Callicott, J.B.   170, *186*
Calow, P.   170, *187*
Canning, D.   222, *255*, 260, *298*
Carson, E.R.   45, *89*
Carson, R.   167, 183, *186*
Casti, J.L.   43, 45, *88*
Catarino, F.   254, *256*, 285, *299*
Cavalier-Smith, T.   56, *88*
Chaitin, C.J.   83, *88*
Chakrabarti, P.P.   119, *124*
Chapin III, F.S.   10, *16*, 175, *187*, 197, 198, *217*, *218*
Charles, A.T.   32, *38*
Chattopadhyay, A.   119, *124*
Chen, J.   30, 31, *38*
Cheslak, E.F.   70, *88*
Chivian, E.   182, *186*, 197, 198, *217*
Choi, J.S.   42, 48, 49, 59, 69, 70, 72, 82, 85–87, *88*, *89*
Chomentowski, W.H.   145, *151*
Chomsky, N.   276, *299*
Christaller, W.   50, *88*
Christensen, N.   168, 170, *187*
Cicero, M.T.   115, *124*
Cincotta, R.P.   198, *217*
Clark, C.   241, *256*
Clark, T.W.   160, *165*
Clench-Aas, J.   162, *166*
Coase, R.H.   250, *256*
Cobb, C.W.   233, *256*, 264, 265, 269, *299*

Cobb, J.B.   31, 32, *38*, 133, *137*, 225, 232, 233, *256*, 267, *299*
Cofala, J.   7, *16*, 162, *166*
Cohen, R.D.   284, *300*
Colborn, T.E.   190, *218*
Collados, C.   224, *256*
Colwell, R.R.   177, *186*
Committee of Scientists   141, *151*
Commoner, B.   193, 196, *217*
Conley, M.   49, 51, *93*
Conley, W.   49, 51, *93*
Consortium for International Earth Science Information Network (CIESIN)   26, *38*
Constitutional Rights Project   277, *299*
Convention on Biological Diversity   181, *186*
Cooperrider, A.Y.   159, 160, *166*
Coops, H.   162, *166*
Cortese, A.D.   102, 113, 114, *124*, 190, *217*
Costanza, R.   1, *3*, 7, 11, *16*, 28, 31, *38*, *39*, 104, 110, *124*, 168, 170, 175, *186*, *187*, 190, *217*, 229, 230, 254, *256*, 267, 279, 285–287, 292, 293, 296, 297, *298*, *299*, *301*, 306, *307*
Couchot, K.   193, 196, *217*
Cousins, S.   49, 50, *88*
Cowen, T.   250, *256*
Crandell, D.R.   141, *151*
Crocker, D.   268, *299*
Crowfoot, J.E.   160, *166*
CSIRO   103, 104, 110, *124*
Curran, P.F.   85, *90*

da Fonseca, G.A.B.   182, *187*
Daily, G.C.   10, 14, *16*, 103, 104, 109, *124*, 129, *137*, 230, 241, *255*, *256*
Dakers, A.J.   135, *137*
Dale, V.H.   142–145, *151*, *152*
Daly, H.E.   1, *3*, 31, 32, *38*, 133, *137*, 225, 230, 232, 233, 247, 250, *256*, *257*, 264, 267, 286, 292, 296, *298*, *299*, *301*
D'Amore, M.   197, *218*
d'Arge, R.   7, 11, *16*, 104, *124*, 190, *217*, 230, *256*
Dasgupta, P.   241, *255*
Davies, T.R.H.   112, *124*
de Groot, R.   7, 11, *16*, 104, *124*, 190, *217*, 230, *256*
de Pires, I.   145, *151*
De Soysa, I.   7, *16*

Author Index

de Vries, B. 12, *16*
DeAngelis, D.L. 85, *92*
DeBonis, J. 160, *165*
DeGennaro, R. 292, *298*
DeHayes, D.H. 190, 191, *217*
Delgado-Acevedo, J. 143, *151*
Denbigh, K.G. 85, *89*
DeSelm, H.R. 143, *151*
Deville, A. 42, *89*
Díaz, S. 10, *16*, 197, 198, *217*
Dieffenbacher-Krall, A.C. 191, *217*
Dietrich, C. 28, *39*
Dirzo, R. 198, *218*
Dixon, B. 7, *16*
Dixon, J. 230, *256*
Dolezal, J. 59, *89*
Doolan, J.M. 28, *38*
Dorland, K. 162, *166*
Dovers, S. 9, *16*
Dowlatabadi, H. 7, 12, *16*, 28, *39*
Downs, P.W. 22, *38*
Draper, D. 169, 176, 178, 182, *186*
Duane, T.P. 224, *256*
Dubourg, W.R. 228, *255*
Dubow, J. 197, *219*
Durning, A.T. 222, 238, *256*, 261, 287, 288, 290, *299*
Dynesius, M. 195, *218*
Dyurgerov, M. 175, *187*

Ehrlich, P.R. 109, *124*
Ehui, S. 279, *299*
Eisl, H. 193, 196, *217*
Ekins, P. 222, *256*, 261, *299*
El Serafy, S. 229, 230, *256*
Engelman, R. 198, *217*
Environment Canada 194, *218*
EPA 177, *186*
Epstein, P.R. 177, *186*
Erlandson, D. 292, *298*
Etnier, C. 116, 117, *124*
European Commission 161, 162, *165*
Eviner, V.T. 10, *16*, 197, 198, *217*

Faber, M. 245, *258*
Faiola, A. 248, *256*
Fajnzylber, P. 284, *299*
Falfán, I. 32, *39*
Farber, S. 7, 11, *16*, 104, *124*, 190, *217*, 229, 230, 250, *256*, *258*
Farley, J. 272, *299*

Farquhar, M. 221, *256*
Fath, B.D. 45, 58, 59, 70, 71, 86, *89–91*
Fearnside, P.M. 145, *151*
Federal, Provincial and Territorial Advisory Committee on Population Health 194, *218*
Feindt, P.H. 75, *91*
Ferreira, G.L. 145, *151*
Ferrer-i-Carbonell, A. 32, *39*
Ferris, D. 292, *298*
Fiorito, E. 263, *301*
Fisher, G. 221, 222, 251, *258*
Fisher, I. 267, *299*
Flavin, C. 168, *186*
Flindt, M.R. 56, *92*
Flood, R.L. 44, 45, *89*
Folke, C. 1, *3*
Fonseca, J.C. 55, 56, 59, *89*
Ford, E.D. 157, *165*
Forsberg, L.E. 119, *125*
Forscher, B.K. 9, 12, *16*
Fox, R.F. 66, *89*
Francis, D.R. 194, *219*
Francis, G. 45, 46, *90*
Frank, R. 251, *256*, 283, 292, *300*
Franklin, J.F. 143, *151*
Freedman, B. 179, 182, *186*
Freitas, A.M. 55, 56, 59, *89*
Frelek, K. 119, *125*
Frenzen, P.M. 143, *151*
Friedman, M. 240, *256*
Friedrich, R. 162, *165*
Frosch, R.A. 102, *125*
Funtonowicz, S. 184, *186*
Futuyma, D.J. 54, 56, *89*
Fyfe, W.S. 175, *187*

Galbraith, J.K. 251, *256*
Gallopin, G.C. 42, *89*
Gan, C. 121, *124*
Gandini, M. 135, *137*
García, J. 32, *39*
Gaskell, P.J. 159, *165*
Gates, J. 238, *256*, 280, *300*
Gaudet, C.L. 170, *187*
Gell-Mann, M. 54, 55, *89*
Gibbons, M. 8, *16*
Gibbs, B. 284, *300*
Gilbert, A. 42, *89*
Gilfedder, M. 10, *17*
Gillin, M. 112, *125*

Giordano, S.   197, *218*
Glansdorff, P.   86, *89*
Gleditsch, N.P.   7, *16*
Glover, J.D.   10, *16*
Goldworthy, P.   27, 28, *38*
Goodland, R.   250, *257*
Goodwin, N.   262, *300*
Goulder, L.   109, *124*
Government of Canada   175, 176, *186*
Graham, B.   222, *255*, 260, *298*
Grasso, M.   7, 11, *16*, 104, *124*, 190, *217*, 229, 230, *256*
Grayson, R.B.   28, *38*
Gregorich, L.J.   197, *217*
Gregory, K.J.   22, *38*
Greiluber, J.   59, *89*
Grimm, M.P.   162, *166*
Groves, R.H.   164, *165*
Gubler, D.   177, *186*
Guerreiro, C.   162, *166*
Guerrero, A.   32, *39*
Guerrero, M.   32, *39*
Gulati, R.D.   162, *166*
Guterstam, B.   118, 119, *125*

Haas, B.K.   221, 222, *257*, 259, 265, *300*
Haedrich, R.L.   160, *165*
Hales, S.   177, *188*
Hamilton, K.   228, 231, *255*, *257*
Hanna, S.   254, *256*, 285, *299*
Hanning, R.M.   196, *218*
Hannon, B.M.   7, 11, *16*, 84, *93*, 104, *124*, 190, *217*, 230, *256*
Hansell, R.I.C.   48, 59, 69, 72, 85, 86, *88*
Hardin, G.   250, *257*
Hare, M.   32, *38*
Harf, J.E.   155, *165*
Harger, J.R.E.   42, *89*
Hargis, B.M.   199, *218*
Harper, C.L.   175, 178, 183, *186*
Harper, J.L.   82, *88*, 108, *125*
Harris, G.P.   13, 14, *16*
Haskell, B.D.   170, *186*, 306, *307*
Hasselgren, K.   117, *125*
Hawken, P.   291, 292, *298*, *300*
Hawley, G.J.   190, *217*
Hayes, D.   182, *186*
Health Canada   193, 194, 196, *218*
Heck, K.E.   284, *300*
Heck, T.   162, *165*
Hect, S.   145, *151*

Heeb, J.   117–119, *125*, 135, *137*
Helmerhorst, T.H.   162, *166*
Hendrayana, H.   79–81, *90*
Hendriksen, J.F.   162, *166*
Henein, K.   49, *91*
Hengeveld, H.G.   175, 176, 178, *186*
Herbst, M.   63, *89*
Herendeen, R.   60, *89*
Herreid, C.F.   160, *165*
Hertel, T.   279, *299*
Heyes, C.   7, *16*
Hicks, J.R.   225, 228, *257*
Hilborn, R.   160, *166*
Hinman, H.R.   10, *16*
Hinz, K.-H.   199, *218*
Hobbie, S.E.   10, *16*, 197, 198, *217*
Holbo, H.R.   61, *91*
Holland, J.H.   13, *16*, 45, *89*
Holland, M.   162, *165*
Hollebeek, P.   162, *166*
Holling, C.S.   46, *89*
Hom, J.   194, *218*
Homer-Dixon, T.F.   169, 174, 179, 180, *186*
Hooper, D.U.   10, *16*, 197, 198, *217*
Horgan, J.   7, 11, *16*
Hornby, A.S.   82, 83, *89*
Horner, J.A.   292, *298*
Horowitz, A.   197, *218*
Horwitz, P.   175, *187*
Hosper, S.H.   162, *165*
Hotelling, H.   241, *257*
Howard, J.   170, *187*
Howarth, R.W.   10, *17*
Hsieh, C.   284, *300*
Huang, S.   50, *89*
Huber, F.   117, *125*
Huber-Sanwald, E.   198, *218*
Huenneke, L.F.   198, *218*
Hueting, R.   231, *257*
Hunphries, H.   49, 51, *93*
Hutchings, J.A.   160, *165*

Ibelings, B.   162, *166*
ICAFS   160, *165*
International Advertising Association   287, *300*
IPENZ   112, *125*
Iverson, L.   191, *217*

Jackson, J.A.   160, *165*

Jackson, R.B.   198, *218*
Jacobson, G.L.   191, *217*
Jagtman, E.   162, *165*
Jakeman, A.   28, *39*
Jana, B.B.   118, 119, *125*
Jansen, H.   162, *166*
Janssen, M.A.   12, *16*
Janssen, W.   27, 28, *38*
Johnson, G.B.   182, *187*
Johnson, L.   69, *89*
Jones, D.L.   170, *187*
Joosse, W.   162, *166*
Jørgensen, S.E.   1, *3*, 36, *39*, 55, 56, 58–62, 64, 69–71, *88–91*, 303, *308*
Josephson, B.   122, *125*

Kandlikar, M.   21, *39*
Kaplan, G.A.   284, *300*
Kaplan, R.   263, *300*
Kaplan, S.   263, *300*
Karagatzides, J.D.   196, *218*
Karnawati, D.   79–81, *90*
Karr, J.R.   168, 170, *187*
Katapatuk, B.   196, *218*
Katchalsky, A.   85, *90*
Kauffman, S.   1, *3*
Kawachi, I.   284, *300*
Kay, J.J.   45, 46, 55, 58, 61, 69, 70, 82, 85, 86, *90*, *92*, *93*
Kellert, S.R.   263, *300*
Kennedy, B.P.   284, *300*
Kent, J.   182, *187*
Kerr, J.B.   194, *219*
Keynes, J.M.   276, *300*
King, A.W.   143, *151*
Kingsley, R.A.   199, *218*
Kinzig, A.   198, *218*
Kiron, D.   266, *300*
Klemer, A.R.   253, *255*, 281, *298*
Klimont, Z.   7, *16*
Klir, G.J.   44, *90*
Koestler, A.   48, 82, 85, 86, *90*
Kokkelenberg, E.C.   229, *257*
Kolmogorov, A.N.   83, *90*
Kraft, M.E.   173, 180, *186*
Krause, D.   174, *186*
Krewitt, W.   162, *165*
Krotschek, C.   123, *125*
Krull, J.L.   174, *187*
Kullback, S.   62, *90*

Lackey, R.T.   159, *165*
Lake, C.   280, *300*
Lamarra, V.A.   70, *88*
Lambert, J.D.   50, *91*
Lammens, E.H.R.R.   162, *166*
Lancelot, J.   292, *298*
Lane, R.E.   237, 238, *257*, 261, 274, 275, *300*
Langis, R.   196, *218*
Langlois, C.   196, *218*
Langridge, J.   12, *16*
Lannigan, R.   170, 175, *187*
Lapham, L.   261, *300*
Larsen, S.   162, *166*
Lasorsa, B.   196, *218*
Laszlo, E.   43, *91*
Lavers, A.   32, *38*
Lavorel, S.   10, *16*, 197, 198, *217*
League of Conservation Voters   253, *257*
Lederman, D.   284, *299*
Leemans, R.   198, *218*
Lehman, S.J.   176, *187*
Leon, W.   203, *217*
Leopold, A.   108, *125*, 170, *186*
Leupelt, M.   57, 69, *91*
Levin, S.   241, *255*
Lewan, L.   65, *94*
Lewin, B.   55, 56, *91*
Li, W.-H.   54, 56, *91*
Likens, G.E.   10, *17*, 169, *186*
Lillebø, A.I.   56, *92*
Limburg, K.   7, 11, *16*, 104, *124*, 190, *217*, 230, 254, *256*, 285, *299*
Limoges, C.   8, *16*
Linares, A.   32, *39*
Lindgren, E.   177, *186*
Lipner, V.   197, *218*
Loayza, N.   284, *299*
Lochner, K.   284, *300*
Lodge, D.M.   198, *218*
Loiselle, S.   135, *137*
Lombardi, M.O.   155, *165*
Lopes, R.J.   56, *91*
Losito, B.D.   263, *301*
Losos, E.   197, *219*
Lotka, A.J.   70, *91*
Lovins, A.   291, *300*
Lovins, H.   291, *300*
Low, B.   254, *256*, 285, *299*
Lubchenco, J.   109, *124*, 190, 197, *219*
Lucretti, S.   59, *89*

Ludwig, D.   160, *166*
Lutz, E.   231, *257*
Luvall, J.C.   61, *91*
Luz, L.M.   145, *151*
Lynch, J.W.   284, *300*
Lysak, M.A.   59, *89*

Ma, S.   114, *126*
Mack, M.C.   10, *16*, 197, 198, *217*
MacMahon, J.   143, *151*
Madeira, V.M.C.   55, 56, 59, *89*
Maister, A.   59, *89*
Mäler, K.-G.   1, *3*, 231, 241, *255*, *257*
Malingreau, J.P.   145, *151*
Mann, L.K.   143, *151*
Månsson, B.Å.   47, *91*
Marchettini, N.   65, 69, 70, *88*, *93*
Marcondes, H.   277, *299*
Margerum, R.D.   22, 28, *39*
Markandya, A.   232, *257*
Marques, J.C.   55, 56, 59, *89*, *91*, *92*
Martinez, A.   286, *300*
Marx, T.   292, *298*
Maskin, E.   241, *255*
Maslow, A.   260, *300*
Matson, P.A.   10, *17*, 109, *124*
Max-Neef, M.   222, 225, 228, 233, 234, 236, *256*, *257*, 260, 261, 266, 279, 288, *300*
Maxwell, T.   28, *39*
Mazumder, A.   48, 59, 69, 72, 85, 86, *88*
McClosky, H.   238, *257*
McCord, R.A.   143, *151*
McCreddin, J.A.   12, *16*
McElroy, C.T.   194, *219*
McGinn, A.P.   277, *300*
McGlade, J.M.   47, *91*
McMichael, A.J.   168, 170, *186*, *187*
McMurtry, R.   170, *187*
Meadows, Dennis   242, *257*
Meadows, Donella   242, *257*
Meffe, G.K.   159, *166*
Meijer, M.L.   162, *166*
Meister, H.P.   75, *91*
Mejer, H.   55, 58, 70, *90*
Melillo, J.M.   190, 197, *219*
Merriam, G.   49, *91*
Metelo, I.   56, *92*
Meyer, F.M.   42, *89*
Mickler, R.A.   194, *218*
Miles, M.A.   263, *301*

Milne, B.   49, *91*
Minsch, J.   75, *91*
Mishan, E.J.   222, 250, *257*
Mitchell, G.   42, *91*
Mitsch, W.J.   102, 114, *125*, 303, *308*
Mittermeier, C.G.   182, *187*
Mittermeier, R.A.   182, *187*
Mobbs, C.   9, *16*
Molitor, M.   254, *256*, 285, *299*
Mooney, H.A.   109, *124*, 190, 197, 198, *218*, *219*
Moran, E.F.   145, *151*
Morgan, M.G.   28, *39*
Morison, J.   175, *187*
Morowitz, H.J.   70, *91*
Morse, S.S.   182, *186*
Moser, A.   73, *91*, 102, *125*
Mulkey, L.A.   30, *38*
Müller, F.   57, 58, 69, *91*
Mullin, D.   174, 183, *187*
Mullineaux, D.R.   141, *151*
Munasinghe, M.   42, *91*, 228, *255*
Munda, G.   32, *39*
Murdock, N.   143, *151*
Myers, I.B.   146, 148, *151*
Myers, N.   182, *187*

Naeem, S.   7, 11, *16*, 104, *124*, 190, *217*, 230, *256*
Nancarrow, B.E.   12, *16*, 26, *39*
Nardi, L.   59, *89*
Narodoslawsky, M.   123, *125*
Navy, H.   180, *185*
Naylor, R.L.   10, *16*, 197, 198, *217*
NCSE   157, 159, 160, *166*
ñeda, B. Casta   229, *256*
Nelson, J.   197, *218*
Nepstad, D.C.   145, *151*
Nickerson, D.   278, *300*
Nicolis, G.   86, *91*
Nicolis, J.S.   83, *92*
Nieboer, E.   196, *218*
Nielsen, S.N.   55, 56, *90*, *91*
Niemi, Å.   134, *138*
Nilsson, R.   190, 195, *218*
Nobre, A.D.   145, *151*
Noordhuis, R.   162, *166*
Norberg-Hodge, H.   183, *187*
Nordhaus, W.D.   229, 232, 234, *257*
Norland, D.   292, *298*
Norton, B.G.   170, *186*, 306, *307*

Noss, R.F.   49, *92*, 159, 160, *166*
Nova Scotia Department of the Environment 209, *218*
Nowotny, H.   8, *16*
NSF   159, *166*
Nussbaum, M.   265, *301*

Oates, W.E.   250, *255*, *257*
Obermayer, R.   59, *89*
Odum, E.P.   69, *92*
Odum, H.T.   49–51, 57, 65, 70, 83, 84, 87, *92*, *93*, 102, 114, *125*
Oechel, W.C.   175, *187*
Oesterheld, M.   198, *218*
Offerman, H.L.   145, *151*
Olson, J.S.   143, *151*
Olsthoorn, X.   162, *166*
O'Neill, R.V.   7, 11, *16*, 49, 51, 85, *92*, *93*, 104, *124*, 144, 145, *151*, *152*, 190, *217*, 230, *256*
Onisto, L.   32, *39*
Onsager, L.   70, 86, *92*
Ophuls, W.   250, *257*
Oreskes, N.   29, *39*
Osterkamp, T.   175, *187*

Pahl-Wostl, C.   32, *38*, 75, *92*
Painter, D.J.   112, *124*
Paiva, A.A.   55, 56, 59, *89*
Palmer, D.   26, *39*
Pamuk, E.R.   284, *300*
Panayatou, T.   279, *301*
Parasuk, C.   279, *301*
Pardal, M.A.   55, 56, *91*, *92*
Park, J.   26, *39*
Parker, P.   32, *39*
Parson, E.A.   13, *16*
Paruelo, J.   7, 11, *16*, 104, *124*, 190, *217*, 230, *256*
Patel, A.   172, *187*
Patil, G.P.   168, 170, 175, *187*
Patten, B.C.   1, *3*, 42, 45–49, 55, 58, 59, 61, 69–72, 82, 84–87, *88*–*90*, *92*, 97, 98, *99*
Patwardhan, A.   21, *39*
Patz, J.A.   177, *187*
Pearce, D.W.   228, 230, *255*, *257*, 292, *301*
Pearson, S.M.   145, *151*
Pedlowski, M.A.   144, 145, *151*
Peet, J.   103, 110, *125*
Peikoff, L.I.   240, *258*
Pereira, J.G.   254, *256*, 285, *299*

Perrings, C.   232, 250, *257*, 293, *299*
Pérusse, M.   196, *218*
Peters, I.   292, *298*
Phelps, E.   241, *257*
Phillips, A.   197, *219*
Pillaityte, R.   119, *125*
Pimentel, D.   197, *218*
Pinkerton, R.C.   49, 70, *92*
Poff, N.L.   198, *218*
Politi, A.   54, *87*
Portielje, R.   162, *166*
Postel, S.   109, *124*, 168, 178, *186*, *187*, 195, 196, *218*
Preckel, P.   279, *299*
Prigogine, I.   57, 70, 86, *89*, *91*, *92*
Primack, R.B.   181, *187*
Proops, J.   245, *258*
Prothrow-Stith, D.   284, *300*
Prugh, T.   296, *301*
Pugh, M.D.   284, *300*

Rabsch, W.   199, *218*
Ramo, J.C.   240, *258*
Rand, A.   240, *258*
Randers, J.   242, *257*
Rapport, D.J.   168, 170–172, 175, *187*
Raskin, R.G.   7, 11, *16*, 104, *124*, 190, *217*, 230, *256*
Raven, P.H.   182, *187*
Ravetz, J.R.   11, *16*, 30, 31, *38*, *39*, 184, *186*
Rawls, J.   239, *258*, 262, 284, *301*
Rayner, S.   254, *256*, 285, *299*
Redzek, L.   119, *125*
Reeders, H.H.   162, *166*
Rees, W.   32, *39*
Rees, W.E.   42, *92*, 128, *138*
Refsgaard, K.   116, 117, *124*
Reganold, J.P.   10, *16*
Regier, H.A.   45, 46, 82, 85, 86, *90*, *92*
Reynolds, D.B.   246, *258*
Reynolds, H.L.   10, *16*, 197, 198, *217*
Rice, P.   197, *218*
Risbey, J.   21, *39*
Risser, P.G.   102, *125*
Roberts, D.V.   102, *125*
Robinson, J.   273, *301*
Romanovsky, V.   175, *187*
Roodman, D.M.   276, 292, *298*, *301*
Rosen, R.   44, *92*
Ross, H.   28, *39*

Rossi, C.   135, *137*
Rothstein, D.   197, *219*
Rotmans, J.   21, 26, *39*
Roux, M.   135, *137*
Roy, S.   135, *137*
Rucevska, I.   119, *125*

Sachs, J.P.   176, *187*
Sadalla, E.K.   174, *187*
Sala, O.E.   10, *16*, 197, 198, *217*, *218*
Salas, W.A.   145, *151*
Salthe, S.N.   50, 83, *92*
Santos, M.   99, *99*
Santos, R.   254, *256*, 285, *299*
Santra, S.C.   119, *124*
Sauvé, L.   173, *187*
Schaberg, P.G.   190, 191, *217*
Scheffer, M.   14, *16*, 162, *166*
Schellnhuber, H.J.   6, 14, *16*, 52, *93*
Schindler, D.W.   10, *17*
Schlesinger, W.H.   10, *17*
Schneider, C.   292, *298*
Schneider, E.D.   55, 58, 69, 70, 85, *93*
Schneider, S.H.   109, *124*
Schneidewind, U.   75, *91*
Schopp, W.   7, *16*
Schreider, S.   28, *39*
Schrödinger, E.   55, 87, *93*
Schrope, M.   14, *16*
Schuessler, K.   221, 222, 251, *258*
Schultes, R.E.   145, *151*
Schultink, G.   42, *93*
Schulz, T.   75, *91*
Schulze, P.C.   102, *125*
Schwartz, M.W.   149, *151*
Schwartzman, S.   8, *16*, 279, *301*
Scoccimarro, M.   28, *39*
Scott, P.   8, *16*
Seaton, R.A.F.   26, *39*
Sen, A.   294, *301*
Seppelt, R.   51, 52, *93*
Serreze, M.C.   175, *187*
Sevilla, J.   222, *255*, 260, *298*
Sforza, M.   248, *258*
Shabecoff, P.   185, *187*
Shearer, W.   42, *91*
Sherman, P.   230, *256*
Shogren, J.   159, 160, *166*
Shue, H.   239, *258*
Shyamsundar, P.   292, *298*
Silva, M.   197, *218*

Simon, H.A.   45, 85, *93*
Simons, J.   162, *166*
Simons, R.F.   263, *301*
Singleton, A.   214, *218*
Skole, D.L.   145, *151*
Smith, A.   248, *258*
Smith, N.J.H.   145, *151*
Snyder, C.   240, *258*
Solow, R.   242, *258*, 273, *301*
Sonnenschein, C.   190, *218*
Sonzogni, W.C.   26, *38*
Soto, A.M.   190, *218*
Southworth, F.   144, 145, *151*, *152*
Spanner, D.C.   48, 85, *93*
Spiegel, J.   175, *187*
Starr, T.B.   50, 82, 85, *87*
Starrett, D.   241, *255*
Stengers, I.   57, *92*
Sterner, T.   241, *255*
Straskraba, M.   55, 59, 61, 69, 71, *90*
Strimbeck, G.R.   190, *217*
Stuart-Smith, K.   49, *91*
Subagyo, P.   79–81, *90*
Sugden, R.   265, *301*
Sukandarrumidi   79–81, *90*
Sullivan, M.F.   67, *88*
Summers, L.   277, *301*
Suter, G.W.   42, *93*
Sutton, P.   7, 11, *16*, 104, *124*, 190, *217*, 230, *256*
Swanson, F.J.   143, *151*
Sykes, M.T.   198, *218*
Syme, G.J.   12, *16*, 26, *39*

Temme, M.M.   52, *93*
Templet, P.H.   250, 253, *255*, *258*, 272, 275, 278, 281, 292, *298*, *301*
The World Commission on Environment and Development   185, *187*
Thom, D.   102, 112, 113, *125*
Tieger, P.   146–148, *152*
Tietenberg, T.   241, *255*
Tilman, D.G.   10, *16*, *17*, 109, *124*, 191, *218*
Time   109, *125*
Tiongco, M.   180, *185*
Tisdell, C.A.   250, *258*
Tobin, J.   232, 234, *257*
Todd, J.   122, *125*
Townsend, C.R.   82, *88*, 108, *125*
Trade Compliance Center   248, *258*

Trow, M.   8, *16*
Trukenmüller, A.   162, *165*
Tschape, H.   199, *218*
Tsiji, L.J.S.   196, *218*
Tsolis, R.M.   199, *218*
Tucker, C.J.   145, *151*
Turner, K.   292, *301*
Turner, S.   49, 51, *93*
Turpin, T.   42, *89*

Ulanowicz, R.E.   70, 84, *93*
Ulgiati, S.   65, *88*, *93*
Ulrich, R.S.   263, *301*
UNESCO   77, *93*
United Nations Development Program   7, *17*, 104, 105, 107–109, 112, *125*, 167, 172, 175, 176, 178, *187*, 198, 208, *218*
United Nations Environment Program   *17*, *125*, 167, *187*, *218*
US Food and Drug Administration   196, *219*

Van Asselt, M.   21, 26, *39*
van Bohemen, H.D.   115, 116, *126*, 135, *138*
van den Belt, M.J.   7, 11, *16*, 99, *99*, 104, *124*, 190, *217*, 230, 254, *256*, 285, *299*
Van den Berg, M.S.   162, *166*
Van den Bergh, J.   32, *39*
Van der Molen, D.T.   162, 163, *165*, *166*
Van Nes, E.H.   162, *166*
van Valen, L.   49, 86, *93*
Van Waveren, R.   28, *39*
Vasseur, L.   171, 180, 181, *188*, 214, *219*
Velinga, P.   6, *17*
Vermij, S.   162, *165*
Vilela Mendes, R.   54, *93*
Villa, F.   28, *39*
Vitousek, P.M.   10, *16*, *17*, 190, 197, 198, *217*, *219*
Voinov, A.   28, *39*
Voinov, H.   28, *39*
Volkenstein, M.V.   63, *93*
von Bertalanffy, L.   57, 84, 86, *93*
Vuthy, L.   180, *185*

Wackernagel, M.   32, *39*, 65, *94*, 128, *138*
Waide, J.B.   85, *92*
Wainger, L.   1, *3*, 28, *39*
Waldrop, M.M.   29, *39*
Walker, A.   28, *39*

Walker, B.H.   12, *16*, 198, *218*
Walker, G.   10, *17*
Walker, M.   198, *218*
Wall, D.H.   198, *218*
Wallach, L.   248, *258*
Walsh, J.E.   175, *187*
Walster, E.   253, *258*
Walster, G.W.   253, *258*
Walters, C.   160, *165*, *166*
Walzer, M.   275, *301*
Wardle, D.I.   194, *219*
Washington-Allen, R.A.   143, *151*
Weinstein, P.   177, *188*
Weitzman, M.   228, *258*
Wetzel, R.G.   66, *94*
Whitford, P.B.   143, *151*
Whitford, W.G.   171, 175, *187*
Wilcove, D.S.   197, *219*
Wilkinson, R.   284, *301*
Williams, J.   10, *17*
Wilson, E.O.   181, 182, *188*, 263, *300*, *301*
Wilson, J.   254, *256*, 285, *299*
Wisnewski, J.   198, *217*
Wondolleck, J.M.   160, *166*
Woods Barlett, P.   193, 196, *217*
Woodward, A.   177, *188*
Woodwell, G.M.   109, *124*
Woodwell, J.   292, *298*
World Bank   *17*, *125*, 167, *187*, *218*, 260, 263, *301*
World Health Organization   168, *188*, 193–196, 199, *219*
World Resources Institute   *17*, *125*, 167, 182, *187*, *188*, *218*
Wright, S.   72, *94*
Wulff, F.   134, *138*
Wurth, A.H.   102, *126*
Wynne, B.   42, *94*
Wyss, P.   117, *125*

Yaffe, S.L.   159, 160, *166*
Yan, J.   114, *126*, 135, *138*
Yassi, A.   175, *187*
Yen, I.H.   284, *300*
Young, C.   228, *255*
Young, M.   254, *256*, 285, *299*

Zaller, J.   238, *257*
Zavaleta, E.S.   10, *16*, 197, 198, *217*
Zelson, M.   263, *301*
Zhang, T.   175, *187*

Subject Index

Aboriginal communities   194
absence of diseases   171
activism   156
adaptation   46, 57, 69, 82
adaptive capacities   43, 49, 86, 95
adaptive hierarchies   49
adaptive landscape   72
adaptive self-organization   43
adaptive self-stabilization   43
advertising   262, 286–290, 296
advocacy   155
agent-based models   32, 76
agriculture   145
agrosystems   178
air pollution effect on forest ecosystems   194
air quality   193, 194
alien species   197
allocation   238–240, 248, 253–255
Amazon (Rondônia State, Brazil)   144–146
anthropocentric valuing   105, 129–131
anthropogenic disturbance   190
anthropogenic ecosystem disruption   190
anthroposphere   73
antibiotics   199
aquaculture   68, 118, 119, 122
Arctic ecosystems   193, 196
Aremark   117
ascendancy   58
assumptions   150
atmospheric CO_2 145
attitude and behaviour   150, 174
AWFUL Theorem   47

barriers   150
bioaccumulation   196
biocomplexity   54, 55
biodiversity   7, 10, 103, 110, 140, 181, 190, 197–200
– hotspots   198
biological function and health   199
biological stress response systems   190

biomass energy fuel   117
biophysical integrity   184
bio-social integrated index   210
biosphere   127–129
biosphere ecosystems   130
blackwater   117
border-lakes   162
bovine spongiform encephalopathy (BSE)   10
brainstorm   147
built capital   263, 264
built environments   108

CAHSystems (Complex Adaptive Hierarchical Systems)   36, 42, 46, 47, 51, 52, 57, 72, 82, 95–97
CAHSystems Theory   41
Calcutta   118
Canada   196
capacity to function   171
carbon release in land clearing   144, 145
case studies   135, 136, 160
– evaluation   121
catchment models   27
catchment scale   22
climate change   144, 145, 175
climate regulation   110
coastal development   77
common good   133
communication   19, 20, 33, 140, 143, 150, 154, 157
community   112
Complex Adaptive Hierarchical Systems, *see* CAHSystems
complex adaptive systems (CAS)   9, 13
complex systems   43
complexity   19, 29, 43–45, 54–56, 73, 74, 78, 83
comprehensive assessment of ecosystem health   184
computer based simulation models   22
computer simulation   21

conceptual frameworks   22
conceptual models   21
coniferous trees   142
consensus building   141, 145
consumerism   222, 261, 295, 297
continental scale   25
control functions   49
control site   142
conventional wastewater systems   122
coordinated modeling   41
cost–benefit studies of environmental policy   162
costs of electricity   161
costs of environmental degradation   190
credibility   141
c-value paradox   56
cycling   47

debris avalanche   142
decentralized wastewater systems   117
decisionmaking   19, 23, 25, 139, 145, 153–155, 157–160, 163, 164
declining carrying capacity   168
deep science   74
deforestation   144, 145
degradation   47, 145
demand-driven research   162
desertification   179, 197
difference principle   239
dimensions of disagreement   155
diseases   177, 183
disorder   55
dissipation   47
dissipative structures   58
distress syndromes   170
disturbances   49, 142
diversity   145
DNA   54, 56
doers   127, 135–137
Donaumoos (Germany)   116
drainage of wetlands   110

Earleaved false foxglove   143
Earth Day   140
earth system modeling (ESM)   6, 13, 25
eco-ethics   41
eco-principles   75
Eco-Social Product   74
eco-technology   114
ecocentric valuing   105
ecological carrying capacity   32

ecological debt   276, 293
ecological economics   96, 110, 307
ecological engineering   96, 102, 114, 115, 128, 132, 133, 306, 307
– case studies   115–122
ecological exergy   55–57
ecological footprint   32
ecological index   210
ecological indicators   306
ecological modeling   96, 305
ecological succession   69
ecological sustainability/integrity   200
ecologically sustainable development (ESD)   6, 14
economic development   145
economic evaluation   32
economic factors   184
economic growth   259, 260, 276, 277, 283–286, 292, 294, 295
economic growth paradigm   24
economic welfare   225–228, 232, 234
economics   160, 161, 164
ecosphere   73
EcoSummit   36
ecosystem capacity   168
ecosystem development   60
ecosystem function   190, 195, 197–199, 267, 272, 278, 291
ecosystem goods   103, 224
ecosystem health   82, 96, 98, 99, 170, 197, 267, 272, 278, 279
– major elements   184
ecosystem services   7, 11, 36, 103–105, 109, 110, 127–137, 140, 190, 191, 195, 198, 199, 201, 224, 226, 240, 249, 267, 270–272, 278, 279, 283–287, 290, 292
– (mis)use of   109, 110
– valuation   5, 229, 230
ecosystem stress   109
ecosystem structure   197
ecosystem theory   71
ecosystem training   108
ecosystems   108, 128–131, 134, 137, 157, 169
– changes affecting human activities   176
– disruption   194
– health indicators   211
edaphic conditions   143
education of scientists   156
educational disconnect   159
embeddedness   105–109

Subject Index

emergent properties 66
emergy 47, 58, 65, 67, 83
emissions 110
empower 47, 58, 65, 68, 69, 84
energy efficiency coefficient 64
energy hierarchies 50, 65
energy production from wastewater 117
engineering 110, 112, 113
– engineers' professional responsibility 112, 123
entropy 47, 62
environ 84
Environmental Action Programme 161
environmental decisionmaking 24, 154, 157, 159, 160, 164
environmental externalities 161
– of energy use 161
environmental impact assessment 51, 52
environmental indicators 42
environmental policymaking 141, 159
– in Europe 161
environmental problems 157
Environmental Protection Agency (USA) 140
environmental science 158, 159
environmental science curriculum 159
environmental scientists 146
equitable development and consumption of resources 204
equity 201, 204, 205
erosion 110, 142
error estimation 27
eruption 141
ethics 184
Ethics for Engineers 113
European Commission 161, 162
eutrophication 56, 57
evolution 57
exergy 47, 54–58, 62, 68, 84
exergy dissipation 58, 61
exergy efficiency 63
exergy efficiency coefficient 62, 64
exergy/empower ratios 67
exergy storage 58, 60, 61
exotic species, influence on diversity 142
expectations 150
external cost estimates 161
externalities 238, 242, 249–251, 253, 255, 270, 272, 278
ExternE 161
extinction 197

farmers 145
fires 144
First Nations 193, 194, 196
fish ponds 119
fish tanks 118
fitness 69, 72
flexibility 147
food-borne disease 196
food chains 197
food resources 196
foot and mouth disease 10, 11
forcing functions 49
Forest Service (USA) 143
fragmentation 145
freshwater resources 195
frontline projects 111
future generations 221–298

general public 172
General Systems Theory 42, 57
genomes 54
genuine progress index (GPI) 35
geographic information systems (GIS) 9, 52
geology 143
global biodiversity 198
global system models 27
goal functions 48, 58, 69–71, 74
grasses 142
greenhouse 118, 119
greywater 117
Gross National Product 74
ground-level ozone 193
ground water 116, 195

habitats 143
– destruction 182
– fragmentation 110, 115
hard problem science (HPS) 1, 2
hard science 97
health 42, 48, 49, 170
healthy ecosystems 184
heterogeneity 73
hierarchy 43, 45, 46, 85
holarchy 43, 45, 47, 85
holistic approach 57, 74, 157
holon 43, 85
human capital 263, 264
human development 222, 227, 265, 266
Human Development Index (HDI) 234
human health 168, 184

human needs 225–297
human welfare 225–228, 234
human well-being 200, 221, 234
hysteresis 14

immigration 145
income inequality 283, 284
increasing human demands 168
Index of Sustainable Economic Welfare (ISEW) 232
indicators 208
– economic 32
– of ecosystem health 208
– of sustainability 31, 32
inequity 201
information 54, 55
information content 64
information overload 41, 47, 85, 95
infrastructure 115
innovative engineering 111
innovative technologies 184
input–output superposition 45
institutionalized model-making 99
insulation 116
integrated applied systems analysis (IASA) 25
integrated assessment and modeling (IAM) 6, 11, 13, 19, 26, 27, 32, 36
integrated assessment (IA) 6, 10, 11, 13, 14, 28–30, 35, 37, 75
– validation 19
integrated environmental management (IEM) 22, 26, 28
integrated frameworks 21
integrated modeling 23, 305
Integrated River Basin Management 23
integrated tools 20
integrating the sciences 112
integrative assessment tools 32
interdisciplinary modeling 95
interdisciplinary research 6
interest groups 140
intergenerational equity 249
intergenerational fairness 242, 247, 248
intergenerational justice 269, 270
international banks 146
International Congress of Systematic and Evolutionary Biologists (Vancouver, Canada) 142
Inuit 193, 194, 196
invasion of exotic species 182

justice theory 239, 240, 247, 269, 270, 272, 273, 276, 278, 284

Kågeröd Recycling Project 117
Kaja 116
Kullback Information 62
Kullback measure 62–64

lack of communication 173
lack of education 173
lack of information 173
lake management 162
land management 140
land use 143
landscape management 15
least specific dissipation 86
legumes 142
link between ecosystem and human health 172
liquid composting 117
living machine system 122
logging 110
logic network 96
long-range atmospheric transport 196

macroscopic pattern analysis 74
major barriers 173
mangrove ecosystems 198
marginal lands 178
marine ecosystems 110
market failures 238, 248, 250, 255, 269, 270, 273, 278
market goods 271, 272, 278, 285, 287, 290
mathematical heterogeneity 51, 52
media 158
mediated modeling 99
meta-population 52
minimax principle 58, 64
Ministry of Transport (Netherlands) 115
model complexity 29
model projections 145
modeling 73, 79, 98
module integration 28
Mount St. Helens (Washington, USA) 141–143
multidisciplinary approach 24, 27, 98
Myers–Briggs psychological types 146

National Science Foundation (USA) 143
native crops 145

Subject Index

natural capital   224–298
natural resource management (NRM)   5, 6
Natural Resources Conservation Service (USA)   143
Nature's services, *see* ecosystem services
network analysis   70
new medicines, food, or fibres   183
nitrogen   116
non-linear behavior   45, 74
non-renewable resources   233, 241, 243, 244, 246
non-scientific audience   158
"normal" metabolic activities   48
nutrient recycling   117, 122

Oak Ridge Cedar Barrens (Tennessee, USA)   143, 144
objectivity   158
order   43, 55
order through fluctuation   86
organization   45
organizational complexity   44, 47, 53
organizational hierarchy   47
organizing principles   70
orientation   69
orientors   57, 58, 61, 69, 71, 72, 74
outbreak   142
overconsumption of resources   173
overuse of ecosystem services   122
overutilization pressure on biodiversity   182
Oxelösund (Våtmark, Sweden)   116
ozone layer   110, 194
ozone levels   194

partnerships   8, 9
pastureland   144
"pathological" state   48
PCBs and mercury   196
peer review   141
persistent organic pollutants   196
pesticides   197
Petri net   51–53
place and culture issues   201
policy priorities   204
policy process   159, 160
policymaking   75, 153, 158–161, 164, 172
political agenda   156, 163
politics   148, 157, 163
pollination   110
pollution   110

polycentric integrated assessment   41, 75, 76
positional wealth   283, 293
pseudononlinearity   46
public acceptance   161
public awareness   184
public goods   270–272, 278, 285, 287–289, 291, 292
public policy   158, 184, 202, 205

quality of life (QOL)   221–255, 259–298
– evaluation   222
– human-development approach   269
– study group   37
quality of soil   197

radiation efficiency coefficient   62, 64
rare plant species   143
Red Queen's Hypothesis   49, 72, 86
reductionism   157, 158
regional water boards   162
regression   29
remote-sensing data   145
renewable resources   240, 241, 243–245
resource managers   149
road development   144
road planning   115
Ruswil (Switzerland)   117

Santa Fe group   36
satisfiers   260–297
scale   21, 27
scenarios   23
– generation   20
science   153
science of sustainability   157
scientific advisory board   142
scientific credibility and neutrality   156
scientific method   140
self-organization   57
self-regulating open holarchic/hierarchic order (Soho) systems   86
simulation   9, 29, 141
social acceptability   184
social capital   263, 264, 284, 296
social learning   41, 76
social systems   201
social traps   287
soft science   97
Soil Conservation Service (USA)   142
soils   143, 145, 197

spatial hierarchy 49, 50, 67
specific exergy 56, 57
stakeholders 19, 22, 27, 31, 34, 127, 132, 133, 137, 141, 162, 163
state space 58
state variables 29
Stensund Aquaculture Centre (Sweden) 119
stewardship of nature 114
stormwater 121
structural asymmetry 45
structural complexity 55
structural exergy 57
succession 49, 142
survival and well-being 170
sustainability 5, 6, 11, 14, 15, 42, 49, 74, 75, 97, 127, 129, 130, 132, 134, 140, 227, 230, 231, 243, 246, 247, 250, 253, 259, 263, 270, 273, 285, 286, 288, 291, 292, 296, 298
sustainability indicators 31, 32
– see also under indicators
– institutional dimension 32
– socioeconomic dimension 32, 35
sustainable and healthy decisions 168
sustainable development 42, 82, 98, 99, 200
sustainable income 225, 228–230, 232
sustainable state 48, 72, 73
sustainable systems 77
sustainable use of ecosystem 109
system-level response 158
systemic complexity 57
systemicity 73
systems ecology 57
systems simulation 13
systems solutions 8
systems theory 43

tall larkspur 143
technocentric approach 113
technological fix 113
technology 112
temporal hierarchy 51
thermodynamic equilibrium 55, 57
thermodynamic principles 47, 53
thermodynamics of irreversible processes 57

thinkers 127, 135–137
throughflow 47
topsoil losses 110
total material requirements (TMR) 35
Toutle River (Washington, USA) 142
trans-boundary pollution 161
transformity 47, 87
transport systems 115
trash dump 144
triple bottom line 5, 6
tropical fruit 118

uncertainty 27, 28, 149
under-use of ecosystem services 122
unpredictable climatic conditions 176
unsustainable practices 179
urban farming 201
urban waterways 120
US Congress 142

vacuum toilets 117
validation 29–31
value system 156
vision 296, 297
volcano 141

waste as fertiliser 118
waste heat 118
wastewater 116, 119
– energy recovery 119
wastewater-fed aquaculture 118, 119
wastewater nutrient cycles 119
water deficits 195
Water Enhancement Programme (Christchurch, New Zealand) 120
water management in the Netherlands 162
water quality 162
water resources 194
water shortages 179, 195
waterborne diseases 194
watershed management 66, 67
watersheds 66
waterways 120
well-being 259, 261, 263, 268, 290
wetlands 116, 118
wholeness 73
world views 148